T0296426

LONDON MATHEMATICAL SOCIETY LECTURE NOTE SERIES

Managing Editor: Professor N.J. Hitchin, Mathematical Institute,
University of Oxford, 24–29 St Giles, Oxford OX1 3LB, United Kingdom

The titles below are available from booksellers, or, in case of difficulty, from Cambridge University Press.

London Mathematical Society Lecture Note Series. 257

An Introduction to Noncommutative Differential Geometry and its Physical Applications

Second Edition

John Madore
Université Paris XI

PUBLISHED BY THE PRESS SYNDICATE OF THE UNIVERSITY OF CAMBRIDGE
The Pitt Building, Trumpington Street, Cambridge, United Kingdom

CAMBRIDGE UNIVERSITY PRESS
The Edinburgh Building, Cambridge, CB2 2RU, UK   www.cup.cam.ac.uk
40 West 20th Street, New York, NY 10011-4211, USA   www.cup.org
10 Stamford Road, Oakleigh, Melbourne 3166, Australia
Ruiz de Alarcón 13, 28014 Madrid, Spain

First published 1995
Second edition 1999
Reprinted with corrections 2000

A catalogue record for this book is available from the British Library

ISBN   0 521 65991 4 paperback

Transferred to digital printing 2002

Contents

1 Introduction

... Einstein was always rather hostile to quantum mechanics. How can one understand this? I think it is very easy to understand, because Einstein had been proceeding on different lines, lines of pure geometry. He had been developing geometrical theories and had achieved enormous success. It is only natural that he should think that further problems of physics should be solved by further development of geometrical ideas. How, to have a × b not equal to b × a is something that does not fit in very well with geometrical ideas; hence his hostility to it. *

If V is a set of points then the set of complex-valued functions on V is a commutative, associative algebra. As a simple example suppose that V has a finite number of elements. Then the algebra is of finite dimension as a vector space. The product of two vectors is given by the product of the components and it satisfies the inequality $\|fg\| \leq \|f\|\|g\|$ with respect to the norm $\|f\| = \max|f|$. Let f^* be the complex conjugate of f. Then obviously the product satisfies also the equality $\|ff^*\| = \|f\|^2$. A normed algebra with an *involution* $f \mapsto f^*$ which satisfies the above two conditions is called a C^*-*algebra*. Conversely any finite-dimensional commutative algebra which is a C^*-algebra can be considered as an algebra of functions on a finite set of points. The number of points is encoded as the dimension of the algebra. It is obviously essential that the algebra be commutative in order that it have an interpretation as an algebra of functions on a set of points. The finite-dimensional example has an interesting extension to infinite sets if they have a topology. If in fact V is a compact space then the subalgebra $C^0(V)$ of continuous functions on V is a C^*-algebra. It can be shown quite generally that conversely any commutative C^*-algebra with a unit element can be considered as an algebra of functions on a compact space. The space can be described using the language of statistical physics as the space of pure states of the algebra and when we pass to noncommutative geometry we shall see that a pure state is a natural generalization of the notion of a point.

If V is a smooth manifold then the algebra of smooth functions $\mathcal{C}(V)$ defined on it is of course a commutative algebra and in this case also there is an intrinsic characterization of the set of all such algebras, an additional structure which can be added to an arbitrary commutative algebra $\mathcal{A}$ which would insure that $\mathcal{A} = \mathcal{C}(V)$ for some V. The manifold V can always be considered as embedded in a euclidean space $\mathbb{R}^n$ of sufficiently high dimension. The coordinates of the embedding space are generators of an algebra of polynomials which is dense in the algebra $\mathcal{C}(\mathbb{R}^n)$ of smooth functions and the equations which define the manifold are relations in $\mathcal{C}(\mathbb{R}^n)$. The quotient of $\mathcal{C}(\mathbb{R}^n)$ by the ideal generated by the relations is equal to the algebra of smooth functions $\mathcal{C}(V)$ on V. Points of V can be again identified with pure states of $\mathcal{C}(V)$ and vector fields on V can be identified with the derivations of

---

*P.A.M. Dirac, as cited in *The Mathematical Intelligencer* **11** (1989) 58

$\mathcal{C}(V)$. It is however not possible in general to recover completely the differentiable and topological structure of V from the algebraic structure of $\mathcal{C}(V)$ alone. Only if the embedding functions are holomorphic or algebraic can the correspondence between the manifold and the algebra be one to one.

The aim of *noncommutative geometry* is to reformulate as much as possible the geometry of a manifold in terms of an algebra of functions defined on it and then to generalize the corresponding results of differential geometry to the case of a noncommutative algebra. We shall refer to this algebra as the *structure algebra*. The main notion which is lost when passing from the commutative to the noncommutative case is that of a point. 'Noncommutative geometry is pointless geometry.' The original algebra which inspired noncommutative geometry is that of the quantized phase space of non-relativistic quantum mechanics. In fact Dirac in his historic papers in 1926 (Dirac 1926a; Dirac 1926b) was aware of the possibility of describing phase-space physics in terms of the quantum analogue of the algebra of functions, which he called the quantum algebra, and using the quantum analogue of the classical derivations, which he called the quantum differentiations. And of course he was aware of the absence of localization, expressed by the Heisenberg uncertainty relation, as a central feature of these geometries. Inspired by work by von Neumann (1955) for several decades physicists studied quantum mechanics and quantum field theory as well as classical and quantum statistical physics giving prime importance to the algebra of observables and considering the state vector as a secondary derived object. This work has much in common with noncommutative geometry. The notion of a pure state replaces that of a point and derivations of the algebra replace vector fields. More recently Connes (1986) introduced an equivalent of the notion of an exterior derivative and generalized the de Rham cohomology of compact manifolds to the noncommutative case.

In Chapter 2 we shall give a brief review of ordinary differential geometry emphasizing those aspects which it is possible to generalize to the noncommutative case. We give as examples the 2-dimensional torus, the sphere and the pseudosphere, chosen because from them by a simple modification noncommutative geometries can be constructed. In Chapter 3 a noncommutative generalization of a metric and a linear connection are proposed. We give some examples of noncommutative geometries using the algebra M_n of $n \times n$ complex matrices as structure algebra. These examples have the advantage of being without any complications due to analysis. The Hodge part of the Hodge-de Rham theory is completely trivial whereas the de Rham part is still of some interest. Matrix geometry has also the advantage of being in some aspects identical to ordinary geometry and constitutes therefore a transition to the more abstract noncommutative geometries which are given in Chapter 4. In Chapter 5 a short review of the theory of vector bundles is given and their noncommutative generalizations are discussed. Most of the mathematical work in noncommutative geometry has been directed towards a generalization of the theory of differential operators on vector bundles over compact manifolds. In Chapter 6 cyclic cohomology is defined and two important results, Morita equivalence and the theorem of Loday and Quillen, are stated but not proven.

The last two chapters are devoted to the suggestion that noncommutative geometry might find interesting applications in high-energy physics. There has been renewed interest recently in the possibility that at very small length scales the structure of space-time is not properly described by a differentiable manifold. A natural alternative is offered by noncommutative geometry. In the conventional formulation of quantum field theory as the theory of formally quantized classical fields on a classical Minkowski space-time, ultraviolet divergences arise when one attempts to measure the amplitude of field oscillations at a precise given point in space-time. One way of circumventing this problem would be to give an additional structure to the point which would render impossible such measurements. For example, one could modify the microscopic structure of space-time with the hypothesis that at a sufficiently small fundamental length the coordinates of a point become noncommuting operators. This means in particular that it would be impossible to measure exactly the position of a particle since the three space coordinates could not be simultaneously diagonalized. We use here the language of quantum mechanics. An observable is an operator on a Hilbert space of states and the result of a measurement of an observable f when the system is in a state ψ of unit norm is given by $(\psi, f\psi)$. The act of measuring f forces the system into an eigenvector of f and so two observables can be simultaneously measured if and only if they commute. The position of a particle would no longer have a well defined meaning. Since we certainly wish this to be so at macroscopic scales, we must require that the fundamental length be not greater than a typical Compton wavelength. In other words, the fuzziness which the noncommutative structure gives a point in space-time could not be greater than the quantum uncertainty in the position of a particle. We can think of space as being divided into *Planck cells*, just as quantized phase space is divided into *Bohr cells*. The cellular structure replaces the point structure in the same way that the Bohr cells replace points in phase space when Planck's constant is not equal to zero. Gravity would in this approach appear as a deformation of the cellular structure and the graviton would become a sort of 'space-time phonon'. In Chapter 7 a geometry is described which is noncommutative at short length scales but which at large scales resembles the geometry of the ordinary 2-sphere. The geometry has the algebra M_n as structure algebra and there are n cells. The cellular structure is uniform as is the curvature. There is at present no consensus on what would be the most satisfactory noncommutative version of Minkowski space.

The question of whether or not space-time has 4 dimensions has been debated for many years. One of the first negative answers was given by Kaluza (1921) and Klein (1926) in their attempt to introduce extra dimensions in order to unify the gravitational field with electromagnetism. Einstein & Bergmann (1938) suggested that at sufficiently small scales what appears as a point will in fact be seen as a circle. Later, with the advent of more elaborate gauge fields, it was proposed that this *internal manifold* could be taken as a compact Lie group or even as a general compact manifold. The great disadvantage of these extra dimensions is that they introduce even more divergences in the quantum theory and lead to an infinite spectrum of new particles. In fact the structure is strongly redundant and most of it has to be discarded. An associated problem is that of localization. We cannot, and indeed do not wish to have to, address the question of the exact position of a particle in the extra dimensions any more than we wish to localize it too exactly in ordinary space-time. We shall take this as motivation for introducing in the last chapter a modification of Kaluza-Klein theory with an *internal structure* which is described by a noncommutative geometry and in which the notion of a point does not exist. As particular examples of such a geometry we shall choose only internal structures which give rise to a finite spectrum of particles.

Quite generally one can address the question of how far it is possible to transcribe all of space-time physics into the language of noncommutative geometry. We shall see in Chapter 6 that a differential calculus can be constructed over an arbitrary associative algebra. This would permit the formulation of gauge theories in any geometry. Matter fields could be incorporated as elements of algebra modules. In a less general setting a sort of Dirac operator has been proposed and a generalized integral (ConnesConnes 1992). This would permit the construction of an (euclidean) action. We shall discuss this briefly in Chapter 5. In this formulation space-time and the bosonic fields are incorporated in the algebra but the fermionic fields appear separately as modules. A more general unification in the spirit of supersymmetry would involve the elimination of the modules in favour of an algebra which is in some sense supersymmetric. Even in the simple matrix models which are discussed in Chapter 3 this has not been done in a completely satisfactory manner. A more serious problem is that of quantization. The 'Central Dogma' of field theory at the present time is that all information is contained in the classical action. The Standard Model is defined by a classical action which is assumed to contain implicitly all of high-energy physics. Quantum corrections are obtained by a standard quantization procedure. This quantization procedure has not been generalized to noncommutative models even in the simplest cases. The examples which have been used to propose classical actions which might be relevant in high-energy physics all involve simple matrix factors. They are quantized by first expanding the noncommutative fields in terms of ordinary space-time components and then quantizing the components. Under quantization the constraints on the model which come from the noncommutative geometry are lost.

Our purpose here is to furnish an elementary introduction to the subject of noncommutative geometry for non-specialists with special emphasis on the matrix case. Some knowledge of ordinary differential geometry including fibre bundles is supposed, although those parts of the subject are recalled which are useful for the understanding of the noncommutative case. We have assumed a few elementary notions and results from the theory of Hilbert and Banach spaces as well as from the theory of rings and algebras. Emphasis has been put on algebraic properties since our prime objects of interest are matrix algebras. This means that subtle, very often important, points of analysis are if at all but briefly mentioned. A certain familiarity with the rudiments of classical field theory has been also supposed since special emphasis has been placed on those aspects which might be useful in field theory and numerous examples have been taken from physics. Although the text is self-contained the examples given in some of the sections can include applications to subjects which require further knowledge.

Important words are placed in italics when first used and sometimes subsequently if the definition is extended. If the definition is not given in the text the word is assumed to be known and the definition can be found in the literature cited in the Notes. Quotes are sometimes used to underline the fact that a word or phrase is ill-defined. In particular, words which are normally only defined in commutative geometry and which are used intuitively in the noncommutative context are placed in quotes. We shall be primarily interested in the 'quasi-commutative' limit and we shall continue to use the word 'function', to designate an element of the algebra generated by the 'coordinates' even though the product of two 'functions' need not commute. We hesitate to use the words 'noncommutative manifold' or 'quantum manifold'. So 'noncommutative geometry' designates both the field of study and the object which is being studied.

We have tried to use conventional notation as far as possible and there is an occasional conflict between the conventions of physics and of algebra and geometry. The upper case H stands for homology, Hermite, Hamilton and Hall; the lower case e denotes an idempotent as well as the electron charge. The symbol δ designates functional variation as well as three different maps of cochains. In Chapter 2 a tilde is placed on quantities which will in subsequent chapters be replaced by noncommutative or 'quantized' equivalents. In quantum mechanics frequently physicists place a hat on an operator to distinguish it from the corresponding classical function. In this respect the tilde is an 'anti-hat'. The tilde is also used, especially in Chapter 8, to designate forms which have been lifted from a manifold to a bundle, as well as for related reasons. Unless otherwise indicated a tensor product is always over the complex numbers. If both factors are of infinite dimension then the symbol '$\otimes$' implies an appropriate completion with respect to the topology used on either factor. The symbol '$\mathcal{A}$' designates an associative algebra over the complex numbers with a unit element. If $\mathcal{A}$ is a formal algebra we shall often use the same symbol or name when referring to a topological completion as a C^*-algebra or a representation thereof as a von Neumann algebra. A commutator $[a, b] = ab \pm ba$ in a graded algebra will always be a graded commutator. This means that the minus sign is used unless the gradings of both a and b are odd. The symbol '$*$' is overworked. In front of a form it indicates the dual form; on a map it indicates an induced map and used as an index, as in H^* and H_*, it stands for a set of natural numbers. In this last usage the symbol '$2*$' stands for a set of even numbers and '$*-1$' a shift by -1. Especially in Section 4.2 it indicates a product in an algebra; it is easier to place a hat on a star than on a point. We use the word 'Example' in a loose sense which indicates anything from a simple particular case to a subject of further research.

In the second edition some errors were corrected and a few more recent results were included, mostly concerning the definition of linear connections. A brief review of von Neumann algebras was also added since there have been some models proposed in which it might be possible to distinguish 'observationally' two representations of the same associative algebra. If one considers the algebra as the analogue of the theory and the representation as a choice of physical state this means that it will be possible to distinguish two states of the same theory. Some of the new material is based on articles written in collaboration with B.L. Cerchiai, Sunggoo Cho, M. Dubois-Violette, G. Fiore, R. Hinterding, Y. Georgelin, T. Masson, J. Mourad, K.S. Park, P. Schupp, S. Schraml, H. Steinacker and J. Wess.

In this new printing some errors were corrected and a few more recent results were included, all concerning the interface with string theory. Some of the new material is based on articles written in collaboration with S. Schraml, P. Schupp and J. Wess.

The author would like to thank A. Čap and M. Dubois-Violette for numerous helpful suggestions as well as F. Constantinescu, C. Duval, J. Gratus, H. Grosse, T. Masson, J. Mourad, F. Müller-Hoissen, A. Nicas, T. Schücker and F. Vanderseypen for pointing out several obscurities and errors in the original manuscript. While preparing the revised text we have benefited from conversations with B.L. Cerchiai, A. Chakrabarti, M. Durdević, G. Fiore, K. Fredenhagen, H. Grosse, A. Kehagias, C. Korthals-Altes, G. Landi, E. Langmann, M. Maceda, D. Madore, C. Ohn, O. Richter, K. Schmüdgen, T. Schücker, A. Turbiner, L. Vainerman, D. Vergara, J. Wess and C. Zachos. Much of the work was done while the author was visiting the Max-Planck-Institut für Physik in München. He would like to thank J. Wess also for his hospitality there.

Allein es steht in einem andern Buch,
Und ist ein wunderlich Kapitel. $*$

$*$Mephistopheles to Faust

2 Differential Geometry

In this chapter we give a brief review of ordinary differential geometry, emphasizing those aspects which it is possible to generalize to the noncommutative case. It is meant to be used only as a reference for the basic definitions which are used in the following chapter on matrix geometry. In the first section we recall the definitions of vector fields and differential forms in such a way that it is evident that they depend directly on the algebra of functions on the manifold. To make this fact more transparent a manifold has been defined first as a submanifold of a higher-dimensional euclidean space and only later in terms of local coordinate charts. The two definitions are equivalent. We recall the definition of a Lie group and Lie algebra because there are many similarities between the geometry of a Lie group and the matrix geometries we shall introduce in the next chapter and also because of course they are essential in the construction of fibre bundles, whose definition is briefly recalled at the end of Section 2.1. The 2-dimensional torus, the sphere and the pseudosphere, with respectively zero, positive and negative Gaussian curvature, have been chosen as examples because from them by a simple modification corresponding noncommutative geometries can be constructed. The sphere furnishes also the simplest example of a manifold defined as a submanifold of a euclidean space. The (flat) torus and the pseudosphere are of interest also for the opposite reason; they cannot be globally embedded in $\mathbb{R}^3$. Symplectic geometry is mentioned as an example because of its importance in the passage from classical to quantum mechanics. In Section 2.2 metrics and connections are defined in terms of local moving frames. The latter are defined first on parallelizable manifolds and then generalized to more general manifolds using principal bundles. The last section contains a very brief survey of cohomology, the minimum necessary to follow the description of vector bundles in Chapter 5.

2.1 Differential manifolds

Let V be a smooth, compact, oriented, real *manifold* without boundary of dimension m and let $\mathcal{C}(V)$ be the commutative, associative *algebra* of smooth, real-valued functions on V. The sum and product of two functions are defined by the sum and product of the value of the function at each point. If $x \in V$ then

$$(f + g)(x) = f(x) + g(x), \qquad (fg)(x) = f(x)g(x).$$

The commutative, associative and distributive rules follow then from those of $\mathbb{R}$. Every such V can be embedded in a euclidean space of sufficiently high dimension $n > m$. It is defined then by $n - m$ relations in the euclidean coordinates x^i and the algebra $\mathcal{C}(V)$ can be considered as a quotient of the algebra of smooth functions on $\mathbb{R}^n$ by the ideal generated by the relations. An *ideal* of an algebra is a subalgebra which is stable under multiplication by a general element of the algebra. An arbitrary intersection of ideals is again an ideal and the ideal generated by a subset of the algebra is the intersection of all ideals which contain it. The algebra $C(V)$ has many ideals, for example the subalgebra of functions which vanish on any closed set in V. Another important algebra, the algebra M_n of all $n \times n$ complex matrices, has no proper ideals.

Let X be a smooth *vector field* on V. The vector space $\mathcal{X}(V)$ of all such X is a left $C(V)$-*module*. That is, if $f \in C(V)$ and $X \in \mathcal{X}(V)$ then $fX \in \mathcal{X}(V)$. We recall that a (left/right) module is a vector space on which there is a (left/right) action of an algebra. The n-dimensional vector space $\mathbb{C}^n$ is a left module for the algebra M_n. Let ∂_i be the *natural basis* of the vectors on the embedding space $\mathbb{R}^n$. The $C(\mathbb{R}^n)$-module $\mathcal{X}(\mathbb{R}^n)$ of smooth vector fields on $\mathbb{R}^n$ is a *free module* of *rank n*. It can be identified as the direct sum of n copies of $C(\mathbb{R}^n)$:

$$\mathcal{X}(\mathbb{R}^n) = \bigoplus_1^n C(\mathbb{R}^n).$$

This means that every element $X \in \mathcal{X}(\mathbb{R}^n)$ can be written uniquely as a linear combination $X = X^i \partial_i$ with $X^i \in C(\mathbb{R}^n)$. In particular every element $X \in \mathcal{X}(V)$ can be written as a linear combination $X = X^i \partial_i$ with $X^i \in C(V)$. As a $C(V)$-module, $\mathcal{X}(V)$ is finitely generated. The expression for X however cannot now be unique since at each point of V there can be at most $m < n$ linearly independent vectors. One can show that there is always a second module $\mathcal{N}$ such that the direct sum $\mathcal{X}(V) \oplus \mathcal{N}$ is a free module of rank n. We have then

$$\mathcal{X}(V) \oplus \mathcal{N} = \bigoplus_1^n C(V).$$

A module is a *projective module* if it is direct summand of a free module. The $C(V)$-module $\mathcal{X}(V)$ is always a projective module. The algebra $C(V)$ and the algebras of matrices are themselves free modules of rank 1. The module $\mathbb{C}^n$ is a projective module over the algebra M_n since the direct sum of itself and a further $n - 1$ copies is isomorphic to M_n. If there is a globally defined *moving frame*, e_α, $1 \leq \alpha \leq m$, on V then $\mathcal{X}(V)$ is a free module of rank m and any element X in $\mathcal{X}(V)$ can be written uniquely as a linear combination $X = X^\alpha e_\alpha$ of the e_α. The manifold V is said to be *parallelizable*. In general a moving frame can be defined only locally on V.

The *Lie bracket* $[X, Y]$ of two elements X and Y of the free module $\mathcal{X}(\mathbb{R}^n)$, defined by

$$[X, Y]f = (XY - YX)f = (X^i \partial_i Y^j - Y^i \partial_i X^j)\partial_j f$$

is also an element of $\mathcal{X}(\mathbb{R}^n)$. Since V is a *submanifold* of $\mathbb{R}^n$ the Lie bracket of two elements of $\mathcal{X}(V)$ is also an element of $\mathcal{X}(V)$. In particular if V is parallelizable we can write the Lie bracket as

$$[e_\alpha, e_\beta] = C^\gamma{}_{\alpha\beta} e_\gamma,$$

where the *structure functions* $C^\gamma{}_{\alpha\beta}$ are elements of $\mathcal{C}(V)$. In general this equation can be written only locally. The converse of the above is known as the *Frobenius theorem*: if m linearly independent elements of $\mathcal{X}(\mathbb{R}^n)$ are closed under the Lie bracket then they are a frame for a manifold V and the $\mathcal{C}(V)$-module generated by them is equal to $\mathcal{X}(V)$.

An important property of a vector field which will interest us here is the fact that it can be defined as a *derivation* of the algebra $\mathcal{C}(V)$, a linear map of $\mathcal{C}(V)$ into itself which satisfies the *Leibniz rule*:

$$X(fg) = (Xf)g + fXg.$$

This permits us to identify $\mathcal{X}(V)$ with the derivations $\mathrm{Der}(\mathcal{C}(V))$ of the algebra $\mathcal{C}(V)$:

$$\mathcal{X}(V) \equiv \mathrm{Der}(\mathcal{C}(V)).$$

It will also enable us to generalize the notion of a vector field to the noncommutative case.

We shall usually choose $\mathcal{C}(V)$ to be an algebra of complex-valued functions. Using complex conjugation one can define an *involution* $f \mapsto f^*$ by $f^*(x) = \overline{f(x)}$. We shall assume that the algebra is closed under the involution. Such an algebra is called a **-algebra*. We shall assume in this case that the elements of $\mathcal{X}(V)$ satisfy a *reality condition*. If $X \in \mathcal{X}(V)$ and $f \in \mathcal{C}(V)$ then

$$(Xf)^* = Xf^*.$$

If $f = f^*$ then f is an *hermitian element* of $\mathcal{C}(V)$. Since the algebra $\mathcal{C}(V)$ is commutative $(fg)^* = f^*g^*$ and the set of real functions is a subalgebra. If f and g are elements of a noncommutative *-algebra then $(fg)^* = g^*f^*$ and in general this is not equal to f^*g^*. The hermitian elements do not form a subalgebra. This is why it is of interest to consider algebras of complex functions even when studying real manifolds.

A *differential form* of order p or *p-form* α is a p-linear completely antisymmetric map of $\mathcal{X}(V)$ into $\mathcal{C}(V)$. In particular if $f \in \mathcal{C}(V)$ and $X_1, \ldots, X_p$ are p vector fields then $\alpha(fX_1, \ldots, X_p) = f\alpha(X_1, \ldots, X_p)$; the value of $\alpha(X_1, \ldots, X_p)$ at a point of V depends only on the values of the vector fields at that point. The set $\Omega^p(V)$ of p-forms is a $\mathcal{C}(V)$-*module*:

$$(f\alpha)(X_1, \ldots, X_p) = (\alpha f)(X_1, \ldots, X_p) = f(\alpha(X_1, \ldots, X_p)).$$

The *exterior product* $\alpha \wedge \beta$ of $\alpha \in \Omega^p(V)$ and $\beta \in \Omega^q(V)$ is an element of $\Omega^{p+q}(V)$ defined by

$$\alpha \wedge \beta(X_1, \ldots, X_{p+q}) = \frac{1}{(p+q)!} \sum \epsilon(i,j)\alpha(X_{i_1}, \ldots, X_{i_p})\beta(X_{j_1}, \ldots, X_{j_q}), \quad (2.1.1)$$

where the summation is taken over all the possible partitions of $(1, \ldots, p+q)$ into $(i_1, \ldots, i_p)$ and $(j_1, \ldots, j_q)$ and $\epsilon(i, j)$ is the signature of the corresponding permutation. It is *graded commutative*, $\alpha \wedge \beta = (-1)^{pq} \beta \wedge \alpha$, because the algebra $\mathcal{C}(V)$ is commutative. In general this will not be the case so we shall accordingly usually write the exterior product of two forms α and β simply as

$$\alpha\beta \equiv \alpha \wedge \beta.$$

Define $\Omega^0(V) = \mathcal{C}(V)$. The set $\Omega^*(V)$ of all $\Omega^p(V)$ for $0 \le p \le m$ with the product (2.1.1) is the *exterior algebra* or *algebra of forms* or algebra of differential forms on V. If V is parallelizable then we can write

$$\Omega^*(V) = \mathcal{C}(V) \otimes \bigwedge{}^*$$

where $\bigwedge^*$ is the exterior algebra over the complex numbers generated by the frame. The algebra of forms is a free $\mathcal{C}(V)$-module of rank 2^m.

The *exterior derivative* $d\alpha$ of $\alpha \in \Omega^p(V)$ is defined by the formula

$$d\alpha(X_0, \ldots, X_p) = \frac{1}{p+1} \sum_{i=0}^{p} (-1)^i X_i(\alpha(X_0, \ldots, \hat{X}_i, \ldots, X_p))$$

$$+ \frac{1}{p+1} \sum_{0 \le i < j \le p} (-1)^{i+j} \alpha([X_i, X_j], X_0, \ldots, \hat{X}_i, \ldots, \hat{X}_j, \ldots, X_p). \tag{2.1.2}$$

The hat on a symbol means here that it is omitted. The case $p = 0$ is especially interesting:

$$df(X) = Xf. \tag{2.1.3}$$

We have written f instead of α since a 0-form is an element of the algebra $\mathcal{C}(V)$. Let now $\alpha \in \Omega^p(V)$ and $\beta \in \Omega^*(V)$. Then one easily sees that d satisfies the conditions

$$d(\alpha \wedge \beta) = d\alpha \wedge \beta + (-1)^p \alpha \wedge d\beta. \tag{2.1.4}$$

Because of the factor $(-1)^p$ in the second term on the right-hand side d does not satisfy Leibniz's rule. It is a *graded derivation* (of degree $+1$) of $\Omega^*(V)$. From (2.1.2) it follows that d is a *differential*:

$$d^2 = 0. \tag{2.1.5}$$

The algebra of forms $\Omega^*(V)$ can also be written as a direct sum

$$\Omega^*(V) = \Omega^{*+}(V) \oplus \Omega^{*-}(V)$$

of even forms and odd forms. The differential takes one into the other. The algebra $\Omega^*(V)$ with the associated differential is a graded *differential algebra*; it is a *calculus* or *differential calculus* over the algebra $\mathcal{C}(V)$. We shall see in Section 6.1 that it is always possible to construct a differential calculus over any associative algebra $\mathcal{A}$; the differential will have however nothing to do with the derivations of $\mathcal{A}$. To distinguish it, the differential calculus we have just constructed over the algebra $\mathcal{C}(V)$, based on the differential (2.1.2), is called the *de Rham calculus*.

The definition of d does not use any properties of $\mathcal{X}(V)$ other than the fact that it has a Lie-algebra structure and that it acts on the algebra $\mathcal{C}(V)$. We can write then $\Omega^*(V) \equiv \Omega^*(\mathcal{C}(V))$. We can use d to give a second definition of a p-form. First we define as above $\Omega^0(V)$ to be equal to $\mathcal{C}(V)$. Then we use (2.1.3) to define df for any $f \in \mathcal{C}(V)$ and we define $\Omega^1(V)$ to be the $\mathcal{C}(V)$-bimodule generated by the df subject to the relations $fdg - dgf = 0$. A p-form is an element of the $\mathcal{C}(V)$-module generated by exterior products of p elements of $\Omega^1(V)$.

Dual to the e_α defined above is a set of 1-forms θ^α which are determined by the equations

$$\theta^\alpha(e_\beta) = \delta^\alpha_\beta. \tag{2.1.6}$$

From (2.1.3) we see that the differential df of a function f can be written as

$$df = e_\alpha f \theta^\alpha.$$

From the commutation relations for the e_α it follows that the exterior derivatives $d\theta^\alpha$ can be expressed in the form

$$d\theta^\alpha = -\frac{1}{2} C^\alpha{}_{\beta\gamma} \theta^\beta \wedge \theta^\gamma. \tag{2.1.7}$$

These equations are referred to as the *structure equations*.

Let ϕ be a smooth map of V into V'. It induces a map ϕ^* of $\mathcal{C}(V')$ into $\mathcal{C}(V)$ given by

$$\phi^* f = f \circ \phi. \tag{2.1.8}$$

This map has a natural extension to a map of $\Omega^*(V')$ into $\Omega^*(V)$ given by

$$\phi^*(df) = d(\phi^* f)$$

on the image of d. If ϕ is a diffeomorphism then one can identify V' with V and consider ϕ^* as an automorphism of the algebras $\mathcal{C}(V)$ and $\Omega^*(V)$. In this case ϕ induces also a map ϕ_* of $\mathcal{X}(V)$ onto itself by the formula

$$(\phi_* X)f = \phi^{*-1} X \phi^* f, \qquad f \in \mathcal{C}(V). \tag{2.1.9}$$

In particular let ϕ_t be a local 1-parameter group of diffeomorphisms of V generated by a vector field X. Then $\phi_t^* f = f + tXf + o(t^2)$ and the *Lie derivative* of a vector field Y with respect to X is given by

$$\phi_{t*} Y = Y - t L_X Y + o(t^2). \tag{2.1.10}$$

A straightforward calculation shows that $L_X Y = [X, Y]$. By requiring that it be a derivation, the Lie derivative can be extended to a general element of the tensor algebra over $\mathcal{X}(V)$.

We define the Lie derivative of a function f to be given by $L_X f = Xf$ and the Lie derivative of $\alpha \in \Omega^p(V)$ to be given by the formula

$$(L_X \alpha)(Y_1, \ldots, Y_p) = X\alpha(Y_1, \ldots, Y_p) - \sum_1^p \alpha(Y_1, \ldots, [X, Y_i], \ldots, Y_p).$$

Again, by requiring that it be a derivation the Lie derivative can be extended to a general element of the tensor algebra over $\Omega^1(V)$.

Define the *interior product* i_X to be a map of $\Omega^{p+1}(V)$ into $\Omega^p(V)$ given for $\alpha \in \Omega^{p+1}(V)$ by the equation

$$(i_X \alpha)(X_1, \ldots, X_p) = (p+1)\alpha(X, X_1, \ldots, X_p).$$

It is a *graded derivation* (of degree -1) of $\Omega^*(V)$. The factor $p+1$ here and in Formula (2.1.2) come from the normalization of the product (2.1.1) which we have chosen. Set $i_X f = 0$. Then it is easy to see that L_X, d and i_X are related by the formula

$$L_X = i_X d + d i_X. \tag{2.1.11}$$

If V is a group then it is a (real, compact) *Lie group*. In this case it is necessarily analytic. Let G be a Lie group of dimension n. Since the left multiplication L_g by an element $g \in G$ defines a diffeomorphism of G it defines also a map L_{g*} of $\mathcal{X}(G)$ onto itself. The Lie algebra $\underline{g}$ of G is the set of left-invariant vector fields X: $L_{g*}X = X$. It is a vector space of dimension n. It is a general result that Lie groups are parallelizable; the left-invariant vector fields constitute a global trivialization of the tangent bundle. If the moving frame e_α is chosen as a basis of $\underline{g}$ then it follows from the left invariance that the structure functions are constants. It follows also from the left invariance that one can identify $\underline{g}$ with the tangent space at the unit element of G and in this case we shall designate the basis by λ_α. The *Maurer-Cartan form* is the Lie-algebra-valued 1-form

$$\theta = \lambda_\alpha \theta^\alpha. \tag{2.1.12}$$

From (2.1.7) follows the *Maurer-Cartan equation*

$$d\theta + \theta^2 = 0.$$

Equation (2.1.7) for the components of θ is also referred to as the Maurer-Cartan equation if V is a group.

The algebra $C(V \times V')$ of smooth functions on the space $V \times V'$ can be identified with the *tensor product* $C(V) \otimes C(V')$ of the two algebras $C(V)$ and $C(V')$:

$$C(V \times V') = C(V) \otimes C(V').$$

This means that every smooth function on $V \times V'$ can be considered as a limit of a sequence of functions which are finite linear combinations of products of a smooth function on V and a smooth function on V'. The 1-forms on $V \times V'$ can be identified with the direct sum of 1-forms on V and V',

$$\Omega^1(V \otimes V') = C(V) \otimes \Omega^1(V') \oplus \Omega^1(V) \otimes C(V')$$

and the differential is the sum of d on V and d' on V'. The exterior algebra $\Omega^*(V \times V')$ is given in terms of the exterior algebra of each factor by

$$\Omega^p(V \times V') = \bigoplus_{i+j=p} \Omega^i(V) \otimes \Omega^j(V'). \tag{2.1.13}$$

We shall return to this in more detail in Section 3.3.

An extension of the definition of the product of two manifolds is a bundle structure. The important special case in which one of the factors is a group is called a *principal bundle*. To define this we need to examine more closely the definition of a manifold, which can also be thought of as a finite set of *coordinate charts*. A compact manifold V of dimension n is a compact topological space which can be covered by a finite set of *open neighbourhoods* $\mathcal{O}_i$ each homeomorphic to $\mathbb{R}^n$:

$$\mathcal{O}_i \xrightarrow{f_i} \mathbb{R}^n.$$

It is a smooth manifold if the induced *coordinate transformations*

$$\mathbb{R}^n \xrightarrow{f_j{}^{-1}} \mathcal{O}_i \cap \mathcal{O}_j \xrightarrow{f_i} \mathbb{R}^n \tag{2.1.14}$$

are smooth. Let G be a Lie group and P a manifold with a set of neighbourhoods $\mathcal{Q}_i$ diffeomorphic to $\mathcal{O}_i \times G$:

$$\mathcal{Q}_i \xrightarrow{\phi_i} \mathcal{O}_i \times G.$$

Suppose that when $x \in \mathcal{O}_i \cap \mathcal{O}_j$ the maps $\phi_i \circ \phi_j{}^{-1}$,

$$(\mathcal{O}_i \cap \mathcal{O}_j) \times G \xrightarrow{\phi_j{}^{-1}} \mathcal{Q}_i \cap \mathcal{Q}_j \xrightarrow{\phi_i} (\mathcal{O}_i \cap \mathcal{O}_j) \times G,$$

are always of the form

$$\phi_i \circ \phi_j{}^{-1}(x, h) = (x, g_{ij}(x)h), \tag{2.1.15}$$

where $h \in G$ and g_{ij} is a smooth function

$$\mathcal{O}_i \cap \mathcal{O}_j \xrightarrow{g_{ij}} G. \tag{2.1.16}$$

Then P is a principal bundle over V with *structure group* G. From the definition of the *transition functions* g_{ij} follow the consistency conditions (with no summation on repeated indices)

$$g_{ik}(x) = g_{ij}(x)g_{jk}(x) \qquad (2.1.17)$$

when $x \in \mathcal{O}_i \cap \mathcal{O}_j \cap \mathcal{O}_k$.

Let g_i be a smooth function from $\mathcal{O}_i$ to G. Then g_i induces a map

$$\mathcal{O}_i \times G \xrightarrow{g_i} \mathcal{O}_i \times G$$

given by $g_i(x, h) = (x, g_i h)$ and called a *change of basis*. The composite maps $\phi_i' = g_i \circ \phi_i$ define a new set of functions $g'_{ij} = g_i g_{ij} g_j^{-1}$ from $\mathcal{O}_i \cap \mathcal{O}_j$ to G which satisfy also the consistency conditions. The bundle is equally well described by the g_{ij} or by the g'_{ij}. If there exist g_i such that the g'_{ij} take their values in a subgroup $G' \subset G$ then the structure group can be *reduced* to G'. If in particular G is reducible to the identity the bundle is *trivial*; the maps ϕ_i' then define a homeomorphism of P with $V \times G$.

There is a projection π of P onto V given in each $\mathcal{Q}_i$ by $\pi(x, h) = x$. By (2.1.15) then π is well defined in $\mathcal{Q}_i \cap \mathcal{Q}_j$. A *section* σ of P is a smooth map of V into P which is such that the composition $\pi \circ \sigma$ is the identity on V. It is defined by a set of functions $g_i(x)$ from $\mathcal{O}_i$ into G such that in $\mathcal{O}_i \cap \mathcal{O}_j$ we have $g_i = g_{ij}g_j$. It follows from the definitions then that if P has a section then it is the trivial bundle. A *local section* can exist over a part of V.

Let $E(V)$ be the *frame bundle* over V. It is a principal $GL(n, \mathbb{R})$-bundle defined in terms of the coordinate transformations (2.1.14). Let $x^\lambda = f_1^\lambda$ and $y^\lambda = f_2^\lambda$ be two coordinate charts. Then in $\mathcal{Q}_1 \cap \mathcal{Q}_2$

$$y^\lambda = f_2 \circ f_1^{-1}(x^\lambda).$$

The element g_{12} of the set (2.1.16) which defines $E(V)$ is given by the matrix-valued function

$$g_{12} = \left(\frac{\partial y^\lambda}{\partial x^\mu}\right).$$

If V is parallelizable then using for example the basis e_α one can identify $E(V)$ with the trivial bundle $V \times GL(n, \mathbb{R})$.

Let F be a vector space, for example $\mathbb{R}^r$ or $\mathbb{C}^r$ for some r. Consider a manifold H with a set of neighbourhoods $\mathcal{Q}_i$ homeomorphic to $\mathcal{O}_i \times F$,

$$\mathcal{Q}_i \xrightarrow{\phi_i} \mathcal{O}_i \times F. \qquad (2.1.18)$$

Suppose that the maps $\phi_i \circ \phi_j^{-1}$ are always of the form (2.1.15) with h replaced by an element $\psi \in F$ and with g_{ij} in some representation of G on F. Then H is an *associated vector bundle* with *fibre* F. Local sections of H can be patched together using local sections of P. If P has a section defined by g_i in $\mathcal{O}_i$ then $\psi_i = g_i\psi$ is a local section of H for any $\psi \in F$. If the principal bundle P is trivial then all associated bundles are also trivial. Two bundles H and H' are said to be *isomorphic* if by a change of basis the transition functions can be chosen to be equal. Although to define it we have made essential use of the manifold and of its definition in terms of local coordinate charts, we shall see that a vector bundle has a natural generalization in noncommutative geometry through its space of sections. The space $\mathcal{H}$ of sections of a vector bundle H is a $\mathcal{C}(V)$-module. The bundle is trivial if and only if the module is free.

Let ϕ be a smooth map from V into V'. A bundle H' over V' is defined by transition functions g'_{ij}. The composite functions $g_{ij} = g'_{ij} \circ \phi$ define a bundle $H = \phi^*H'$ over V which is the *pull-back* of H'.

The *tangent bundle* TV to V is a vector bundle with structure group $GL(n, \mathbb{R})$ and with fibre $F = \mathbb{R}^n$. The transformation (2.1.18) from the vector X to the fibre coordinates $X^\lambda \in F$ is given in terms of the natural basis ∂_λ of the vectors on $\mathbb{R}^n$. Let X be a vector field. Then in the coordinate chart $\mathcal{Q}_1$

$$\phi_1(X) = (x, X_1^\lambda(x)\partial_\lambda)$$

and similarly for $\mathcal{Q}_2$. In the intersection $\mathcal{Q}_1 \cap \mathcal{Q}_2$ the map g_{12} defined by (2.1.15) is given by

$$X_2^\lambda = \frac{\partial y^\lambda}{\partial x^\mu}X_1^\mu.$$

That is, the tangent bundle is an associated $E(V)$-bundle. Each element of the moving frame can be expressed in terms of the local natural basis:

$$e_\alpha = e_\alpha^\lambda \partial_\lambda.$$

The map from ∂_λ to e_α is an example of a change of basis. The space of sections of TV is the space $\mathcal{X}(V)$ of vector fields on V. If V is parallelizable then using the basis e_α one can identify TV with the trivial bundle $V \times \mathbb{R}^n$. We have already noticed that in this case $\mathcal{X}(V)$ is a free $\mathcal{C}(V)$-module of rank n.

The *cotangent bundle* T^*V to V is defined similarly. It is also a vector bundle with structure group $GL(n, \mathbb{R})$ and with fibre $F = \mathbb{R}^n$. If we designate an element of F by α_λ then the transformation (2.1.18) from the 1-form α to the fibre coordinates α_λ is given in terms of the *natural basis* dx^λ dual to ∂_λ. Let α be a form. Then in $\mathcal{Q}_1$

$$\phi_1(\alpha) = (x, \alpha_{1\lambda}(x)dx^\lambda)$$

and similarly for $\mathcal{Q}_2$. In the intersection $\mathcal{Q}_1 \cap \mathcal{Q}_2$ the map g_{12} is given by

$$\alpha_{1\mu} = \frac{\partial y^\lambda}{\partial x^\mu}\alpha_{2\lambda}.$$

The cotangent bundle is also an associated $E(V)$-bundle. The dual moving frame θ^α can be expressed in terms of the dual natural basis

$$\theta^\alpha = \theta^\alpha_\lambda dx^\lambda.$$

The space of sections of T^*V is the space $\Omega^1(V)$ of 1-forms on V. If V is parallelizable then using the basis θ^α dual to e_α one can identify T^*V with the trivial bundle $V \times \mathbb{R}^n$. In this case $\Omega^1(V)$ is a free $\mathcal{C}(V)$-module of rank n.

Consider the set $\mathrm{Vect}(V)$ of all vector bundles over V. To be specific we suppose that they are real vector bundles although we shall frequently use also complex vector bundles. An operation on vector spaces defines in a natural way a corresponding operation on $\mathrm{Vect}(V)$. Let H_1 and H_2 be two vector bundles over V. The *direct sum* or *Whitney sum* is the vector bundle $H_1 \oplus H_2$ over V whose fibre is the direct sum of the corresponding fibres. If $GL(l_1, \mathbb{R})$ and $GL(l_2, \mathbb{R})$ are the structure groups of H_1 and H_2 respectively then the structure group of the direct sum is $GL(l_1 + l_2, \mathbb{R})$. The *direct product* is the vector bundle $H_1 \otimes H_2$ over V whose fibre is the direct product of the corresponding fibres. The structure group of the direct product is $GL(l_1 l_2, \mathbb{R})$.

Let V be a compact manifold of dimension m embedded in $\mathbb{R}^n$ for some n. Let N be the normal bundle to V. This is a vector bundle with fibre $\mathbb{R}^{n-m}$ consisting of the vectors in $\mathbb{R}^n$ which are normal to V with respect to the standard euclidean inner product of $\mathbb{R}^n$. In general it is nontrivial. From TV and N we can construct the direct-sum bundle $TV \oplus N$. The module $\mathcal{N}$ introduced at the beginning of the section is the space of sections of N and the fact that $\mathcal{X}(V) \oplus \mathcal{N}$ is a free module is equivalent to the fact that $TV \oplus N$ is a trivial $GL(n, \mathbb{R})$-bundle:

$$TV \oplus N \simeq V \times \mathbb{R}^n. \tag{2.1.19}$$

Example 2.1 Consider $\mathbb{R}^3$ with coordinates $\tilde{x}^a$, $1 \le a \le 3$, and standard euclidean metric given by the components $g_{ab} = \delta_{ab}$. Let S^2 be the 2-sphere defined by

$$g_{ab}\tilde{x}^a\tilde{x}^b = r^2. \tag{2.1.20}$$

Consider the algebra $\mathcal{P}$ of polynomials in the $\tilde{x}^a$ and let $\mathcal{I}$ be the ideal generated by the relation (2.1.20). That is, $\mathcal{I}$ consists of elements of $\mathcal{P}$ with $g_{ab}\tilde{x}^a\tilde{x}^b - r^2$ as factor. Then the quotient algebra $\mathcal{A} = \mathcal{P}/\mathcal{I}$ is dense in the algebra $\mathcal{C}(S^2)$. Any element of $\mathcal{A}$ can be represented as a finite multipole expansion of the form

$$\tilde{f}(\tilde{x}^a) = f_0 + f_a\tilde{x}^a + \frac{1}{2}f_{ab}\tilde{x}^a\tilde{x}^b + \cdots, \tag{2.1.21}$$

where the $f_{a_1\ldots a_i}$ are completely symmetric and trace-free. The vector space of the $f_{a_1\ldots a_i}$ is of dimension $2i + 1$ and so we obtain a vector space of dimension $\sum_{i=0}^{n-1}(2i + 1) = n^2$ if we consider only polynomials of order $n - 1$. Later we shall redefine the product of the $\tilde{x}^a$ to make this vector space into the algebra of $n \times n$ matrices. Logically the algebras $\mathcal{P}$ and $\mathcal{A}$ should have a tilde on them here and without the tilde they should designate the matrix algebra.

To construct a set of generators of the derivations of $\mathcal{A}$ we use the natural action of SO_3 on S^2. For $1 \leq a \leq 3$ let λ_a be a basis of $\underline{su}_2$, the Lie algebra of SU_2. The adjoint action $\lambda_a \mapsto g^{-1}\lambda_a g = \Lambda^b_a \lambda_b$ maps the element $g \in SU_2$ onto the element $\Lambda^a_b \in SO_3$ and so induces a projection π of SU_2 onto S^2. The basis of $\underline{su}_2$ can be considered then as a set $\tilde{e}_a$ of vector fields on S^2. It has a simple action on the generators $\tilde{x}^a$ of $\mathcal{A}$ given by

$$\tilde{e}_a \tilde{x}^b = C^b{}_{ca} \tilde{x}^c. \tag{2.1.22}$$

The $C^b{}_{ca}$ are the $\underline{su}_2$ structure constants: $C_{abc} = r^{-1}\epsilon_{abc}$. The $\tilde{e}_a$ satisfy the commutation relations:

$$[\tilde{e}_a, \tilde{e}_b] = C^c{}_{ab}\, \tilde{e}_c. \tag{2.1.23}$$

Only two of them are independent at each point of S^2. They satisfy the relation $\tilde{x}^a \tilde{e}_a f = 0$ for $f \in \mathcal{C}(S^2)$. This is an expression of the fact that the 2-sphere is not parallelizable and that the module $\mathcal{X}(S^2)$ is not a free module.

Let $d\tilde{x}^a$ be the exterior derivatives of the coordinates of $\mathbb{R}^3$. Restricted to S^2 they satisfy the relation $\tilde{x}_a d\tilde{x}^a = 0$. The module $\Omega^1(S^2)$ is also not a free module. We shall use the projection π to lift the $d\tilde{x}^a$ to 1-forms $\pi^*(d\tilde{x}^a)$ on SU_2. Dual to the basis $\tilde{e}_a$ are the components $\tilde{\theta}^a$ of the Maurer-Cartan form on SU_2. If we identify $d\tilde{x}^a$ with its image in SU_2 then we have the relations

$$d\tilde{x}^a = C^a{}_{bc}\tilde{x}^b \tilde{\theta}^c, \qquad \tilde{\theta}^a = -C^a{}_{bc}\tilde{x}^b d\tilde{x}^c - iA\tilde{x}^a. \tag{2.1.24}$$

The 1-form A is the potential of the *Dirac monopole* of unit magnetic charge. Although it is not in $\Omega^1(S^2)$, its exterior derivative

$$F = dA = \frac{i}{2r^2}C_{abc}\tilde{x}^a\tilde{\theta}^b \wedge \tilde{\theta}^c = \frac{i}{2r^2}C_{abc}\tilde{x}^a d\tilde{x}^b \wedge d\tilde{x}^c \tag{2.1.25}$$

can be considered as an element of $\Omega^2(S^2)$. We see then that when lifted to SU_2 the $d\tilde{x}^a$ can be expressed in terms of globally defined 1-forms. As a manifold, SU_2 is parallelizable and the module $\Omega^1(SU_2)$ is a free module. □

Example 2.2 Let $\mathbb{T}^2$ be the 2-dimensional torus, which we identify with the points $(\tilde{x}, \tilde{y})$ in the plane such that $0 \leq \tilde{x}, \tilde{y} \leq 2\pi r$. Consider the functions

$$\tilde{u} = e^{i\tilde{x}/r}, \qquad \tilde{v} = e^{i\tilde{y}/r}$$

defined on $\mathbb{T}^2$. Let $\mathcal{P}$ be the algebra of polynomials in $\tilde{u}$ and $\tilde{v}$. This algebra is dense in the algebra $\mathcal{C}(\mathbb{T}^2)$ of smooth complex-valued functions on $\mathbb{T}^2$. It is obviously commutative:

$$\tilde{u}\tilde{v} = \tilde{v}\tilde{u}.$$

Consider the derivations $\tilde{e}_1$ and $\tilde{e}_2$ of $\mathcal{C}(\mathbb{T}^2)$ defined by the partial derivatives with respect to $\tilde{x}$ and $\tilde{y}$:

$$\tilde{e}_1\tilde{f} = \frac{\partial \tilde{f}}{\partial \tilde{x}}, \qquad \tilde{e}_2\tilde{f} = \frac{\partial \tilde{f}}{\partial \tilde{y}}.$$

Then

$$\begin{aligned} \tilde{e}_1\tilde{u} = ir^{-1}\tilde{u}, \quad & \tilde{e}_1\tilde{v} = 0, \\ \tilde{e}_2\tilde{u} = 0, \quad & \tilde{e}_2\tilde{v} = ir^{-1}\tilde{v}. \end{aligned} \qquad (2.1.26)$$

The torus is parallelizable and the $\tilde{e}_\alpha$ form a moving frame. That is, $\mathcal{X}(\mathbb{T}^2)$ is a free $\mathcal{P}$-module of rank 2. Since $\tilde{e}_1$ and $\tilde{e}_2$ commute the structure functions vanish. The de Rham forms dual to these derivations are given by

$$\tilde{\theta}^1 = -ir\tilde{u}^{-1}d\tilde{u}, \qquad \tilde{\theta}^2 = -ir\tilde{v}^{-1}d\tilde{v}. \qquad (2.1.27)$$

They are closed forms. If one use them as a moving frame to define a metric (as described in the following section) one finds that the corresponding curvature vanishes. The flat torus cannot be embedded in $\mathbb{R}^3$ but it does have an embedding in $\mathbb{R}^4$. It is possible to define $\mathcal{P}$ as an abstract algebra with two commuting generators but without topology it is not possible to speak of a closure. One can define two derivations by Formulae (2.1.26). The action on arbitrary elements of the algebra is defined by the Leibniz rule. The most general derivation lies in the module generated by $\tilde{e}_\alpha$.

The algebra $\mathcal{P}$ is infinite dimensional. If we consider not the entire torus but only, for example, a finite regular lattice of points then $\tilde{u}$ and $\tilde{v}$ will satisfy relations $\tilde{u}^n = 1$, $\tilde{v}^n = 1$ for some integer n and the dimension of the resulting algebra $\mathcal{P}_n$ will be equal to n^2, the number of points on the lattice. It is easy to see that $\mathcal{P}_n$ has no derivations. Suppose in fact X to be one. Then the equalities

$$0 = X(1) = X(\tilde{u}^n) = n\tilde{u}^{n-1}X\tilde{u}$$

imply that $X\tilde{u} = 0$. A similar argument leads easily to the conclusion that the algebra of functions on any finite set has no derivations. We shall see however later in Section 6.1 that any algebra has an extension to a graded differential algebra with a natural graded derivation.   □

Example 2.3 The *Poincaré half-plane* or *Lobachevsky half-plane* or *pseudosphere* is defined to be the upper half of the real plane $\mathbb{R}^2$ endowed with a metric of constant negative curvature. It is a complete, parallelizable manifold. Let $\tilde{u}$ and $\tilde{v} > 0$ be the coordinates and introduce the derivations defined by

$$\begin{aligned} \tilde{e}_1\tilde{u} = \tilde{v}, \quad & \tilde{e}_1\tilde{v} = 0, \\ \tilde{e}_2\tilde{u} = 0, \quad & \tilde{e}_2\tilde{v} = -\tilde{v}. \end{aligned} \qquad (2.1.28)$$

These derivations form a (solvable) Lie algebra with commutation relation

$$[\tilde{e}_1, \tilde{e}_2] = \tilde{e}_1.$$

If one compares (2.1.26) with (2.1.28) one sees that in the latter a length scale has been set equal to one. The dual frame is given by

$$\tilde{\theta}^1 = \tilde{v}^{-1}d\tilde{u}, \qquad \tilde{\theta}^2 = -\tilde{v}^{-1}d\tilde{v}.$$

The $\tilde{\theta}^1$ is not a closed form: $d\tilde{\theta}^1 = -\tilde{\theta}^1\tilde{\theta}^2$. It follows from the general formulae of the following section that the Gaussian curvature is equal to -1. The Poincaré half-plane cannot be embedded in $\mathbb{R}^3$ but it can be immersed in $\mathbb{R}^5$.   □

Example 2.4 If the dimension of V is even then V can possess a *symplectic form*, a non-degenerate 2-form ω which is closed:

$$d\omega = 0.$$

We can use ω to define a map $f \mapsto X_f$ of $\mathcal{C}(V)$ into $\mathcal{X}(V)$ by the formula

$$\omega(X_f, Y) = -Yf, \qquad Y \in \mathcal{X}(V). \tag{2.1.29}$$

This can be equivalently written as $i_{X_f}\omega = -2df$ from which it follows that $L_{X_f}\omega = 0$. Conversely if $L_X\omega = 0$ then, at least locally, $X = X_f$ for some $f \in \mathcal{C}$. The *Poisson bracket* is related to ω by the equation

$$\omega(X_f, X_g) = \{f, g\}. \tag{2.1.30}$$

It follows that $X_f g = \{f, g\}$ and the hamiltonian flow is given by the differential equation

$$\frac{dg}{dt} = \{f, g\}.$$

This is Hamilton's equation with f as hamiltonian.   □

Example 2.5 Let V be a 3-dimensional manifold which is the product $V = \mathbb{R} \times \Sigma$ of the real line and a compact Riemann surface Σ. Let t be the coordinate of $\mathbb{R}$. Let $A \in \Omega^1(V)$ and set $F = dA$. Consider A as a classical field with action S given by

$$S = \int_V AdA. \tag{2.1.31}$$

The integral is the integral of a 3-form over a 3-surface. The action is obviously invariant under the change of field variable $A \mapsto A + df$, provided that F tends to zero for large values of $|t|$. Variation of S yields the classical field equations $F = 0$. Using a change of field variable it is possible to set the t-component A^0 of A equal to to zero. In general however A itself cannot be set to zero because of the non-trivial topology of Σ. Let g be the genus of Σ and let the closed curves c_i and c_i', for $1 \leq i \leq g$ be the basis of the first homology group of Σ. It is easy to see that there is a finite number of independent dynamical variables given by the $2g$ functions

$$\phi_i = \int_{c_i} A, \qquad \phi_i' = \int_{c_i'} A,$$

which satisfy the equations

$$\frac{d\phi_i}{dt} = 0, \qquad \frac{d\phi_i'}{dt} = 0.$$

The classical dynamics of this field are quite trivial but nevertheless the action has the interesting property of having been defined without the use of a metric. The only property which distinguishes the t direction is the topological triviality. This is the simplest example of what has come to be known as a topological field theory. In more elaborate versions of the theory which involve a 1-form A with values in the Lie algebra of a compact group there is an additional term on the right-hand side of (2.1.31). If V is the boundary of a 4-dimensional manifold W then by Stokes' theorem (2.3.5) S can be written in the form

$$S = \int_W F^2. \tag{2.1.32}$$

□

A commutative algebra need not of course be an algebra of smooth functions on a manifold. Another important example is the $*$-algebra $\mathcal{A} = C^0(V)$ of all continuous complex-valued functions on an arbitrary topological space V. The algebra $C^0(V)$ has many important subalgebras which we shall define and discuss below. It is not true in general that a commutative $*$-algebra over the complex numbers can be considered as the algebra of all continuous functions on some topological space.

Consider a second space V' and a map ϕ from V' to V. Then one can define a map ϕ^* from $\mathcal{A}$ to $\mathcal{A}' = C^0(V')$ by Equation (2.1.8). There are two limit cases of particular interest. The map ϕ is a projection of V' onto V if every element of V is the image of an element of V'. In this case ϕ^* defines an injection of $\mathcal{A}'$ into $\mathcal{A}$. In fact $\phi^* f = \phi^* g$ means that $f(\phi(x')) = g(\phi(x'))$ which in turn implies that $f = g$. The algebra $\mathcal{A}'$ can be considered as a subalgebra of $\mathcal{A}$. The map ϕ is an injection of V' into V if $\phi(x) = \phi(x')$ implies that $x = x'$. In this case ϕ^* is a projection of $\mathcal{A}$ onto $\mathcal{A}'$. In fact suppose f' is an element of $\mathcal{A}'$ and define $f \in \mathcal{A}$ by $f(x) = f'(x')$ if x is in the image of ϕ and $f(x)$ arbitrary if not. Then obviously $f' = \phi^* f$. One can define the subalgebra $\mathcal{I}$ of functions which vanish on $\phi(V')$. It is clear that if f vanishes on $\phi(V')$ and g is an arbitrary element of $\mathcal{A}$ then $fg = gf$ vanishes on

$\phi(V')$. The subalgebra $\mathcal{I}$ is stable under multiplication by an arbitrary element of $\mathcal{A}$; it is an *ideal* of $\mathcal{A}$. The set $\mathcal{A}/\mathcal{I}$ of equivalence classes of elements of $\mathcal{A}$ modulo elements of $\mathcal{I}$ is thus an algebra, the *quotient algebra*. It can be obviously identified with the algebra of all continuous functions on V'. That is, $\mathcal{A}' = \mathcal{A}/\mathcal{I}$.

The algebra $C^0(V \cup V')$ of all continuous functions on the union $V \cup V'$ of V with some other space V' can be identified with the *direct sum* $C^0(V) \oplus C^0(V')$ of $C^0(V)$ and the algebra $C^0(V')$. Since a set V can always be considered as the union of its points and since the algebra of functions over a single point is the algebra of complex numbers we can write in particular

$$C^0(V) = \bigoplus_1^n \mathbb{C}$$

in the case when V consists of a finite set of n points with the *discrete topology*.

The algebra $C^0(V \times V')$ of continuous functions on the space $V \times V'$ can be identified with the *tensor product* $C^0(V) \otimes C^0(V')$ of the two algebras $C^0(V)$ and $C^0(V')$:

$$C^0(V \times V') = C^0(V) \otimes C^0(V').$$

This means that every continuous function on $V \times V'$ can be considered as a limit of a sequence of functions which are finite linear combinations of products of a continuous function on V and a continuous function on V'.

In studying differential geometry one is especially interested in algebras with derivations. A *derivation* of an algebra $\mathcal{A}$ is a linear map X of $\mathcal{A}$ into itself which satisfies the *Leibniz rule*: $X(fg) = (Xf)g + fXg$. If X is a derivation then so is fX for any element of $\mathcal{A}$; the space $\text{Der}(\mathcal{A})$ of derivations of $\mathcal{A}$ is an $\mathcal{A}$-*module*. This property makes essential use of the fact that $\mathcal{A}$ is commutative. Algebras of functions on finite sets, algebras of continuous functions and algebras of measurable functions are examples of important algebras which have no derivations. Suppose the algebra has an involution. For an arbitrary derivation X we define

$$X^\dagger f = (Xf^*)^*. \tag{2.1.33}$$

We shall impose in general a *reality condition* $X^\dagger = X$ on the derivations. It is to be noticed that $X^\dagger$ is not an adjoint of an operator X. It is defined here uniquely in terms of the involution of $\mathcal{A}$ whereas X acts on this algebra as a vector space.

Let $\mathcal{P}$ be the algebra of polynomials on a finite set of n elements x^i with a commutative product, $x^i x^j = x^j x^i$ but otherwise no relations. Consider the ideal $\mathcal{I} \subset \mathcal{P}$ generated by a finite set of polynomials $R_p(x^i)$ with $1 \le p \le n - m$. The quotient algebra $\mathcal{A} = \mathcal{P}/\mathcal{I}$ is defined by a set of *generators* x^i and a set of *relations*

$$R_p(x^i) = 0.$$

One can think of the x^i as the coordinates of the euclidean space $\mathbb{C}^n$ and the algebra $\mathcal{A}$ as the algebra of all polynomials on the subset $V \subset \mathbb{C}^n$ on which all elements of $\mathcal{I}$ vanish. Generically the dimension of V will be equal to m. The exact correspondence between sets of points V defined by polynomial relations and ideals $\mathcal{I}$ of $\mathcal{P}$ is the subject of algebraic geometry. The construction of an algebra by a set of generators and relations is useful even if $x^i x^j \neq x^j x^i$. There are n derivations ∂_i of the algebra $\mathcal{P}$ defined on the generators by the equations $\partial_i x^j = \delta_i^j$ and extended to the algebra by the Leibniz rule. The set $\mathrm{Der}(\mathcal{P})$ of all derivations of $\mathcal{P}$ is a free $\mathcal{P}$-module with basis ∂_i; every derivation X can be uniquely written in the form $X = X^i \partial_i$ with $X^i \in \mathcal{P}$. A derivation X which leaves $\mathcal{I}$ invariant can be identified with an element of $\mathrm{Der}(\mathcal{A})$. In general $\mathrm{Der}(\mathcal{A})$ is not free; if it is a *projective module* there is a second module $\mathcal{N}$ such that

$$\mathrm{Der}(\mathcal{A}) \oplus \mathcal{N} = \bigoplus_1^n \mathcal{P}.$$

If the ideal $\mathcal{I}$ is generated by a set of smooth functions then generically V will be a smooth complex manifold. If V is compact then $\mathcal{A}$ is dense in the algebra $\mathcal{C}(V)$. If one considers algebras over the real numbers instead of the complex numbers one obtains a smooth real manifold.

Some basic definitions of the algebraic theory of commutative *rings* can be briefly summarized since we shall have little occasion to use them. A *maximal ideal* of a ring is such that the only ideals which contain it are itself and the entire ring. The intersection of all maximal ideals is known as the (Jacobson) *radical* $\mathrm{Rad}(\mathcal{A})$ of $\mathcal{A}$. A *prime ideal* is such that if the product of two elements belongs to it then at least one of the two factors belongs also. A maximal ideal is always prime. An element is *nilpotent* if raised to some power it vanishes. An algebra is a ring which is also a vector space over some field.

Example 2.6 For each point $x \in V$ one can consider the ideal $\mathcal{I}_x$ of $\mathcal{C}^0(V)$ of all functions which vanish at x. It is a maximal ideal. The quotient algebra $\mathcal{A}/\mathcal{I}_x = \mathbb{C}$ can be identified with the algebra of functions on the point x. In general not every maximal ideal is of the form $\mathcal{I}_x$ for some point $x \in V$. Those which do not, correspond to 'points at infinity'. Adding them to V is a sort of *compactification* (Stone-Čech compactification). The description of a space V in terms of a set of maximal ideals of $\mathcal{C}^0(V)$ is one of the basic tools of algebraic geometry; it has however no satisfactory generalization to noncommutative algebras. $\square$

Example 2.7 Consider the algebra of smooth functions defined in some open neighbourhood of the origin of $\mathbb{R}^n$ and define an equivalence relation by the condition that $f \equiv g$ if f and g are equal on some open sub-neighbourhood. Then the set of equivalence classes is a commutative algebra $\mathcal{A}$, called the algebra of *germs* of functions at the origin. Let $[f]$ be the equivalence class of f and let $\mathcal{I}_0$ be the ideal generated by [0]. Then $\mathcal{I}_0$ is the unique maximal ideal and $\mathcal{A}/\mathcal{I}_0 = \mathbb{C}$. Although there is only one maximal ideal and therefore, according to a definition given above, only one point, the algebra is of infinite dimension. It also has an infinite-dimensional vector space of derivations. In fact any derivation X of the algebra of smooth functions defines a derivation of $\mathcal{A}$ by the formula $X[f] = [Xf]$. Because of the lack of any relation in examples like this one between the number of maximal ideals and the dimension of the algebra, one also in algebraic geometry defines points using prime ideals. One can show that $\mathcal{A}$ has an infinite number of prime ideals.   □

Example 2.8 For any fixed integer n consider the subring T_n of the ring of $n \times n$ matrices consisting of upper-triangular matrices which have equal values along each diagonal. A typical element $f \in T_3$ for example would be then of the form

$$ f = \begin{pmatrix} a & b & c \\ 0 & a & b \\ 0 & 0 & a \end{pmatrix}. $$

It is easy to see that T_n is a commutative ring. The ideal $\mathcal{R}$ consisting of those elements with zero along the diagonal is the unique prime ideal. It is nilpotent with $\mathcal{R}^n = 0$ and so $\mathrm{Rad}(\mathcal{A}) = \mathcal{R}$. In the language of algebraic geometry T_n describes a space with but one point. The value of f at the point is equal to a. The fact that the number of points is independent of n is related to the fact that there are nilpotent elements.   □

Example 2.9 Return to the algebra $\mathcal{P}$ of polynomials in n commuting elements x^i and consider the ideal $\mathcal{I}$ generated by the n polynomials $(x^i)^2$. Then the quotient algebra $\mathcal{A} = \mathcal{P}/\mathcal{I}$ is generated as a vector space by the elements

$$ 1, \quad x^i, \quad x^i x^j, \, i < j, \quad \cdots \quad x^1 x^2 \cdots x^{n-1} x^n. $$

It is a vector space of dimension 2^n but there is only one prime ideal and therefore only one point.   □

2.2   Metrics and connections

A *metric* g can be defined as a $\mathcal{C}(V)$-linear, nondegenerate, symmetric map

$$ \Omega^1(V) \otimes_{\mathcal{C}(V)} \Omega^1(V) \xrightarrow{\ g\ } \mathcal{C}(V). $$

Let $g^{\alpha\beta}$ be the components of the standard euclidean (or lorentzian) metric in $\mathbb{R}^n$. If V is parallelizable one can define a metric on V by the condition that the moving frame θ^α be orthonormal:

$$ g(\theta^\alpha \otimes \theta^\beta) = g^{\alpha\beta}. $$

Using the moving frame one can introduce a *linear connection* $\omega^\alpha{}_\beta$, a 1-form on V with values in the Lie algebra of SO_n. It satisfies the *structure equations*:

$$\Theta^\alpha = d\theta^\alpha + \omega^\alpha{}_\beta \wedge \theta^\beta, \tag{2.2.1}$$

$$\Omega^\alpha{}_\beta = d\omega^\alpha{}_\beta + \omega^\alpha{}_\gamma \wedge \omega^\gamma{}_\beta. \tag{2.2.2}$$

The *torsion form* Θ^α and the *curvature form* $\Omega^\alpha{}_\beta$ satisfy the *Bianchi identities*

$$d\Theta^\alpha + \omega^\alpha{}_\beta \wedge \Theta^\beta = \Omega^\alpha{}_\beta \wedge \theta^\beta, \tag{2.2.3}$$

$$d\Omega^\alpha{}_\beta + \omega^\alpha{}_\gamma \wedge \Omega^\gamma{}_\beta - \Omega^\alpha{}_\gamma \wedge \omega^\gamma{}_\beta = 0. \tag{2.2.4}$$

We shall suppose that the torsion form vanishes,

$$\Theta^\alpha = 0,$$

in which case there is a unique solution to the resulting equation (2.2.1) and the $\omega^\alpha{}_\beta$ can be expressed in terms of the structure functions:

$$\omega^\alpha{}_\beta = -\frac{1}{2}(C^\alpha{}_{\beta\gamma} - C_\gamma{}^\alpha{}_\beta + C_{\beta\gamma}{}^\alpha)\theta^\gamma.$$

This is the relation between Equation (2.1.7) and Equation (2.2.1).

Using the metric a map of $\Omega^1(V)$ onto $\mathcal{X}(V)$ can be defined. If $\xi = \xi_\alpha\theta^\alpha \in \Omega^1(V)$ then $\xi \mapsto X$ where $X = g^{\alpha\beta}\xi_\alpha e_\beta$. The map is equivalent to raising the index on ξ_α. The inverse map of $\mathcal{X}(V)$ onto $\Omega^1(V)$, using the inverse $g_{\alpha\beta}$ of the matrix $g^{\alpha\beta}$, is equivalent to lowering the index on X^α. The metric is often expressed using local coordinates in terms of a *line element* $ds^2 = g_{\mu\nu}dx^\mu dx^\nu$. If one writes $\theta^\alpha = \theta^\alpha_\mu dx^\mu$ then

$$g_{\mu\nu} = g_{\alpha\beta}\theta^\alpha_\mu\theta^\beta_\nu.$$

Using a metric one can identify the cotangent bundle with the tangent bundle:

$$T^*V \simeq TV.$$

We shall also have occasion to use the *unit-ball bundle* BV, defined to be all covectors (or vectors) over V of length less than or equal to 1 with respect to some metric, and its boundary the *unit-sphere bundle* SV.

If the manifold V is not parallelizable then a metric can still be defined using the frame bundle. The formula is the same except for the fact that now a moving frame need not be globally defined. If there is a metric then an important subbundle of $E(V)$ is the bundle $O(V)$ of *orthonormal frames*. It is a principal SO_n-bundle. If a metric has been given on V then one can choose a moving frame e_α to be orthonormal. Conversely, given an arbitrary moving frame one can define a metric by the condition that the frame be orthonormal. The bundle $E(V)$ can always be reduced to $O(V)$ and each reduction is equivalent to a metric on V.

Using the bundle $O(V)$ (or $E(V)$) the linear connection can be generalized to the case where the manifold V is not necessarily parallelizable. The construction is identical except that one needs to consider the possibility that the bundle $O(V)$ have no sections and therefore that the (orthonormal) moving frame and the 1-forms $\omega^\alpha{}_\beta$ cannot be globally defined on V. This is done by defining them on $O(V)$ instead. Let X be tangent to $O(V)$ at some frame e_α and write $\pi_*(X) = X^\alpha e_\alpha$ where π is the projection of $O(V)$ onto V. Then the *canonical form* θ^α with values in $\mathbb{R}^n$ is defined at the frame e_α by the equation

$$\theta^\alpha(X) = X^\alpha. \tag{2.2.5}$$

If V is parallelizable this definition of θ^α is equivalent to the one given above in (2.1.6). The manifold $O(V)$ itself is always parallelizable. The canonical form and any linear connection constitute a global basis of $\Omega^1(O(V))$.

In general the forms on $O(V)$ are related to forms on V using local sections. Let σ (σ') be a local section over a region $\mathcal{O}$ ($\mathcal{O}'$). Then the canonical 1-form θ^α on $O(V)$ defines a local moving frame $\sigma^*\theta^\alpha$ ($\sigma'^*\theta^\alpha$) which we shall more conveniently write simply also as θ^α (θ'^α). In the intersection $\mathcal{O} \cap \mathcal{O}'$ of two regions the transformations (2.1.15) take the form

$$\theta^\alpha = \Lambda^\alpha_\beta \theta'^\beta,$$

where the Λ^α_β is a function from $\mathcal{O} \cap \mathcal{O}'$ into SO_n equivalent to (2.1.16).

A (torsion-free) *linear connection* is also now a 1-form $\omega^\alpha{}_\beta$ on $O(V)$ which satisfies the Bianchi identities but considered as equations on $O(V)$. Each local section of $O(V)$ defines a set of 1-forms on V which we shall designate by the same symbol and which we shall also call a linear connection. Using the same notation as for the canonical form it follows directly from Equation (2.2.1) that

$$\omega'^\alpha{}_\beta = \Lambda^{-1\alpha}_\gamma \omega^\gamma{}_\delta \Lambda^\delta_\beta + \Lambda^{-1\alpha}_\gamma d\Lambda^\gamma_\beta, \tag{2.2.6}$$

by again identifying $\omega^\alpha{}_\beta$ ($\omega'^\alpha{}_\beta$) with its image $\sigma^*\omega^\alpha{}_\beta$ ($\sigma'^*\omega^\alpha{}_\beta$).

Let e_α be a (local, orthonormal) moving frame and let X be a vector field on V, which we expand using e_α as $X = X^\alpha e_\alpha$. Then the *covariant derivative* DX of X can be defined by the local expression $DX = (DX^\alpha) \otimes e_\alpha$ where

$$DX^\alpha = dX^\alpha + \omega^\alpha{}_\beta X^\beta.$$

The transformation properties of the connection under a change of local section assure that D is a well defined map from $\mathcal{X}(V)$ to $\Omega^1(V) \otimes_{\mathcal{C}(V)} \mathcal{X}(V)$. The divergence of X is defined as $\mathrm{div}(X) = D_\alpha X^\alpha$; it is a map of $\mathcal{X}(V)$ into $\mathcal{C}(V)$. It is not however a 1-form since it is not $\mathcal{C}(V)$-linear.

The covariant derivative $D\xi$ of an element $\xi \in \Omega^1(V)$ can be similarly defined. If $\xi = \xi_\alpha \theta^\alpha$ then it is given by the local expression $D\xi = (D\xi_\alpha) \otimes \theta^\alpha$ where

$$D\xi_\alpha = d\xi_\alpha - \omega^\beta{}_\alpha \xi_\beta.$$

Again the transformation properties of $\omega^\alpha{}_\beta$ assure that D is a well defined map from $\Omega^1(V)$ into $\Omega^1(V) \otimes_{C(V)} \Omega^1(V)$.

A covariant derivative can also be defined as a linear map

$$\Omega^1(V) \xrightarrow{D} \Omega^1(V) \otimes_{C(V)} \Omega^1(V)$$

which satisfies

$$D(f\xi) = df \otimes \xi + f D\xi.$$

It is given by

$$D\theta^\alpha = -\omega^\alpha{}_\beta \otimes \theta^\beta$$

where $\omega^\alpha{}_\beta$ is the associated linear connection. There is an immediate extension of D to $\Omega^*(V) \otimes_{C(V)} \Omega^1(V)$ by requiring that it be a graded derivation of degree 1. If $\alpha \in \Omega^p(V)$ the covariant derivative of $\alpha \otimes \theta^\alpha$ is given by

$$D(\alpha \otimes \theta^\alpha) = d\alpha \otimes \theta^\alpha + (-1)^p \alpha \wedge D\theta^\alpha.$$

It follows that the curvature can be expressed as

$$D^2\theta^\alpha = -\Omega^\alpha{}_\beta \otimes \theta^\beta.$$

Let π be the projection of $\Omega^1(V) \otimes_{C(V)} \Omega^1(V)$ onto $\Omega^2(V)$ defined by the exterior product. Then the torsion form is given by

$$\Theta^\alpha = (d - \pi \circ D)\theta^\alpha.$$

Let σ be the action on $\Omega^1(V) \otimes_{C(V)} \Omega^1(V)$ defined by the *ordinary flip*:

$$\sigma(\theta^\alpha \otimes \theta^\beta) = \theta^\beta \otimes \theta^\alpha.$$

Obviously the equation

$$\pi \circ (1 + \sigma) = 0$$

is identically satisfied. It can be considered as the definition of π.

Let g be a metric. Introduce the notation $\sigma_{12} = \sigma \otimes 1$ and $g_{23} = 1 \otimes g$, a notation which we shall often use. There is a natural extension of D to the tensor product $\Omega^1(V) \otimes_{C(V)} \Omega^1(V)$ given by

$$D_2(\theta^\alpha \otimes \theta^\beta) = D\theta^\alpha \otimes \theta^\beta + \sigma_{12}(\theta^\alpha \otimes D\theta^\beta).$$

Both $g_{23} \circ D_2(\theta^\alpha \otimes \theta^\beta)$ and $dg(\theta^\alpha \otimes \theta^\beta)$ are elements of $\Omega^1(V)$. A linear connection is *metric-compatible* if they are equal:

$$g_{23} \circ D_2(\theta^\alpha \otimes \theta^\beta) = dg(\theta^\alpha \otimes \theta^\beta) = dg^{\alpha\beta} = 0. \qquad (2.2.7)$$

This is so written for later convenience.

A vector field X is a *Killing vector field* if $L_X g = 0$. Let ξ be the associated element of $\Omega^1(V)$ obtained using the metric to lower the index. Then this condition can be equivalently written using σ as

$$(\sigma + 1) \circ D\xi = 0.$$

Using the metric one can also define the *Hodge duality* map $*$ of $\Omega^p(V)$ into $\Omega^{n-p}(V)$. For each $p \leq n$, introduce the $(n-p)$-forms

$$*\theta_{\alpha_1 \dots \alpha_p} = \frac{1}{(n-p)!} \epsilon_{\alpha_1 \dots \alpha_p \alpha_{p+1} \dots \alpha_n} \theta^{\alpha_{p+1}} \wedge \dots \wedge \theta^{\alpha_n}.$$

Then $*$ is defined by the map

$$\theta^{\alpha_1} \wedge \dots \wedge \theta^{\alpha_p} \mapsto *\theta^{\alpha_1 \dots \alpha_p} \qquad (2.2.8)$$

where the indices on the right-hand side have been raised with $g^{\alpha\beta}$. If $\alpha \in \Omega^p(V)$ then

$$* * \alpha = (-1)^{(n+1)p} \alpha.$$

Using $*$ one can define a map δ of $\Omega^p(V)$ into $\Omega^{p-1}(V)$. The image $\delta\alpha$ of $\alpha \in \Omega^p(V)$ is given by

$$\delta\alpha = (-1)^{np+n+1} * d * \alpha. \qquad (2.2.9)$$

The map δ is not a (graded) derivation. However from (2.1.5) we have $\delta^2 = 0$.

Using d and δ one can define the *Laplace operator* or laplacian Δ,

$$\Delta = d\delta + \delta d, \qquad (2.2.10)$$

which maps $\Omega^p(V)$ into itself. This can be written also in the form

$$\Delta = (d + \delta)^2.$$

The map $d + \delta$ from $\Omega^*(V)$ into itself is called the *Kähler-Dirac operator*. It takes $\Omega^{*\pm}(V)$ into $\Omega^{*\mp}(V)$. A *harmonic form* α satisfies the equation $\Delta\alpha = 0$.

There is a natural inner product on the algebra of forms. Let α and β be two elements of $\Omega^*(V)$. Then one can define (α, β) to be given by

$$(\alpha, \beta) = \int \alpha \wedge *\beta \qquad (2.2.11)$$

if α and β have the same grading and to be zero otherwise. With respect to this inner product the operator δ is the adjoint of d:

$$(\delta\alpha, \beta) = (\alpha, d\beta).$$

The *Clifford algebra* associated to the vector space $\mathbb{R}^n$ is the complex algebra generated by n self-adjoint elements γ^α which satisfy the relations

$$\gamma^\alpha \gamma^\beta + \gamma^\beta \gamma^\alpha = 2g^{\alpha\beta}.$$

On the right we have suppressed the identity element of the algebra. The γ^α can be represented by $N \times N$ Dirac matrices with $N = 2^{[n/2]}$. If n is even the Clifford algebra is the algebra M_N; otherwise it can be identified with two copies of M_N. The spinor representation of the group SO_n is defined by the map $\Lambda \mapsto S(\Lambda)$ of the element $\Lambda \in SO_n$ to the $N \times N$ matrix $S(\Lambda)$ defined by the equations

$$\Lambda^\alpha_\beta \gamma^\beta = S^{-1}(\Lambda)\gamma^\beta S(\Lambda), \qquad \det S(\Lambda) = 1. \qquad (2.2.12)$$

This is a double-valued map since if $S(\Lambda)$ is a solution then so is $-S(\Lambda)$. If Λ is near the identity in SO_n, that is $\Lambda^\alpha_\beta \simeq \delta^\alpha_\beta + \lambda^\alpha{}_\beta$, then there is a unique solution to Equations (2.2.12) given by

$$S(\Lambda) \simeq 1 + \frac{1}{4}\lambda_{\alpha\beta}\gamma^\alpha\gamma^\beta.$$

The set of all $S(\Lambda)$ is the compact Lie group Spin(n). The map $S(\Lambda) \mapsto \Lambda$ defines an epimorphism

$$\text{Spin}(n) \to SO_n \qquad (2.2.13)$$

of Spin(n) onto SO_n which covers it twice. For $n \geq 3$, Spin(n) is simply connected. We saw in Example 2.1 that Spin(3) $= SU_2$.

Using the (orthonormal) moving frame e_α we can identify the tangent space at each point of V with $\mathbb{R}^n$ and so at each point we have a representation $S(\Lambda)$ of SO_n. A spin structure on V is a Spin(n)-bundle over V with a projection onto the orthonormal frame bundle given at each point by (2.2.13). The construction of a spin structure on a manifold involves therefore lifting an SO_n structure and this imposes certain topological conditions. A necessary but not sufficient condition is that V be orientable. If a spin structure does exist it is not necessarily unique.

A section ψ of an associated Spin(n)-bundle H with fibre $\mathbb{C}^N$ is a *Dirac spinor*. Let $\mathcal{H}$ be the space of smooth sections of H. If V is parallelizable then ψ can be considered as a function with values in $\mathbb{C}^N$, that is, an element of the $\mathcal{C}(V)$-module $\mathcal{H} = \mathcal{C}(V) \otimes_\mathbb{R} \mathbb{C}^N$. The covariant derivative $D\psi$ of ψ is given as

$$D\psi = d\psi + \frac{1}{4}\omega_{\alpha\beta}\gamma^\alpha\gamma^\beta\psi. \qquad (2.2.14)$$

The covariant derivative of the adjoint ψ^* of ψ is defined by $D\psi^* = (D\psi)^*$. The covariant derivative of a section of the tensor product of 2 bundles is defined using the Leibniz rule. A Dirac matrix can be considered as an element of $\mathcal{X}(V) \otimes_{C(V)} \mathcal{H} \otimes_{C(V)} \mathcal{H}^*$. The subscript $C(V)$ in the *tensor product* $\mathcal{H} \otimes_{C(V)} \mathcal{H}^*$ indicates that $\psi f \otimes \psi^*$ and $\psi \otimes f\psi^*$ are to be identified. The covariant derivative of γ^α is given therefore by

$$D\gamma^\alpha = \omega^\alpha{}_\beta \gamma^\beta + \frac{1}{4}\omega_{\beta\delta}[\gamma^\beta\gamma^\delta, \gamma^\alpha] \equiv 0.$$

Suppose that n is an even integer. Consider the differential operator $D\psi$ defined in (2.2.14), which we write in a local moving frame as $D\psi = D_\alpha \psi \theta^\alpha$. Then the *Dirac operator* $\slashed{D}$ is defined by replacing θ^α by $i\gamma^\alpha$:

$$\slashed{D}\psi = i\gamma^\alpha D_\alpha \psi. \tag{2.2.15}$$

It is a self-adjoint, first-order, differential operator acting on $\mathcal{H}$. Consider the element $\epsilon = i^{n/2}\gamma^1 \cdots \gamma^n$ of the Clifford algebra. The Dirac matrices can be chosen so that ϵ takes the form

$$\epsilon = \begin{pmatrix} 1 & 0 \\ 0 & -1 \end{pmatrix}. \tag{2.2.16}$$

The unit here is the identity matrix in dimension $n/2$. A general Spin(n)-bundle H can be written as the direct sum $H = H^+ \oplus H^-$ corresponding to the eigenvalues of ϵ. The space $\mathcal{H}$ can be accordingly written also as a direct sum

$$\mathcal{H} = \mathcal{H}^+ \oplus \mathcal{H}^- \tag{2.2.17}$$

of spinors with positive and negative *helicity*. The Dirac operator then takes the form

$$\slashed{D} = \begin{pmatrix} 0 & D^- \\ D^+ & 0 \end{pmatrix}, \tag{2.2.18}$$

with $D^- = (D^+)^*$. For $\psi \in \mathcal{H}$ we have $\slashed{D}\psi = D^+\psi^+ + D^-\psi^-$ and $D^\pm\psi^\pm \in \mathcal{H}^\mp$. The Dirac operator anticommutes with ϵ. The algebra of operators on $\mathcal{H}$ has a $\mathbb{Z}_2$ grading and $\slashed{D}$ is an odd element. It changes the helicity of a spinor. Each component $\psi^\pm$ of the Dirac spinor ψ is called a *Weyl spinor*. The Dirac spinor is real if and only if each Weyl spinor is *conjugate* to its partner.

Example 2.10 Let $\mathcal{H}$ be the module of sections of a general bundle of spinors over V and let $f \in C(V)$. Then we have $\slashed{D}(f\psi) = (ie_\alpha f)\gamma^\alpha\psi + f\slashed{D}\psi$ where e_α is a local moving frame. By left multiplication we have an embedding $f \mapsto \hat{f}$ of $C(V)$ into the *even operators* on $\mathcal{H}$ and we can write

$$\widehat{e_\alpha f}\gamma^\alpha = -i[\slashed{D}, \hat{f}]. \tag{2.2.19}$$

Suppose the dimension of V is even. Then the exterior algebra and the Clifford algebra have the same dimension and are isomorphic as graded vector spaces; we can map the latter onto the former by the replacement $\gamma^\alpha \mapsto \theta^\alpha$. We can identify then $e_\alpha f \gamma^\alpha$ with df. Define $\hat{d}$ by $\hat{d}\hat{f} = \widehat{e_\alpha f \gamma^\alpha}$ and rewrite (2.2.19) as

$$\hat{d}\hat{f} = -i[\not{D}, \hat{f}]. \tag{2.2.20}$$

The embedding of $\mathcal{C}(V)$ can be thus extended to an embedding of $\Omega^1(V)$ into the *odd operators* on $\mathcal{H}$ and thereby to an embedding of $\Omega^*(V)$ which respects the $\mathbb{Z}_2$ grading. If the commutator is taken to be graded we can use (2.2.20) to extend $\hat{d}$ to all operators on $\mathcal{H}$. In particular we have

$$\hat{d}^2 \hat{f} = -[\not{D}^2, \hat{f}] \tag{2.2.21}$$

and so $\hat{d}^2 \neq 0$. The Clifford algebra and the exterior algebra have the same dimension but they are not isomorphic as algebras. The construction of a differential follows closely the construction of d_η given later in Section 3.2 and will be given also in Section 6.1 in a more general context.   □

Let H be the trivial $\mathbb{C}^N$-bundle. Since an element of $\mathcal{H}$ is equivalent to N complex functions the ordinary Laplace operator Δ can also be defined on it. One sees after a short calculation that the difference $\not{D}^2 - \Delta$ is a multiplicative term which depends only on the curvature of the linear connection. The Dirac operator can be considered therefore as a square root of the Laplace operator. This is how it was introduced initially by Dirac. The Laplace operator can also be defined on a general spinor bundle and is again equal to the square of the Dirac operator to within terms which vanish with the curvature. We saw above that another square root of the Laplace operator is the Kähler-Dirac operator $d + \delta$.

With the use of the right action of G one can define quite generally a *gauge connection* or a *Yang-Mills connection* ω on an arbitrary principal bundle P. To simplify we shall suppose that G is a matrix group. Let X be a vector field on P. The right action of an element g of the structure group G defines a new vector field $R_{g*}X$ on P by (2.1.9). Let $\underline{g}$ be the Lie algebra of G. Every element $Y \in \underline{g}$ induces a vector field on P which we shall designate also Y. These vector fields satisfy the condition $R_{g*}Y = g^{-1}Yg$. The right-hand side is the adjoint action of the group on its Lie algebra. Let ω be a 1-form on P with values in $\underline{g}$. If the two conditions

$$\omega(Y) = Y, \qquad \omega(R_{g*}X) = g^{-1}\omega(X)g \tag{2.2.22}$$

are satisfied for every vector field X on P then ω is a connection on P. These conditions can be written equivalently, using (2.1.10), as

$$i_Y\omega = Y, \qquad L_Y\omega = [\omega, Y].$$

On the left of both of these equations, Y is considered as a vector field on P and on the right as an element of $\underline{g}$.

If P is the trivial bundle $P = V \times G$ the conditions (2.2.22) imply that ω is uniquely determined by a *gauge potential* or a *Yang-Mills potential*, a 1-form A on V with values in $\underline{g}$. Let (x, h) be coordinates of P and let $\sigma(x) = (x, 1)$ be a global section. We define $A = \sigma^* \omega$. Using the identity $(\pi^* A)(x, 1) = A(x)$ we can write ω as

$$\omega = h^{-1} A h + h^{-1} dh,$$

We have here written the Maurer-Cartan form $\theta = h^{-1} dh$ in terms of the differential dh of the element h of G. With the identifications we have made, $\theta(Y) = Y$ for arbitrary $Y \in \underline{g}$. This is to be compared with (2.2.5).

In general, using a local section σ of P over a region $\mathcal{O}$ of V, the form ω defines a local gauge potential $A = \sigma^* \omega$. Let σ' be a local section over a second region $\mathcal{O}'$ and define $A' = \sigma'^* \omega$. In the intersection $\mathcal{O} \cap \mathcal{O}'$, it follows from (2.2.22) that A and A' are related by the *gauge transformation*

$$A' = g^{-1} A g + g^{-1} dg. \tag{2.2.23}$$

Here g is a function from $\mathcal{O} \cap \mathcal{O}'$ into G which defines P over $\mathcal{O} \cup \mathcal{O}'$. A change of basis is also called a gauge transformation. The ambiguity in the gauge potential A can be removed by imposing a *gauge condition*. This determines, at least locally, a section of P. Formula (2.2.23) is to be compared with (2.2.6).

There is no notion of torsion in general but a *curvature form* Ω can be defined by the equation

$$\Omega = d\omega + \omega^2 \tag{2.2.24}$$

analogous to (2.2.2). It satisfies also the *Bianchi identities*

$$d\Omega + \omega \wedge \Omega - \Omega \wedge \omega = 0.$$

The curvature form is a 2-form on P but it follows from the conditions (2.2.22) that it can be considered as a form on V. Using again the section σ over the region $\mathcal{O}$ and the coordinates (x, h) in the region $\pi^{-1}(\mathcal{O})$, we can write Ω in terms of the *gauge field strength* or *Yang-Mills field strength* $F = \sigma^* \Omega$ as

$$\Omega = h^{-1} F h.$$

It follows that F is related to $F' = \sigma'^* \Omega$ by

$$F' = g^{-1} F g.$$

The field strength is related to the gauge potential by the formula

$$F = dA + A^2. \tag{2.2.25}$$

Since there is no risk of ambiguity we shall identify ω and Ω with their local images $\sigma^*\omega$ and $\sigma^*\Omega$ and reserve the symbols A and F as well as the designations gauge potential and field strength for applications to gauge theory.

In terms of a moving frame on V we can write

$$F = \frac{1}{2}F_{\alpha\beta}\theta^\alpha \wedge \theta^\beta, \qquad D\psi = D_\alpha\psi\theta^\alpha.$$

The functions $\mathrm{Tr}(F_{\alpha\beta}F^{\alpha\beta})$ and $(D_\alpha\psi)^*D^\alpha\psi$ are *gauge-invariant*. Their values remain unchanged under a gauge transformation. It is not possible to construct a gauge-invariant expression which is quadratic in the potential A. Comparing with Example 2.14 below we conclude that a gauge potential describes a massless field. This is fine in the case of the U_1-potential, which can be identified with the electromagnetic field, but can be a problem for the U_r-potentials for $r \geq 2$. A way to give a mass to a gauge field without breaking gauge invariance will be explained in Chapter 8.

Let H be an hermitian vector bundle associated to P and let $\mathcal{H}$ be the space of sections. The *covariant derivative $D\psi$* of an element $\psi \in \mathcal{H}$ is given locally in $\mathcal{O}$ by the formula

$$D\psi = d\psi + A\psi. \tag{2.2.26}$$

The product $A\psi$ is defined by the action of the structure group on the fibre. It is also a tensor product since A is a 1-form. Let ψ and ψ' be two elements of $\mathcal{H}$. Then they are necessarily connected over $\mathcal{O} \cap \mathcal{O}'$ by a gauge transformation

$$\psi \mapsto \psi' = g^{-1}\psi.$$

The covariant derivative of ψ' is given in $\mathcal{O}'$ by $D'\psi' = d\psi' + A'\psi'$. From (2.2.23) we find that

$$D'\psi' = g^{-1}D\psi. \tag{2.2.27}$$

We see then that D is a well defined map of $\mathcal{H}$ into $\Omega^1(V) \otimes_{C(V)} \mathcal{H}$.

A covariant derivative can also be defined as a linear map

$$\mathcal{H} \xrightarrow{D} \Omega^1(V) \otimes_{C(V)} \mathcal{H} \tag{2.2.28}$$

which satisfies the condition

$$D(f\psi) = df \otimes \psi + fD\psi. \tag{2.2.29}$$

It follows that the difference between two covariant derivatives is an algebra morphism of $\mathcal{H}$. We assume also that D respects the hermitian structure of $\mathcal{H}$:

$$d(\psi_1^*\psi_2) = (D\psi_1)^*\psi_2 + \psi_1^*D\psi_2.$$

Let P be the trivial U_1-bundle $V \times U_1$ and let $\mathcal{U}_1$ be the global sections of P. We can identify $\mathcal{U}_1$ with the unitary elements of $\mathcal{C}(V)$. With $f = g^{-1}$, $g \in \mathcal{U}_1$, Equations (2.2.27) and (2.2.29) are the same equation written differently.

There is an immediate extension of D to $\Omega^*(V) \otimes_{\mathcal{C}(V)} \mathcal{H}$ by requiring that it be a graded derivation of degree 1. Let $\alpha \in \Omega^p(V)$. The covariant derivative of $\alpha \otimes \psi$ is given by

$$D(\alpha \otimes \psi) = d\alpha \otimes \psi + (-1)^p \alpha \wedge D\psi.$$

The field strength (2.2.25) can be defined therefore directly in terms of the covariant derivative. If $\psi \in \mathcal{H}$ we have

$$D^2\psi = F\psi. \tag{2.2.30}$$

The definition of the covariant derivative on an arbitrary left (or right) module is conceptually simpler than it is on the bimodule $\Omega^1(V)$ because the two factors on the right-hand side of (2.2.28) are distinct.

The field strength itself is a section of a vector bundle associated to P. The fibre is the Lie algebra and the action of the group is the adjoint action. The covariant derivative DE of an element E of the corresponding space of sections is given by

$$DE = dE + [A, E].$$

Using this covariant derivative the *Bianchi identities* can be written as

$$DF = 0.$$

Let P be a $(\mathrm{Spin}(n) \times U_r)$-bundle over V. A section ψ of the associated line bundle with fibre $\mathbb{C}^N \otimes \mathbb{C}^r$ is a spinor with *internal symmetry*. The covariant derivative of ψ is obtained using the Leibniz rule and Equations (2.2.14) and (2.2.26):

$$D\psi = d\psi + A\psi + \frac{1}{4}\omega_{\alpha\beta}\gamma^\alpha\gamma^\beta\psi. \tag{2.2.31}$$

In particular then the Dirac operator can be defined on the space of smooth sections of any $(\mathrm{Spin}(n) \times U_r)$-bundle over V.

Let ψ be a section of an arbitrary vector bundle H over V associated to a principal bundle with structure group G and let $x(t)$ be a curve in V. The *parallel transport* of $\psi(x(0))$ to the point $x(t)$ is the unique solution to the differential equation

$$\frac{dx^\mu}{dt}D_\mu\psi = 0$$

with initial value $\psi(0) = \psi(x(0))$. The value of $\psi(t)$ depends on the curve. In particular if the curve is closed with, for example, $x(1) = x(0)$ then in general $\psi(x(1)) \neq \psi(x(0))$. But, of course, $\psi(x(1)) = g\psi(x(0))$ for some $g \in G$. The set of g such that this is true for some closed curve is a closed subgroup of G called the *holonomy group*. It is easy to see that the structure group can always be reduced to the holonomy group of any connection.

Example 2.11 Let P be the trivial U_1-bundle $V \times U_1$ and let $\mathcal{U}_1$ be the global sections of P. We can identify $\mathcal{U}_1$ with the unitary elements of $\mathcal{A} = \mathcal{C}(V) \otimes \mathbb{C}$. Let H be the trivial line bundle $V \times \mathbb{C}$. Then $\mathcal{H}$ is equal to $\mathcal{A}$ considered as a vector space. A connection ω on P can be identified with an anti-hermitian element of $\Omega^1(\mathcal{A}) = \Omega^1(V) \otimes \mathbb{C}$. More generally let P be the trivial U_r-bundle $V \times U_r$ and let $\mathcal{U}_r$ be the global sections of P. We can identify $\mathcal{U}_r$ with the unitary elements of $\mathcal{A} = \mathcal{C}(V) \otimes M_r$, where M_r is the algebra of $n \times n$ complex matrices. Let H be the trivial vector bundle $V \times \mathbb{C}^r$. Then $\mathcal{H}$ is the free $\mathcal{C}(V)$-module of rank r and a projective $\mathcal{A}$-module. It is also a $\mathcal{U}_r$-module. A gauge potential is an anti-hermitian element of $\Omega^1(\mathcal{A}) = \Omega^1(V) \otimes M_r$.   □

Example 2.12 The group SU_2 is a U_1-bundle over S^2 and the projection of the bundle onto its base space is called the *Hopf fibration*. Using the notation of Example 2.1 we see that

$$A = \frac{i}{r^2} \tilde{x}_a \tilde{\theta}^a \qquad\qquad (2.2.32)$$

is a connection with curvature $F = dA$. Every other connection A' on SU_2 is obtained from (2.2.32) by the addition of a 1-form which is the image by π^* of a 1-form on S^2. A direct calculation, using (2.1.25) shows that

$$\frac{1}{2\pi i} \int_{S^2} F' = \frac{1}{2\pi i} \int_{S^2} F = 1,$$

where $F' = dA'$ is the curvature of A'.

Since the frame bundle of SU_2 is parallelizable a connection on it can be identified as a set of 1-forms on SU_2. A torsion-free connection is given by

$$\tilde{\omega}^a{}_b = -\frac{1}{2} C^a{}_{bc} \tilde{\theta}^c.$$

There are of course others, obtained using different moving frames. From (2.1.24) or the above expression for A we see that the connection on SU_2 can be considered as a component of the moving frame. The corresponding curvature can be thus considered as part of a linear connection on the bundle. We shall see this in a more general context in the next example.   □

Example 2.13 Let P be a principal SU_r-bundle with a projection π onto V. Let $\tilde{\theta}^a - A^a$ be the components of a connection form A on P written in some fixed basis of $\underline{su}_r$ orthonormal with respect to the Killing metric (as introduced in Section 3.1). Let θ^α be a local moving frame on V and lift it to a set of 1-forms $\tilde{\theta}^\alpha = \pi^*(\theta^\alpha)$ on P. Then the set

$$\tilde{\theta}^i = (\tilde{\theta}^\alpha, \tilde{\theta}^a)$$

is a moving frame on P. The index i takes the values (α, a). We can define a metric on P by the requirement that the frame be orthonormal. Let $\omega^\alpha{}_\beta$ be a torsion-free connection form on V. Then a torsion-free connection form $\tilde{\omega}^i{}_j$ on P is given by

$$\tilde{\omega}^\alpha{}_\beta = \omega^\alpha{}_\beta - \frac{1}{2}F_a{}^\alpha{}_\beta\tilde{\theta}^a,$$

$$\tilde{\omega}^\alpha{}_a = -\frac{1}{2}F_a{}^\alpha{}_\beta\tilde{\theta}^\beta, \qquad\qquad (2.2.33)$$

$$\tilde{\omega}^a{}_b = -\frac{1}{2}C^a{}_{bc}\tilde{\theta}^c.$$

The set of local connection forms $\tilde{\omega}^i{}_j$ defined on each coordinate patch of P can be used to construct a connection form on $O(P)$. $\square$

We end this section with a short list of fields which are commonly used in physics. They are of respectively spin 0 (Klein-Gordon), spin 1/2 (Dirac), spin 1 (Yang-Mills and Maxwell) and spin 2 (Einstein). We have skipped the spin 3/2 (Rarita-Schwinger) because it involves technicalities which do not interest us here. A supersymmetric version of gravity would contain a combination of a massless spin-3/2 field and the gravitational field.

Example 2.14 A real *scalar field* is an element $\phi \in C(V)$. If it is a scalar field with an *internal symmetry* group G then it is a section of a G-bundle H. It is a *free massless* scalar field if the equations of motion follow from the classical *Klein-Gordon action*

$$S_0[\phi] = \frac{1}{2}\int D_\alpha\phi D^\alpha\phi dx = \frac{1}{2}\int \phi\Delta\phi dx.$$

Here dx is the invariant volume element of V with respect to some metric. The derivative is the covariant derivative with respect to a connection on the principal G-bundle to which H is associated. The *Laplace operator* (2.2.10) becomes the gauge-covariant one

$$\Delta = -D_\alpha D^\alpha.$$

If V has a boundary we suppose that ϕ vanishes smoothly on it. The normalization of S is a convention. It has been chosen so that in the non-relativistic limit of a classical point particle the action takes the form of the free kinetic energy. If the field has a *mass* μ then there is an additional term in the action of the form

$$S_\mu[\phi] = \frac{1}{2}\mu^2\int \phi^2 dx. \qquad\qquad (2.2.34)$$

An *interaction* can be introduced by including in the action the integral of an extra function $U(\phi)$ of ϕ of higher degree than quadratic. The resulting field equations are then nonlinear and can usually only be solved by approximation methods. If the theory is to be stable then U must be bounded below. If the theory is to be renormalizable then U cannot be arbitrary; it can for example be at most a polynomial of degree four if V is of dimension four. If there is an *external source* J then the action acquires an additional interaction term of the form

$$S_J[\phi] = \int J\phi dx.$$

Let S be the total action, including the source term:

$$S[\phi, J] = \frac{1}{2} \int \phi(\Delta + \mu^2)\phi dx + \int U(\phi)dx + \int J\phi dx. \qquad (2.2.35)$$

Consider the case of a simple scalar field. The classical field equations are obtained by setting the functional derivative of S with respect to the ϕ equal to zero:

$$\frac{\delta S}{\delta \phi} = 0. \qquad (2.2.36)$$

The *functional derivative* is defined as the derivative on $\mathcal{C}(V)$ considered as an infinite-dimensional manifold:

$$S[\phi(x) + \epsilon(x)] = S[\phi] + \int \frac{\delta S[\phi]}{\delta\phi(x)}\epsilon(x)dx + o(\epsilon^2).$$

A straightforward calculation yields

$$\frac{\delta S}{\delta \phi} = (\Delta + \mu^2)\phi + U' + J, \qquad U' \equiv \frac{dU}{d\phi}$$

and so the classical field equation is the nonlinear *Klein-Gordon equation*:

$$(\Delta + \mu^2)\phi + U' + J = 0. \qquad (2.2.37)$$

From the definition we see also that the field ϕ can be considered as the functional derivative of S with respect to J:

$$\phi = \frac{\delta S}{\delta J}. \qquad (2.2.38)$$

There is a missmatch of physical dimensions in this and similar equations. It should be written

$$\phi(x) = \frac{\delta S}{\delta J(y)}\delta^{(n)}(x - y)$$

and the Dirac distribution has dimensions of mass.   □

Example 2.15 The free *Dirac action* is the gauge-covariant functional

$$S[\psi, \psi^*] = \int \psi^*(\slashed{D} + \mu)\psi dx$$

where $\not{D}$ is the operator defined in Equation (2.2.15) but possibly with a covariant derivative as defined in (2.2.28). If V is without boundary then the Dirac action is real; otherwise one must subtract a surface term. Variation with respect to ψ^* yields the *Dirac equation*

$$(\not{D} + \mu)\psi = 0.$$

□

Example 2.16 The functional derivative of the *Yang-Mills action*

$$S[A] = \frac{1}{2}\text{Tr}\int F \wedge *F = \frac{1}{4}\text{Tr}\int F_{\alpha\beta}F^{\alpha\beta}dx$$

yields the vacuum *Yang-Mills equations*

$$*D * F = 0, \qquad D_\alpha F^{\alpha\beta} = 0.$$

If the gauge group is the abelian group U_1 then the action is called the *Maxwell action*.

□

Example 2.17 If the dimension of the manifold is odd then gauge-invariant quantities can be formed which do not depend on a choice of metric. For example, consider the case of Example 2.5. The action (2.1.31) has a generalization to an action for an arbitrary gauge potential:

$$S_{CS}[A] = \text{Tr}\int_V (AdA + \frac{2}{3}A^3).$$

This action is known as a *Chern-Simons action*. It is gauge-invariant under gauge transformations which tend to the identity for large values of $|t|$. It plays a useful role, either alone or as a supplementary term, in certain problems in surface physics and in 3-dimensional field theories. If V is the boundary of a 4-dimensional manifold W then by Stokes' theorem (2.3.5) S can be written in the form

$$S[A] = \text{Tr}\int_W F^2.$$

□

Example 2.18 One can define the *Einstein tensor* $G_{\alpha\beta}$ by the equation

$$-\frac{1}{2}\Omega^{\beta\gamma} \wedge *\theta_{\alpha\beta\gamma} = G_{\alpha\beta} *\theta^\beta. \qquad (2.2.39)$$

The $(n-2)$-form

$$\sigma_\alpha = -\frac{1}{2}\omega^{\beta\gamma} \wedge *\theta_{\alpha\beta\gamma}$$

can be used when $n = 4$ to define the energy of a gravitational system. Because of the transformation properties of $w^\alpha{}_\beta$, the expression for σ_α is frame dependent and can be made to vanish at any given point. This is a manifestation of the equivalence principle. There is no meaningful definition of local gravitational energy. Let μ_P^2 be the square of the Planck mass, given in terms of Newton's constant G by the formula

$$\mu_P^2 = \frac{1}{16\pi G}. \tag{2.2.40}$$

Suppose that the torsion form vanishes. Then the *Einstein-Hilbert action*

$$S[\theta^\alpha] = \mu_P^2 \int \Omega^{\alpha\beta} \wedge *\theta_{\alpha\beta} \tag{2.2.41}$$

depends on θ^α alone. The *Einstein field equations*

$$G_{\alpha\beta} = 0$$

are obtained by variation of S with respect to θ^α. This follows immediately from the second of the Bianchi identities and the expression (2.2.39) for the Einstein tensor. □

2.3   Cohomology

The embedding ϵ of the real numbers into $\Omega^0(V)$ as the constant functions and the set of differentials d can be written out in the form of a sequence

$$0 \to \mathbb{R} \xrightarrow{\epsilon} \Omega^0(\mathcal{C}(V)) \xrightarrow{d} \Omega^1(\mathcal{C}(V)) \xrightarrow{d} \cdots$$
$$\xrightarrow{d} \Omega^{n-1}(\mathcal{C}(V)) \xrightarrow{d} \Omega^n(\mathcal{C}(V)) \xrightarrow{d} 0. \tag{2.3.1}$$

We have here supposed that $\dim(V) = n$. We shall suppose further that V is compact, oriented and without boundary. If a p-form is in the kernel of d it is a *closed form* or a *p-cocycle*; if it is in the image of d it is an *exact form* or a *p-coboundary*. From the identity (2.1.5) it is clear that every exact form is closed. If we replace V by a sufficiently small open neighbourhood around one of its points then, by the *Poincaré lemma*, every closed form is exact. The kernel of the first differential is equal to the image of ϵ and the kernel of each succeeding one is equal to the image of the predecessor. The sequence is an *exact sequence*. In general this is not the case. The p-th *cohomology group* $H^p(V;\mathbb{R})$ in the sense of de Rham is the set of p-cocycles modulo the p-coboundaries. We shall use the same symbol to designate a closed p-form and the corresponding equivalence class in $H^p(V;\mathbb{R})$. The groups $H^p(V;\mathbb{R})$ are a measure of the non-exactness of the sequence (2.3.1). Under the product induced from the wedge product they form a ring $H^*(V;\mathbb{R})$. Although $\Omega^p(V)$ is an infinite-dimensional vector space the dimension of $H^p(V;\mathbb{R})$ is always finite.

If a form is in the image of δ, the adjoint of d, then it is a *coexact* form. Using the inner product (2.2.11) one obtains the *Hodge decomposition* of $\Omega^*(V)$ into three orthogonal subspaces,

$$\Omega^*(V) = d\Omega^*(V) \oplus \delta\Omega^*(V) \oplus \Omega_\Delta^*(V),$$

where $d\Omega^*(V)$ and $\delta\Omega^*(V)$ are respectively the exact and coexact forms and $\Omega_\Delta^*(V)$ are the harmonic forms. The central result of the *Hodge-de Rham theory* is the fact that the dimension of the space $H^p(V;\mathbb{R})$ is equal to the dimension of $\Omega_\Delta^p(V)$. This fact has no general noncommutative equivalent since there is no general definition of an harmonic form. There is however an analogous statement for the matrix geometries which we shall study in the next chapter.

For each $p \geq 0$ we define the *standard simplex* $[\sigma_0, \cdots, \sigma_p]$ to be the set

$$[\sigma_0, \cdots, \sigma_p] = \{x^i \in \mathbb{R}^{p+1} : x^i \geq 0, \ \sum x^i = 1\}.$$

Any image of the standard simplex in V under a diffeomorphism will be called a simplex. A *simplicial complex* or *chain* is a formal sum of simplices. If we define the negative of a given simplex to be the same simplex with the opposite orientation then the set of all simplicial complexes is a free abelian group $C_p(V)$. For each p-simplex a $(p-1)$-face $[\sigma_0, \cdots, \hat{\sigma}_i, \cdots, \sigma_p]$ is obtained by suppressing the i-th vertex. There is a *boundary operator* ∂ which maps $C_p(V)$ into $C_{p-1}(V)$. The boundary of a p-simplex is defined by the formula

$$\partial[\sigma_0, \cdots, \sigma_p] = \sum_0^p (-1)^i [\sigma_0, \cdots, \hat{\sigma}_i, \cdots, \sigma_p]. \tag{2.3.2}$$

It obviously satisfies the relation

$$\partial^2 = 0. \tag{2.3.3}$$

The maps defined by (2.3.2) can be written out in the form of a sequence

$$0 \xrightarrow{\partial} C_n(V) \xrightarrow{\partial} C_{n-1}(V) \xrightarrow{\partial} \cdots$$
$$\xrightarrow{\partial} C_1(V) \xrightarrow{\partial} C_0(V) \xrightarrow{\partial} 0. \tag{2.3.4}$$

If a p-complex is in the kernel of ∂ it is a *p-cycle*; if it is in the image of ∂ it is a *p-boundary*. From the identity (2.3.3) it is clear that every boundary is a cycle but in general the converse of this is not true. The p-th *homology group* $H_p(V;\mathbb{Z})$ is the set of p-cycles modulo the p-boundaries. The set $H_*(V;\mathbb{Z})$ of all the groups $H_p(V;\mathbb{Z})$ is a measure of the non-exactness of the sequence (2.3.4). From the assumptions we have placed on V it follows that $H_n(V;\mathbb{Z}) = \mathbb{Z}$ and that $H_p(V;\mathbb{Z}) = 0$ for all $p > n$. Homology can also be defined using an arbitrary ring as coefficients. Using chains with real coefficients yields

$$H_*(V;\mathbb{R}) = H_*(V;\mathbb{Z}) \otimes_\mathbb{Z} \mathbb{R}$$

as homology. Information can be lost in this process. For example the first homology group $H_1(SO_3; \mathbb{Z})$ of the rotation group SO_3 is equal to $\mathbb{Z}_2$ whereas the group $H_1(SO_3; \mathbb{R})$ vanishes.

We have already used the integral of an n-form over an n-dimensional manifold. A p-form α can be integrated over a p-chain. In fact if ϕ is the smooth map which takes the standard p-simplex σ into the given simplex then the *integral* of α over $\phi(\sigma)$ is defined to be the ordinary Riemann integral of $*\phi^*(\alpha)$ over σ. The duality map here is within $\Omega^*(\sigma)$ and so $*\phi^*(\alpha)$ is a function on σ. The integral of an n-form is a particular case. Let α be a p-form and σ an arbitrary $(p+1)$-chain. Then *Stokes'* *theorem* states that

$$\int_\sigma d\alpha = \int_{\partial\sigma} \alpha. \qquad (2.3.5)$$

Because of this equality a p-form is also called a p-*cochain* and the differential d is also called a *coboundary operator*.

Let α be a p-form and σ a p-cycle. Then there is a natural map of $\Omega^p(V) \times C_p(V)$ into the real numbers given by

$$(\alpha, \sigma) \mapsto \int_\sigma \alpha.$$

In particular the integral defines a linear map of $\Omega^n(V)$ into the real numbers. Using Stokes' theorem one can show that there is therefore a natural pairing,

$$H^*(V; \mathbb{R}) \times H_*(V; \mathbb{R}) \to \mathbb{R}, \qquad (2.3.6)$$

of the cohomology groups with the homology groups and that the former can be considered as the dual of the latter:

$$H^*(V; \mathbb{R}) = \mathrm{Hom}(H_*(V; \mathbb{R}), \mathbb{R}).$$

A smooth map ϕ from V into V' induces by composition a homomorphism from $H_*(V; \mathbb{R})$ into $H_*(V'; \mathbb{R})$. The map ϕ^* defined in (2.1.8) induces a homomorphism from $H^*(V'; \mathbb{R})$ into $H^*(V; \mathbb{R})$.

Example 2.19 We have defined the dual to the chains in terms of differential forms but this can be done in a more abstract way which makes no explicit use of the differential structure of V. Chains can be defined using homeomorphisms instead of smooth maps and an abstract group of cochains $C^p(V; \mathbb{Z})$ can be defined as homomorphisms of the chains $C_p(V)$ into the integers: $C^p(V; \mathbb{Z}) = \mathrm{Hom}(C_p(V), \mathbb{Z})$. A *cup product* can be defined which is analogous to the exterior product of forms and $C^*(V; \mathbb{Z})$ can be made into a ring. Dual to the boundary operator a coboundary operator can be defined and a cohomology ring $H^*(V; \mathbb{Z})$ can be constructed. Alternatively, the group $C_p(V)$ can be made into a vector space by taking combinations of simplices with arbitrary real coefficients; as such it has a dual vector space $C^p(V;\mathbb{R})$ which can be thought of as a set of abstract differential forms. The de Rham theorem states that the cohomology of $C^*(V;\mathbb{R})$ is isomorphic to the previous one constructed using differential forms.

Designate by (α,σ) the result of the pairing of an element $\alpha \in H^*(V;\mathbb{Z})$ with an element $\sigma \in H_*(V;\mathbb{Z})$. Let $\alpha \cup \beta$ be the cup product of $\alpha \in H^{q-p}(V;\mathbb{Z})$ and $\beta \in H^p(V;\mathbb{Z})$ and let $\sigma \in H_q(V;\mathbb{Z})$. Then $(\alpha\cup\beta,\sigma)$ is an integer. If real coefficients are used then α and β can be identified as forms. In this case $\alpha \cup \beta = \alpha \wedge \beta$ and

$$(\alpha \cup \beta, \sigma) = \int_\sigma \alpha \wedge \beta.$$

One can also define a pairing,

$$H^p(V;\mathbb{Z}) \times H_q(V;\mathbb{Z}) \to H_{q-p}(V;\mathbb{Z}), \tag{2.3.7}$$

of β and σ given by the *cap product* $\alpha \cap \sigma$ which is defined by the equality

$$(\alpha, \beta \cap \sigma) = (\alpha \cup \beta, \sigma).$$

If σ is the generator of $H_n(V;\mathbb{Z})$ then (2.3.7) induces the *Poincaré duality* isomorphism:

$$H^p(V;\mathbb{Z}) \simeq H_{n-p}(V;\mathbb{Z}).$$

In the case $p = q$ Equation (2.3.7) implies (2.3.6).   □

Example 2.20 There are several ways of defining cohomology, one of which we have presented in some detail. Another equivalent construction is called *Čech cohomology*. As an example we shall show how it can be used to classify principal G-bundles over a manifold. Let $\mathcal{O}_i$ be a covering of V as in Section 2.1. A 1-cochain with values in G is a set of functions

$$\mathcal{O}_i \xrightarrow{g_i} G$$

A 2-cochain is a set of functions g_{ij} of the form (2.1.16). A 2-cochain is a cocycle if the consistency conditions (2.1.17) are satisfied. It is a coboundary if every g_{ij} is of the form

$$g_{ij} = g_i g_j{}^{-1}.$$

We see then that to each cocycle there is associated a principal G-bundle over V and that the bundle is trivial if the cocycle is a coboundary. The G-bundles over V are given then by the cohomology group $H^1(V;G)$. This version of Čech cohomology is unusual in that the coefficients lie in a general Lie group which is not necessarily abelian. To define higher-dimensional Čech cohomology groups the group G must be abelian.   □

Example 2.21 An important variation of (co)homology is relative (co)homology. If W is a submanifold of V then there is an obvious inclusion of $C_*(W)$ into $C_*(V)$ which is compatible with the boundary operator. The relative homology is the homology of the quotient $C_*(V)/C_*(W)$. Define $C^*(V, W)$ to be the homomorphism of $C_*(V)$ into $\mathbb{R}$ which vanish on $C_*(W)$. Obviously $C^*(V, W) \subset C^*(V)$. The *relative cohomology* $H^*(V, W)$ is the cohomology of $C^*(V, W)$. As an example we mention the cohomology ring $H^*(BV, SV; \mathbb{Z})$ of the unit-ball bundle relative to the unit-sphere bundle. An important result which we shall use in Section 5.1 is the Thom isomorphism:

$$H^*(V; \mathbb{Z}) \simeq H^{*+n}(BV, SV; \mathbb{Z}). \qquad (2.3.8)$$

$\square$

Example 2.22 Cohomology can also be defined for algebraic systems. Let v be a $\underline{g}$-module and let $\bigwedge^* \underline{g}$ be the exterior algebra of $\underline{g}$. Define

$$C^0(\underline{g}, v) = v, \qquad C^p(\underline{g}, v) = \operatorname{Hom}(\textstyle\bigwedge^p \underline{g}, v).$$

Then Formula (2.1.2) defines a map of C^p into C^{p+1} which is conventionally designated by δ instead of d:

$$C^p(\underline{g}, v) \xrightarrow{\delta} C^{p+1}(\underline{g}, v). \qquad (2.3.9)$$

This sequence of maps can be written out as in (2.3.1):

$$0 \longrightarrow C^0(\underline{g}, v) \xrightarrow{\delta} C^1(\underline{g}, v) \xrightarrow{\delta} \cdots$$
$$\xrightarrow{\delta} C^p(\underline{g}, v) \xrightarrow{\delta} C^{p+1}(\underline{g}, v) \xrightarrow{\delta} \cdots .$$

The cohomology $H^*(\underline{g}; v)$ of the Lie algebra $\underline{g}$ is defined to be the cohomology of this sequence. If G is a compact Lie group with Lie algebra $\underline{g}$ then $H^*(G; \mathbb{R})$ and $H^*(\underline{g}; \mathbb{R})$ are isomorphic. Here $\mathbb{R}$ is the trivial $\underline{g}$-module. We noticed that the set $\mathcal{X}(V)$ of smooth vector fields is a (infinite-dimensional) Lie algebra. However since it uses only smooth forms the de Rham cohomology is not equal to the *Lie-algebra cohomology* of $\mathcal{X}(V)$. The latter is known as Gelfand-Fuchs cohomology and is much more difficult to compute. In Section 6.2 we shall define the corresponding map δ for associative algebras.

Using an element $s \in H^2(\underline{g}; v)$ the vector space $\underline{g} \oplus v$ can be made into a Lie algebra. We define the Lie bracket of $X_1 + Y_1$ and $X_2 + Y_2$ to be

$$[X_1 + Y_1, X_2 + Y_2] = [X_1, X_2] + X_1 Y_2 - X_2 Y_1 + s(X_1, X_2). \qquad (2.3.10)$$

It satisfies the Jacobi identity because $\delta s = 0$. If s is a coboundary, $s = 2\delta t$, then we can rewrite (2.3.10) in the form

$$[X_1 - t(X_1) + Y_1, X_2 - t(X_2) + Y_2] = [X_1, X_2] - t([X_1, X_2]) + X_1 Y_2 - X_2 Y_1.$$

The map $X \mapsto X - t(X)$ permits us to identify this again with (2.3.10) but with $s \equiv 0$. □

Example 2.23 Choose the v of the previous example to be equal to g itself. A cochain s can be used to define a new product $[X, Y]' = [X, Y] + \epsilon s(X, Y)$ in g which is a first-order deformation of the original bracket. If the cochain is a cocycle the Jacobi identity remains valid to first order and we have constructed an infinitesimal deformation of the original Lie algebra. If the cocycle is a coboundary, $s = 2\delta t$, then the deformation is trivial. It is equivalent to a change of basis $X \to X + \epsilon t(X)$. The first-order deformations of a Lie algebra are in one-to-one correspondence with the elements of $H^2(g; g)$. Semi-simple Lie algebras are rigid because they have vanishing 2-cohomology. They have also vanishing 1-cohomology, which implies that all finite-dimensional representations are completely reducible. □

Example 2.24 If the action of a classical field theory has a Lie group G as invariance group then there exists a set of conserved quantities which form a representation of the Lie algebra g of G. Let $X \in g$ and let $\tilde{q}(X)$ be the associated conserved quantity. Then since $\tilde{q}$ is a representation of g we have $[\tilde{q}(X), \tilde{q}(Y)] - \tilde{q}([X, Y]) = 0$. If the theory is quantized then $\tilde{q}$ becomes a map q of g into the algebra $\mathcal{A}$ of quantum observables and $\mathcal{A}$ becomes a g-module. In general the 2-cochain s with values in $\mathbb{C}$ defined by

$$s(X, Y) = \mathrm{Tr}([q(X), q(Y)] - q([X, Y]))$$

does not vanish. We shall suppose that it is finite. It is known as a *Schwinger term*. Since $\mathbb{C}$ is a trivial g-module we have from (2.1.2)

$$-3\delta s(X, Y, Z) = s([X, Y], Z) + s([Y, Z], X) + s([Z, X], Y)$$

and this vanishes because of the Jacobi identities. The Schwinger term defines then a 2-cocycle of g with values in $\mathbb{C}$. If the $q(X)$ are all of trace class then $s(X, Y) = -\mathrm{Tr}(q([X, Y]))$. We have then $s = 2\delta t$ where the 1-cochain t is defined by $t(X) = \mathrm{Tr}(q(X))$. We shall return again to Schwinger terms in Example 6.13. □

Notes

More details of the contents of this chapter are to be found in the books by Kobayashi & Nomizu (1969), Greub *et al.* (1976), Choquet-Bruhat & DeWitt-Morette (1989) and Bredon (1993). Differential geometry was first described in the more algebraic language which is necessary for noncommutative geometry by Koszul (1960). More details on the structure of fibre bundles can be found in Husemoller (1974). For a review of topological field theory we refer to Birmingham *et al.* (1991). As an introduction to the theory of commutative algebras and algebraic geometry we cite the textbook by Eisenbud (1995).

Section 2.2 contains a short review of gauge theory in the language of fibre bundles. There have been several detailed studies of the geometric aspects of gauge theories at various levels of mathematical sophistication. We refer, for example, to Eguchi *et al.* (1980), Madore (1981), Nash & Sen (1982), Göckeler & Schücker (1987), Nakahara (1989) or Isham (1999). The three examples of geometries we have given are useful because of their simple noncommutative generalizations, due to the fact that they have constant Gaussian curvature. Noncommutative deformations of the algebra $\mathcal{P}$ of Example 2.2 have been studied by Weyl (1950) and by Rieffel (1990). We shall mention these in Chapter 4. A modification of the algebra of Example 2.1 will be given in Section 7.2. A discussion of the embedding of the Poincaré half-plane is to be found for example in Gamkrelidze (1991) or in Mleko & Sterling (1993). A noncommutative version will be given in Section 4.1. For more details on the Chern-Simons action and its relation to topological field theory and to knot theory we refer to the book by Guadagnini (1993). A more sophisticated mathematical treatment is given in the monograph by Turaev (1994). The content of Example 2.18 is due to Thirring (1979). See also Dubois-Violette & Madore (1987). A description of the peculiarities of the Rarita-Schwinger equation in interaction with the electromagnetic and gravitational field can be found in Madore (1975). For more details on the calculations of the homology and cohomology of Lie algebras we refer to Greub *et al.* (1976).

3   Matrix Geometry

In the introduction we mentioned that one can formulate much of the ordinary differential geometry of a manifold in terms of the algebra of smooth functions defined on it. It is possible to define finite noncommutative geometries by replacing this algebra with the algebra M_n of $n \times n$ complex matrices. Since M_n is of finite dimension all calculations reduce to pure algebra. The Hodge-de Rham theorem, for example, becomes an easy result on the range and the kernel of a symmetric matrix. Matrix geometry is also interesting in being similar to the ordinary geometry of compact Lie groups. It constitutes therefore a transition to the more abstract formalism of general noncommutative geometry which is given in the next chapter. We shall show in Chapter 7 that there is a sense in which the ordinary geometry of the torus and the sphere can be considered as the limit of a sequence of matrix geometries. In this chapter we shall be exclusively interested in matrix algebras. When however a result has nothing to do with the dimension of the algebra and is valid in more general contexts we shall often designate the algebra by $\mathcal{A}$ in order to avoid an unnecessary repetition of formulae in the next chapter. It is worth emphasizing also that, as we shall see in Section 4.3, many interesting algebras with involutions can be obtained as limits of matrix algebras. The present chapter can be read and understood as algebra without any knowledge of geometry but a certain familiarity with this subject is necessary as motivation. We gave in Chapter 2 a brief introduction.

In the first section we shall define vectors and differential forms more or less as this is normally done in geometry. To begin with we shall use the Lie algebra of all derivations, an algebra which corresponds to the Lie algebra of all smooth vector fields on a manifold. We shall then see that there are interesting subalgebras of this algebra which could be considered as the analogues of the set of moving frames. The distinction is not of interest in ordinary geometry because the most general vector field can always be expressed at least locally in terms of a moving frame. When in Chapter 7 we consider a matrix geometry as a finite approximation to an ordinary differential manifold the dimension of the latter will be determined by the choice of subalgebra. An important difference with the commutative case is the fact that the derivations no longer form a module over the algebra. The product of a 'vector' by a 'function' is no longer a 'vector'. This fact is connected with an ambiguity concerning the algebra of derivations which it is appropriate to use. In Section 3.2 we shall introduce a different type of differential form which has the interesting property that it need not rely on derivations for its definition. The existence of several different types of differential forms is a characteristic property of noncommutative geometry. We shall recapitulate what we have learned about 1-forms and 2-forms at the end of Section 3.1 in terms of an arbitrary algebra $\mathcal{A}$ with unit element. The construction will be completed in Section 3.2 and Section 6.1. In Section 3.4 we propose a general

definition of a metric. In Section 3.5 we define and discuss briefly the properties of
a left-module connection, which we refer to as a Yang-Mills connection, using the
terminology of physicists. We could of course equally well have chosen to define a
Yang-Mills connection as a connection on a right module. İn the following section we
discuss in more detail linear connections and in Section 3.7 we discuss the problem of
the definition of the curvature of linear connections. There is no completely general
satisfactory definition in the noncommutative case.

3.1   Differential forms I

In Chapter 2 we recalled some elementary results of the differential geometry of
a manifold V expressed in terms of the algebra $\mathcal{C}(V)$ of smooth functions on V.
We shall now show that to a certain extent matrices furnish finite analogues of the
geometries of compact parallelizable manifolds. We shall need some basic facts about
modules over a noncommutative (associative) algebra $\mathcal{A}$. Consider some subalgebra
$\mathcal{A} \subset \mathcal{L}(\mathcal{H})$ of the algebra of all linear operators on an arbitrary vector space $\mathcal{H}$. In
this chapter $\mathcal{A}$ will always be some subalgebra of a matrix algebra and $\mathcal{H}$ a finite-
dimensional vector space. We shall use the notation f, g etc. for the elements of $\mathcal{A}$
when we wish to emphasize their role as 'functions'. Let ψ be an element of $\mathcal{H}$. The
sum and product of two operators are defined in the obvious way

$$(f + g)\psi = f\psi + g\psi, \qquad (fg)\psi = f(g\psi)$$

and the associative and distributive rules follow from the linearity assumption. The
important difference with functions is of course the fact that in general

$$fg \neq gf.$$

The *center* $\mathcal{Z}(\mathcal{A})$ of a general noncommutative algebra $\mathcal{A}$ is the subalgebra of
elements of $\mathcal{A}$ which commute with all other elements. The algebra is commutative
then if and only if $\mathcal{Z}(\mathcal{A}) = \mathcal{A}$. We shall be most interested in the case where the
center is trivial: $\mathcal{Z}(\mathcal{A}) = \mathbb{C}$.

Suppose that the vector space $\mathcal{H}$ has an inner product and suppose the equality

$$(f^*\phi, \psi) = (\phi, f\psi)$$

defines an *involution* $f \mapsto f^*$. Then f^* is called the *adjoint* of f. As in the
commutative case an algebra with an involution is called a *-algebra*. It is not
true that a general noncommutative *-algebra over the complex numbers can be
considered as the algebra of all linear operators on some vector space with inner
product. An important difference with the commutative case is the fact that whereas
the hermitian elements of a commutative algebra form a subalgebra the hermitian

elements of $\mathcal{L}(\mathcal{H})$ do not. If f and g are hermitian then in general $(fg)^* = g^* f^* = gf \neq fg$. For this reason even when considering noncommutative versions of real manifolds we shall be forced to consider the noncommutative analogues of algebras of complex-valued functions.

Since $\mathcal{A}$ acts on $\mathcal{H}$ from the left we say that $\mathcal{H}$ is a left $\mathcal{A}$-*module* (or $\mathcal{A}$-left-module). If $\mathcal{A}$ were to act on the right we would call $\mathcal{H}$ a right $\mathcal{A}$-module. If it acts from both sides we say that $\mathcal{H}$ is an $\mathcal{A}$-*bimodule*. For example $\mathcal{A}$ itself is a vector space and as an algebra it can be identified as a subalgebra of $\mathcal{L}(\mathcal{A})$. Using the notation introduced in Section 2.2, to each $f \in \mathcal{A}$ we associate the operator $\hat{f} \in \mathcal{L}(\mathcal{A})$ defined by $\hat{f} : g \mapsto fg$ for arbitrary $g \in \mathcal{A}$. As a vector space $\mathcal{A}$ is obviously an $\mathcal{A}$-bimodule. It is possible for a module to be a left $\mathcal{A}$-module and a right $\mathcal{A}'$-module for some other algebra $\mathcal{A}'$. Let now $\mathcal{A}$ be an arbitrary fixed subalgebra of $\mathcal{L}(\mathcal{H})$. A left $\mathcal{A}$-*submodule* of $\mathcal{H}$ is a subspace $\mathcal{H}'$ of $\mathcal{H}$ which is stable under the (left) action of $\mathcal{A}$. Let $\mathcal{I}$ be the subalgebra of $\mathcal{A}$ which annihilates such an $\mathcal{H}'$:

$$\mathcal{I} = \{f \in \mathcal{A} : f\mathcal{H}' = 0\}.$$

Then obviously $\mathcal{I}$ is stable under left multiplication by an arbitrary element of $\mathcal{A}$; it is a left *ideal* of $\mathcal{A}$. It is easy to see also that since $\mathcal{H}'$ is stable under the action of $\mathcal{A}$ then $\mathcal{I}$ is a right ideal; it is thus a 2-sided ideal. In this case one can define, as in the commutative case, the *quotient algebra* $\mathcal{A}/\mathcal{I}$ as the space of equivalence sets of elements of $\mathcal{A}$ modulo elements of $\mathcal{I}$. Suppose that $\mathcal{H}'$ has a complement in $\mathcal{H}$ in the sense that $\mathcal{H}$ can be written as a *direct sum*

$$\mathcal{H} = \mathcal{H}' \oplus \mathcal{H}''$$

of two left $\mathcal{A}$-modules. One can identify then the quotient algebra $\mathcal{A}/\mathcal{I}$ as a subalgebra of $\mathcal{L}(\mathcal{H}'')$.

From a left $\mathcal{A}$-module $\mathcal{H}$ and a right $\mathcal{A}$-module $\mathcal{H}'$, one can form the *tensor product* $\mathcal{H} \otimes_{\mathcal{A}} \mathcal{H}'$. It has in general no interesting module structure. As in the commutative case, the subscript $\mathcal{A}$ means that $\psi f \otimes \psi'$ and $\psi \otimes f \psi'$ are to be identified. The tensor product $\mathcal{H}' \otimes \mathcal{H}$ is both a left $\mathcal{A}$-module and a right $\mathcal{A}$-module. The tensor product $\mathcal{A} = \mathcal{A}' \otimes \mathcal{A}''$ over the complex numbers of two algebras $\mathcal{A}'$ and $\mathcal{A}''$ is an algebra. Here one considers both $\mathcal{A}'$ and $\mathcal{A}''$ as (bi)modules with respect to the algebra of complex numbers. There is, for example, a natural identification $M_m \otimes M_n = M_{mn}$. The *opposite* $\mathcal{A}^{\mathrm{op}}$ of an algebra $\mathcal{A}$ is the algebra with the same structure as $\mathcal{A}$ as vector space but with a new multiplication $\circ$ defined by $f \circ g = gf$. A left $\mathcal{A}$-module can be considered as a right $\mathcal{A}^{\mathrm{op}}$-module. For any algebra $\mathcal{A}$ the *enveloping algebra* $\mathcal{A}^{\mathrm{e}}$ is defined to be the tensor product $\mathcal{A}^{\mathrm{e}} = \mathcal{A} \otimes \mathcal{A}^{\mathrm{op}}$. An $\mathcal{A}$-bimodule can also be considered then as a left $\mathcal{A}^{\mathrm{e}}$-module. If we had defined $\mathcal{A}^{\mathrm{e}} = \mathcal{A}^{\mathrm{op}} \otimes \mathcal{A}$ then an $\mathcal{A}$-bimodule would have been the same as a right $\mathcal{A}^{\mathrm{e}}$-module.

We saw above that the algebra itself is both a left and a right module. A left $\mathcal{H}$-module $\mathcal{H}$ is said to be a *free module* if it can be identified as the direct sum of n

copies of $\mathcal{A}$ for some integer n known as the *rank* of the module. An $\mathcal{A}$-bimodule is free if it can be written as the direct sum of a finite number of copies of the enveloping algebra $\mathcal{A}^e$. A left module $\mathcal{H}'$ is said to be a *projective module* if there exists a second left module $\mathcal{H}''$ such that the direct sum $\mathcal{H} = \mathcal{H}' \oplus \mathcal{H}''$ is free. Similarly of course right modules and bimodules can be free or projective. With rare exceptions all modules which we shall consider will be projective.

Example 3.1 Let $\mathbb{C}^n$ be the n-dimensional euclidean vector space with the standard inner product. Then $\mathcal{L}(\mathbb{C}^n) = M_n$, the algebra of all $n \times n$ complex matrices. The involution takes a matrix into its hermitian adjoint. For some integer $m < n$ we write $\mathbb{C}^n = \mathbb{C}^m \oplus \mathbb{C}^{n-m}$ and decompose the algebra M_n as the direct sum

$$M_n = M_n^+ \oplus M_n^-$$

of the *even operators* M_n^+ which respect the decomposition and the *odd operators* M_n^- which do not. Obviously we can identify

$$M_n^+ = M_m \times M_{n-m} \subset M_n$$

with the subalgebra of block-diagonal matrices. The M_n^- is an M_n^+-bimodule but not an algebra. The vector space $\mathbb{C}^m$ is a left M_n^+-module which is annihilated by the ideal M_m of M_n^+. The quotient algebra can be identified with the algebra M_{n-m}. Obviously $\mathbb{C}^{n-m}$ is a complement of $\mathbb{C}^m$ as a left M_n^+-module.

The vector space $\mathcal{H}_{mn}$ of all $m \times n$ matrices is a left M_m-module and a right M_n-module. In particular M_n itself is an M_n-bimodule and as an algebra M_n can be considered as a subalgebra of $\mathcal{L}(M_n)$. We consider $\mathbb{C}^n$ as a column vector and write the corresponding row vector as $\mathbb{C}^{nT}$. Then we have the direct-sum decompositions

$$\mathcal{H}_{mn} = \bigoplus_1^n \mathbb{C}^m, \qquad \mathcal{H}_{mn} = \bigoplus_1^m \mathbb{C}^{nT}$$

of $\mathcal{H}_{mn}$ in terms of left and right M_n-modules. The matrix algebra M_n has no 2-sided ideals but it has a decomposition

$$M_n = \bigoplus_1^n \mathbb{C}^n, \qquad M_n = \bigoplus_1^n \mathbb{C}^{nT}$$

as a direct sum of n left and right ideals. Since as a left module we can identify

$$M_n = \mathbb{C}^n \oplus \bigoplus_1^{n-1} \mathbb{C}^n$$

we see that $\mathbb{C}^n$ is a projective left M_n-module. Similarly $\mathbb{C}^{nT}$ is a projective right M_n-module and $\mathbb{C}^n \otimes \mathbb{C}^{nT}$ is a projective M_n-bimodule. $\square$

Let X be a *derivation* of the algebra M_n and let $\mathrm{Der}(M_n)$ be the vector space of all derivations of M_n. It is an elementary fact of algebra that a derivation of M_n is necessarily an *inner derivation*. This means that every element X of $\mathrm{Der}(M_n)$ is of the form $X = \mathrm{ad}\, f$ for some f in M_n. We recall that the *adjoint action* is defined by

$$Xg = \mathrm{ad}\, f(g) = [f, g], \qquad g \in M_n.$$

We introduced a reality condition on derivations in Section 2.1. The derivation X is real if and only if f is anti-hermitian. The main difference with the commutative case lies in the fact that $\mathrm{Der}(M_n)$ is not a left M_n-module. If $X \in \mathrm{Der}(M_n)$ and $f \in M_n$ then in general fX is not in $\mathrm{Der}(M_n)$. It is however a free module over the center of M_n, that is, over the complex numbers. It is a vector space of dimension $n^2 - 1$. This is the analogue of the fact that the vector fields on a parallelizable manifold V form a free $\mathcal{C}(V)$-module.

We shall see in Section 6.1 that over an arbitrary algebra, even a commutative algebra, there are many differential calculi in the sense in which this expression is used in noncommutative geometry. The de Rham calculus is not the only calculus over the algebra of smooth functions on a manifold; it is certainly however the most interesting one. As in the preceding chapter we shall be interested in this section in differential calculi based on derivations. We shall first treat the case where the calculus is based on the Lie algebra of all derivations of M_n and then consider two important cases where the derivations form a subalgebra. In the first case the subalgebra is the Lie algebra $\underline{su}_2$, of dimension three and in the second case it is an abelian subalgebra of dimension two.

Let λ_a, for $1 \le a \le n^2 - 1$, be an anti-hermitian basis of the Lie algebra of the special unitary group SU_n. The product $\lambda_a \lambda_b$ can be written in the form

$$\lambda_a \lambda_b = \frac{1}{2} C^c{}_{ab} \lambda_c + \frac{1}{2} D^c{}_{ab} \lambda_c - \frac{1}{n} g_{ab}. \tag{3.1.1}$$

The g_{ab} are the components of the *Killing metric*, a positive, nondegenerate, invariant inner product on the Lie algebra. We have suppressed the unit element of M_n which should appear in the above formula as a factor of g_{ab}. The *structure constants* $C^c{}_{ab}$ are real. The Killing metric can be defined in terms of them by the equation

$$g_{ab} = -\frac{1}{2n} C^c{}_{ad} C^d{}_{bc}.$$

We shall lower and raise indices with g_{ab} and its inverse g^{ab}. The tensor C_{abc} is completely antisymmetric and D_{abc} is completely symmetric and trace-free.

The set of λ_a is a set of generators of M_n. It is not a minimal set but it is convenient because of the fact that the derivations

$$e_a = \mathrm{ad}\, \lambda_a \tag{3.1.2}$$

form a basis over the complex numbers for $\text{Der}(M_n)$. Any element X of $\text{Der}(M_n)$ can be written as a linear combination $X = X^a e_a$ of the e_a where the X^a are complex numbers. The vector space $\text{Der}(M_n)$ has a Lie-algebra structure. In particular the derivations e_a satisfy the commutation relations

$$[e_a, e_b] = C^c{}_{ab} e_c. \tag{3.1.3}$$

We define the *algebra of forms* $\Omega^*(M_n)$ over M_n just as we did in the commutative case. First we define $\Omega^0(M_n)$ to be equal to M_n. Then we use (2.1.3) to define df for $f \in M_n$. We have then

$$df(e_a) = e_a f. \tag{3.1.4}$$

This means in particular that

$$d\lambda^a(e_b) = [\lambda_b, \lambda^a] = -C^a{}_{bc}\lambda^c. \tag{3.1.5}$$

We define the set of 1-forms $\Omega^1(M_n)$ to be the set of all elements of the form $f dg$ with f and g in M_n. So $\Omega^1(M_n)$ is a left M_n-module. We could have also defined $\Omega^1(M_n)$ to be the set of all elements of the form $(dg)f$ which would define $\Omega^1(M_n)$ as a right module. Since M_n is not a commutative algebra one has to distinguish between left and right multiplication. Although $f dg$ and $(dg)f$ are not equal the two definitions of $\Omega^1(M_n)$ coincide as a bimodule because of the relation $d(fg) = f(dg) + (df)g$. A *differential form* of order p or *p-form* is defined exactly as in Section 2.1 and the product is given by (2.1.1). With the *differential d* defined by the Equation (2.1.2), which satisfies again Equations (2.1.4) and (2.1.5), the algebra $\Omega^*(M_n)$ is a graded *differential algebra*. It is easy to see however that if $\alpha, \beta \in \Omega^*(M_n)$ then in general

$$\alpha\beta \neq \pm\beta\alpha.$$

The set of $d\lambda^a$ constitutes a system of generators of $\Omega^1(M_n)$ as a left or right module but it is not a convenient one. The left and right module structures of $\Omega^1(M_n)$ are defined by the equations

$$(\lambda^a d\lambda^b)(e_c) = \lambda^a d\lambda^b(e_c) = \lambda^a[\lambda_c, \lambda^b], \qquad (d\lambda^b)\lambda^a(e_c) = d\lambda^b(e_c)\lambda^a = [\lambda_c, \lambda^b]\lambda^a.$$

Obviously then

$$\lambda^a d\lambda^b \neq d\lambda^b \lambda^a.$$

There is a better system of generators completely characterized by the equations

$$\theta^a(e_b) = \delta^a_b. \tag{3.1.6}$$

We have suppressed the unit matrix which should appear as a factor of δ^a_b on the right-hand side of this equation. The θ^a are related to the $d\lambda^a$ by the equations

$$d\lambda^a = C^a{}_{bc} \lambda^b \theta^c, \tag{3.1.7}$$

and their inverse

$$\theta^a = \lambda_b \lambda^a d\lambda^b. \tag{3.1.8}$$

They are the matrix analogue of the *dual basis* of the 1-forms defined in (2.1.6). Equation (3.1.7) follows immediately from (3.1.5) and (3.1.6) but the proof of the inverse Equation (3.1.8) uses the identities

$$C^a{}_{bc} C^c{}_{de} C^e{}_{fa} = -n C_{bdf},$$

$$C^a{}_{bc} C^c{}_{de} D^e{}_{fa} = -n D_{bdf},$$

$$C^a{}_{bc} D^c{}_{de} D^e{}_{fa} = -\frac{1}{n}(n^2 - 4) C_{bdf}$$

between the structure constants and the tensor D_{abc}.

Because of the relation (3.1.6) we have $f\theta^b = \theta^b f$ for arbitrary $f \in M_n$; the θ^a commute with the elements of M_n. This describes the structure of $\Omega^1(M_n)$ as a right or left module. It follows also from the definition of the product that the θ^a anticommute: $\theta^a \theta^b = -\theta^b \theta^a$. This describes the structure of $\Omega^*(M_n)$ as an algebra. The θ^a satisfy the same structure equations as the components of the Maurer-Cartan form on the special unitary group SU_n:

$$d\theta^a = -\frac{1}{2} C^a{}_{bc}\, \theta^b \theta^c. \tag{3.1.9}$$

The product on the right-hand side of this formula is the product in $\Omega^*(M_n)$. Although this product is not in general antisymmetric, the subalgebra $\bigwedge^* \subset \Omega^*(M_n)$ generated by the θ^a is an exterior algebra. Formula (3.1.9) means that it is a differential subalgebra. We can identify then $\Omega^*(M_n)$ with the tensor product of M_n and $\bigwedge^*$. In particular it follows that as a left or right M_n-module $\Omega^1(M_n)$ is free of rank $(n^2 - 1)$.

From the generators θ^a we can construct a 1-form

$$\theta = -\lambda_a \theta^a \tag{3.1.10}$$

in $\Omega^1(M_n)$. From (3.1.8) we see that θ can be written as

$$\theta = -\frac{1}{n}\lambda_a d\lambda^a = \frac{1}{n} d\lambda_a \lambda^a.$$

Using θ we can rewrite (3.1.8) as

$$\theta^a = C^a{}_{bc} \lambda^b d\lambda^c - n\lambda^a \theta. \tag{3.1.11}$$

Apart from the second term on the right-hand side this equation is related to (3.1.7) by an interchange of $d\lambda^a$ and θ^a. From (3.1.7) and (3.1.9) one sees that θ satisfies the condition

$$d\theta + \theta^2 = 0. \tag{3.1.12}$$

It satisfies with respect to the algebraic exterior derivative the same equation which
the Maurer-Cartan form (2.1.12) satisfies with respect to ordinary exterior derivation
on the group SU_n. Notice however the change in sign between the definition of the
Maurer-Cartan form and the definition (3.1.10) of the present form θ. The former
is a Lie-algebra-valued form on a group. The exterior derivative acts on the θ^α but
not on the basis λ_α whereas the coefficients λ_a in the definition (3.1.10) of θ are
elements of M_n and their exterior derivative is not zero. It follows directly from
the definitions that the exterior derivative df of an element of M_n can be written in
terms of a commutator with θ:

$$df = -[\theta, f]. \tag{3.1.13}$$

This is not true however for an arbitrary element of $\Omega^*(M_n)$. It follows from (3.1.13)
that as a bimodule $\Omega^1(M_n)$ is generated by θ; we shall see below in Example 3.40
that it is a projective bimodule. There is a map of the trace-free elements of M_n
onto the derivations of M_n given by $f \mapsto X_f = \mathrm{ad}\, f$. The 1-form θ can be defined,
without any reference to the θ^a, as the inverse map:

$$\theta(X_f) = -f. \tag{3.1.14}$$

The complete set of all derivations of M_n is the natural analogue of the space
of all smooth vector fields $\mathrm{Der}(\mathcal{C}(V))$ on a manifold V. If V is parallelizable then
$\mathrm{Der}(\mathcal{C}(V))$ is a free module over $\mathcal{C}(V)$ with a set of generators e_α which is closed
under the Lie bracket and which has the property that if $e_\alpha f = 0$ for all e_α then f is
a constant function. Consider the Lie algebra (3.1.3) with the e_a defined by (3.1.2).
We have supposed that the λ_a are elements of the fundamental representation of
$\underline{su}_n$, the Lie algebra of SU_n. We could equally well have assumed that they lie in an
n-dimensional representation of $\underline{su}_m$ for $m < n$. The corresponding set of derivations
$\mathrm{Der}_m(M_n)$ will be a Lie subalgebra of the Lie algebra $\mathrm{Der}(M_n)$ of all the derivations
of M_n and if the representation is irreducible the λ_a will generate M_n as an algebra.
This freedom of representation of the algebra $\mathrm{Der}_2(M_n)$ with $n \to \infty$ will allow us
in Chapter 7 to consider the geometry of the 2-sphere as a limit of the appropriately
modified matrix geometries.

When we speak of a *matrix geometry* we mean a choice of matrix algebra M_n and
a choice of generators of M_n such that the associated derivations form a Lie algebra,
which can be a proper subalgebra of the Lie algebra $\mathrm{Der}(M_n)$ of all derivations. This
extra freedom of choice is connected to the fact that $\mathrm{Der}(M_n)$ is not a module over
M_n. We shall call a basis of the derivations $\mathrm{Der}_m(M_n)$ a *frame* or *Stehbein* and we
shall view it as the matrix equivalent of a moving frame. The smallest value of m is
$m = 2$. The matrices λ_a generate then the irreducible n-dimensional representation
of $\underline{su}_2$. The most general element of M_n is a polynomial in the λ_a. Consider the
derivations $e_a = \mathrm{ad}\, \lambda_a$ with $1 \le a \le 3$. The equations

$$e_a f = 0, \tag{3.1.15}$$

imply that f is proportional to the unit element. The set e_a is not the minimal set with this property. One need choose only e_1 and e_2 for example since the third derivation appears as the commutator of these two. Let e_r for $1 \le r \le n^2 - 1$ be a basis of $\mathrm{Der}(M_n)$ and let e_a for $1 \le a \le 3$ be a basis of $\mathrm{Der}_2(M_n)$. Then for some real $3 \times (n^2 - 1)$ matrix Λ^{-1} we have

$$e_a = \Lambda^{-1r}_{a} e_r.$$

One can order the e_r so that the first 3 elements satisfy the same commutation relations as the e_a. They form then a reducible representation of $\underline{su}_2$.

If a basis J_a of the n-dimensional representation of $\underline{su}_2$ satisfies the commutation relations $[J_a, J_b] = i\epsilon_{abc} J^c$ then the value of the *Casimir operator* $C(J_a) = J_a J^a$ is given by

$$R(J_a) = C(J_a) - \frac{1}{4}(n^2 - 1) = 0.$$

This is the Casimir relation. Because of it the set $\mathrm{Der}_2(M_n)$ could be considered as the natural analogue of the set of moving frames on a 2-dimensional manifold. We introduce 3 matrices

$$x^a = kr^{-1} J^a \tag{3.1.16}$$

such that the Casimir relation becomes $g_{ab} x^a x^b - r^2 = 0$. The parameter k must be therefore related to r by the equation

$$4r^4 = (n^2 - 1)k^2.$$

It has units of (length)^2. The two parameters r and k are related then, for large n, by the condition

$$n = \frac{4\pi r^2}{2\pi k}. \tag{3.1.17}$$

The x^a satisfy the commutation relations

$$[x_a, x_b] = ik C^c_{ab} x_c, \qquad C_{abc} = r^{-1} \epsilon_{abc}. \tag{3.1.18}$$

As a basis for $\mathrm{Der}_2(M_n)$ we chose the derivations

$$e_a = \frac{1}{ik}\mathrm{ad}\, x_a. \tag{3.1.19}$$

This will coincide with the previous expression (3.1.2) if we normalize the λ_a so that

$$x^a = ik\lambda^a.$$

The derivative $e_a x^b$ of x^b is given then as

$$e_a x^b = C^b_{ca} x^c. \tag{3.1.20}$$

The derivative of the remaining elements of M_n, higher-order polynomials in the x^a, is given by the Leibniz rule. The e_a satisfy the commutation relations (2.1.23)

$$[e_a, e_b] = C^c{}_{ab} e_c. \tag{3.1.21}$$

With a restricted Lie algebra of derivations, one can define the differential exactly as before using Equation (3.1.4). However now the set of e_a is a basis of $\mathrm{Der}_m(M_n) \subseteq \mathrm{Der}(M_n)$. The derivations are taken, so to speak, only along the preferred directions. If we choose the complete set $\mathrm{Der}(M_n)$ of derivations then $1 \leq a \leq n^2-1$ and each θ^a is an $(n^2-1) \times n^2$ matrix. It takes the vector space $\mathrm{Der}(M_n)$ into M_n. If we choose for example $\mathrm{Der}_2(M_n)$ as the derivations then $1 \leq a \leq 3$ and each θ^a is a $3 \times n^2$ matrix. It takes the vector space $\mathrm{Der}_2(M_n)$ into M_n. Since M_n is an irreducible representation of $\mathrm{Der}_2(M_n)$ however the θ^a have a unique extension to all of $\mathrm{Der}(M_n)$. Except for the change of structure constants, Equation (3.1.7) remains unchanged but (3.1.8) will in general be modified.

It is obvious also that the θ^a dual to the e_a can be written as a linear combination

$$\theta^a = \Lambda^a_r \theta^r$$

of the θ^r dual to the e_r with coefficients Λ^a_r which can be general elements of M_n. Both sides of the above equation are to be considered as $(n^2 - 1) \times n^2$ matrices, maps of $\mathrm{Der}(M_n)$ into M_n. The Λ is a one-sided inverse,

$$\Lambda^a_r \Lambda^{-1r}{}_b = \delta^a_b,$$

of the Λ^{-1}. Since the θ^a have a unique extension to $\mathrm{Der}(M_n)$ one can in principle calculate the Λ^a_r from the identity

$$\theta^a(e_r) = \Lambda^a_r.$$

Dual to the derivations (3.1.19) are the set of 1-forms

$$\theta^a = -C^a{}_{bc} x^b dx^c - \frac{ik}{r^2} \theta x^a, \qquad \theta = \frac{i}{k} x_a \theta^a = \frac{r^2}{k^2} x_a dx^a. \tag{3.1.22}$$

The similarity between (2.1.24) and (3.1.22) will be exploited in Chapter 7.

The algebra of forms constructed from M_n using $\mathrm{Der}_m(M_n)$ as derivations will be designated $\Omega^*_m(M_n)$ or simply $\Omega^*(M_n)$. Let $\bigwedge^*$ be the exterior algebra generated by the forms θ^a defined using the derivations $\mathrm{Der}_m(M_n)$. Then $\Omega^*_m(M_n)$ can be identified with the tensor product of M_n and $\bigwedge^*$:

$$\Omega^*_m(M_n) = M_n \otimes \bigwedge{}^*.$$

In Chapter 7 we shall discuss a sense in which $\Omega^*_2(M_n)$ tends as $n \to \infty$ to the algebra of differential forms $\Omega^*(S^2)$ on the 2-sphere. We have considered the group

SU_n as acting on the algebra M_n and we have defined differential forms as dual to the Lie algebra of derivations induced by the group action. This construction can be generalized to include the action of any Lie group acting freely on an arbitrary associative algebra. See Example 4.12 below.

The formalism has an immediate extension to the case of a matrix algebra which is the product of two matrix algebras M and M'. The simple algebra M_n possesses certain analogies with the algebra of smooth functions on a completely parallelizable compact manifold. The product algebra $M \times M'$ is to be compared with the algebra of functions on the disjoint union of two manifolds. We have $\text{Der}(M \times M') = \text{Der}(M) \oplus \text{Der}(M')$ and therefore $\Omega^1(M \times M') = \Omega^1(M) \oplus \Omega^1(M')$. Since the product of an element in M and an element of M' vanishes we have also

$$\Omega^*(M \times M') = \Omega^*(M) \times \Omega^*(M').$$

Example 3.2 Since the matrix algebras do not commute, the corresponding geometry resembles a quantized version of phase space. In fact it resembles also the unquantized version. There is a symplectic form ω defined as

$$\omega = -2id\theta.$$

From the Equation (3.1.12) and the relation (3.1.14) it follows that for each $f \in M_n$, Equation (2.1.29) has $X_f = \text{ad } f$ as solution. In particular we have

$$\omega(X_f, X_g) = i[f, g]. \tag{3.1.23}$$

This is to be compared with Equation (2.1.30). □

Example 3.3 The sequence of matrix algebras M_n has the *Heisenberg algebra* $\mathcal{A}$ as an interesting formal limit. The Heisenberg algebra is the formal *-algebra generated by a single element a and its adjoint a^* which satisfy the commutation relation $[a, a^*] = 1$. Let $J_\pm = J_1 \pm iJ_2$, J_3 be the basis of the n-dimensional irreducible representation of the Lie algebra $\underline{sl}_2(\mathbb{C})$ of $SL(2, \mathbb{C})$ with $J_+^* = J_-$, $J_3^* = J_3$ and with the commutation relations $[J_3, J_\pm] = \pm J_\pm$, $[J_+, J_-] = 2J_3$. The $(J_\pm, J_3)$ generate M_n as an associative algebra. Let f_k be the eigenvectors of J_3: $J_3 f_k = k f_k$ for $-s \leq k \leq s$ where s is an integer or half-integer with $n = 2s + 1$. Consider the change of basis

$$a_\pm = \frac{1}{\sqrt{2s}} J_\pm, \qquad a_3 = J_3 + s.$$

The commutation relations become

$$[a_3, a_\pm] = \pm a_\pm, \qquad [a_-, a_+] = 1 - \frac{1}{s} a_3.$$

For integer $k \geq 0$ set

$$|k\rangle = \lim_{s \to \infty} f_{k-s}$$

and define the operators a, a^* by

$$a|k\rangle = \lim_{s\to\infty} a_- f_{k-s}, \quad a^*|k\rangle = \lim_{s\to\infty} a_+ f_{k-s}.$$

Then we have

$$[a, a^*]|k\rangle = \lim_{s\to\infty} [a_-, a_+] f_{k-s} = \lim_{s\to\infty} (1 - \frac{1}{s} a_3) f_{k-s} = |k\rangle.$$

We have also $a|0\rangle = 0$. If we define $N|k\rangle = \lim_{s\to\infty} a_3 f_{k-s}$ then we have $N|k\rangle = k|k\rangle$ and from the Casimir relation

$$R(J_a) = J_+ J_- + J_3(J_3 - 1) - s(s + 1) = 2s(a_+ a_- - a_3) + a_3(a_3 - 1) = 0$$

we can conclude that

$$(N - a^* a)|k\rangle = \lim_{s\to\infty} (a_3 - a_+ a_-) f_{k-s} = \lim_{s\to\infty} \frac{1}{2s} k(k - 1) f_{k-s} = 0.$$

When M_n tends to $\mathcal{A}$, the commutator in M_n tends to the commutator in $\mathcal{A}$. But we saw above that the commutator in M_n is given by the symplectic form $d\theta$. On the other hand we shall see also in Chapter 7 that the form $d\theta$ tends in the commutative limit to a multiple of the field strength of the Dirac monopole, which can be used to define a phase-space structure on S^2. Under quantization, the Poisson bracket with respect to this structure goes over into the commutator in $\mathcal{A}$. Since M_n is a simple algebra it follows that all derivations of M_n are hamiltonian. We shall see in Chapter 7 that they generate in the commutative limit symplectomorphisms of the sphere.   □

We shall now construct an algebra of forms over M_n based on two commutating derivations. Choose an orthonormal basis $|j\rangle_1$, $0 \le j \le n - 1$, of $\mathbb{C}^n$, set $|n\rangle_1 \equiv |0\rangle_1$ and introduce unitary elements u and v of M_n defined by

$$u|j\rangle_1 = q^j |j\rangle_1, \quad v|j\rangle_1 = |j + 1\rangle_1. \tag{3.1.24}$$

One deduces immediately the relations

$$uv = qvu, \quad u^n = 1, \quad v^n = 1, \quad q^n = 1. \tag{3.1.25}$$

There is also an orthonormal basis $|j\rangle_2$ in which the v is diagonal. The two bases are related by the *Fourier transformation*

$$|j\rangle_1 = \frac{1}{\sqrt{n}} \sum_{l=0}^{n-1} q^{+jl} |l\rangle_2, \quad |l\rangle_2 = \frac{1}{\sqrt{n}} \sum_{j=0}^{n-1} q^{-jl} |j\rangle_1.$$

Introduce the hermitian matrices x and y by

$$x|j\rangle_1 = \frac{k}{r} j |j\rangle_1, \quad y|l\rangle_2 = \frac{k}{r} l |l\rangle_2.$$

One finds that

$$u = e^{ix/r}, \qquad v = e^{iy/r}, \qquad q = e^{-i k/r^2} \tag{3.1.26}$$

and the two parameters r and k must be related by the condition

$$n = \frac{(2\pi r)^2}{2\pi k}. \tag{3.1.27}$$

This is the same as (3.1.17) given above, with the area of the sphere replaced by the area of the torus. The matrices x and y or u and v generate the entire matrix algebra.

The commutation relations

$$[x, v] = \frac{k}{r} v(1 - nP_2), \qquad [y, u] = -\frac{k}{r} u(1 - nP_1) \tag{3.1.28}$$

are easily derived. We have here introduced the projectors

$$P_2 = |n-1\rangle_1 \langle n-1|, \qquad P_1 = |0\rangle_2 \langle 0|.$$

As every element of the algebra, they can be expressed as polynomials in the generators:

$$P_2 = \frac{1}{n} \sum_0^{n-1} q^l u^l, \qquad P_1 = \frac{1}{n} \sum_0^{n-1} v^l.$$

It follows that the action of the derivations

$$e_1 = -\frac{1}{ik} \operatorname{ad} y, \qquad e_2 = \frac{1}{ik} \operatorname{ad} x$$

of the algebra is given by

$$e_1 u = i r^{-1} u(1 - nP_1), \quad e_1 v = 0,$$
$$e_2 u = 0, \qquad\qquad e_2 v = i r^{-1} v(1 - nP_2). \tag{3.1.29}$$

They form a Lie algebra with commutation relation

$$[e_1, e_2] = 0.$$

The projector terms on the right-hand side of Equations (3.1.29) are due to the constraints (3.1.25). It is because of these terms that we find $e_1 u^n = 0$ and $e_2 v^n = 0$ as we must. The 1-forms dual to the derivations (3.1.29) are easily seen to be given by

$$\theta^1 = -ir(1 - \frac{n}{n-1} P_1) u^{-1} du, \qquad \theta^2 = -ir(1 - \frac{n}{n-1} P_2) v^{-1} dv. \tag{3.1.30}$$

The similarity between (2.1.27) and (3.1.30) will be exploited in Chapter 7. With the choice of derivations (3.1.29) the matrix algebra M_n can be made to look then like a noncommutative version of the algebra of functions on a torus. The relations (3.1.25) define a finite-dimensional version of what is known as the Weyl algebra.

Example 3.4 The sequence of matrix algebras M_n has a representation of the *Weyl algebra* $\mathcal{A}$ as an interesting limit. The algebra $\mathcal{A}$ is a $*$-algebra generated by two unitary elements u and v which satisfy the commutation relations $uv = qvu$ with q a complex number of unit modulus. If we write q as in (3.1.26) then the condition (3.1.27) is not necessarily satisfied. If we impose it then u^n and v^n commute and $\mathcal{A}$ can be identified with a matrix algebra over a commutative algebra; the matrix entries are functions of u^n and v^n. If one sets $q = e^{2\pi i l/n}$ for some integer $l < n$ then $q^n = 1$ and u and v still generate M_n. As l/n tends to an irrational number then $\mathcal{A}$ tends to an algebra $\mathcal{M}$ of the type described later in Example 4.41.   □

In this section we have used the same matrix algebra to describe a noncommutative version of a sphere and a noncommutative version of a torus, two surfaces which have different topologies. This opens the possibility that at small length scales, of the order of the Planck length, topology is an ill-defined quantity in physics. We shall return to this in Example 7.29.

The construction we have just given does not essentially use the fact that the derivations formed a Lie algebra and it can also be formulated so as to be valid for the more general algebras which we shall introduce in the following chapter. Let n be an arbitrary integer and let λ_i be n linearly independent elements of a noncommutative algebra $\mathcal{A}$. Introduce the derivations e_i of $\mathcal{A}$ by Formula (3.1.2). For an arbitrary element $f \in \mathcal{A}$ then we set $e_i f = [\lambda_i, f]$. In general the e_i do not form a Lie algebra but they do however satisfy commutation relations as a consequence of the commutation relations of $\mathcal{A}$. We shall suppose that if $e_i f = 0$ then $f \in \mathcal{Z}(\mathcal{A})$. Since we shall usually suppose that the center of the algebra we are considering is trivial this condition means that there are 'enough' derivations; it implies that if all the derivatives of a 'function' vanish then the 'function' is a constant. It means also that at least in some formal sense the λ_i generate $\mathcal{A}$.

We repeat some of the formulae given in the case $\mathcal{A} = M_n$ to underline the fact that they have nothing to do with matrices and that they remain valid even when the differential is based on a family of n derivations e_i which do not necessarily close to form a Lie algebra. We define the differential df of an arbitrary element $f \in \mathcal{A}$ by

$$df(e_i) = e_i f \qquad\qquad (3.1.31)$$

The commutation relations of the algebra fix the relations between fdg and dgf for all f and g. The major difference with the commutative case lies in the fact that because of the lack of an interesting module structure for the derivations one can construct a large number of different differential calculi over a given algebra. We shall see this in the examples given in the next chapter. The bimodule of 1-forms $\Omega^1(\mathcal{A})$ is generated by the elements of the form fdg or dgf with f and g in $\mathcal{A}$.

If $\mathcal{A}$ is a $*$-algebra we suppose that the λ_i are anti-hermitian so that the associated derivations e_i satisfy the standard reality condition given in Section 2.1. If we define

$(df)^*$ by

$$(df)^*(X) = (df(X^\dagger))^* = Xf^* = df^*(X)$$

then the *reality condition*

$$(df)^* = df^* \tag{3.1.32}$$

on the differential is also satisfied. From a similar sequence of identities one finds that $(fdg)^* = dg^*f^*$ and for general $f \in \mathcal{A}$ and $\xi \in \Omega^1(\mathcal{A})$ one has

$$(f\xi)^* = \xi^*f^*, \qquad (\xi f)^* = f^*\xi^*.$$

Such a relation need not be satisfied if the differential is not based on real derivations.

As in the matrix case it is often possible to construct a *frame*, a set of 1-forms dual to the derivations e_i, characterized by the equations

$$\theta^i(e_j) = \delta^i_j. \tag{3.1.33}$$

We shall find below necessary conditions on the λ_i for these equations to have a solution. It follows directly from this that

$$\theta^i f = f\theta^i. \tag{3.1.34}$$

Using the frame we can set

$$df = e_i f\theta^i \tag{3.1.35}$$

from which it follows that the module structure of $\Omega^1(\mathcal{A})$ is given by

$$fdg = (fe_ig)\theta^i, \qquad dgf = (e_ig)f\theta^i.$$

If a frame exists the module $\Omega^1(\mathcal{A})$ is free of rank n as a left or right module. It can therefore be identified with the direct sum

$$\Omega^1(\mathcal{A}) = \bigoplus_1^n \mathcal{A} \tag{3.1.36}$$

of n copies of $\mathcal{A}$. In this representation θ^i is given by the element of the direct sum with the unit in the ith position and zero elsewhere. We shall refer to the integer n as the *dimension* of the geometry. Because of (3.1.36) the complete algebra of forms $\Omega^*(\mathcal{A})$ has a simple structure in terms of $\mathcal{A}$ which we mention in Section 6.1.

From the generators θ^i we can construct a 1-form

$$\theta = -\lambda_i\theta^i \tag{3.1.37}$$

in $\Omega^1(\mathcal{A})$ which will play an important role in the definition of connections in Section 3.5. It follows directly from the definitions that the exterior derivative df of an element of $\mathcal{A}$ can be written in terms of a commutator with θ:

$$df = -[\theta, f]. \tag{3.1.38}$$

This formula is to be compared with (2.2.20). It follows from it that as a bimodule $\Omega^1(\mathcal{A})$ is generated by one element. If the algebra is a $*$-algebra then it follows from the definitions that the elements of the frame are hermitian:

$$(\theta^i)^* = \theta^i, \qquad \theta^* = -\theta.$$

In Section 6.1 we shall show how to construct an entire differential algebra consistent with the module structure of the 1-forms.

Let π be the product map

$$\Omega^1(\mathcal{A}) \otimes_{\mathcal{A}} \Omega^1(\mathcal{A}) \xrightarrow{\pi} \Omega^2(\mathcal{A}). \tag{3.1.39}$$

The product of two elements ξ and η in $\Omega^1(\mathcal{A})$ is the element $\xi\eta = \pi(\xi\otimes\eta)$ in $\Omega^2(\mathcal{A})$. It is easy to see that in general $\xi\eta \neq -\eta\xi$. In fact if $f\,dg \neq dg\,f$ then necessarily $df\,dg \neq -dg\,df$. We shall suppose that the map π has a right inverse, a map

$$\Omega^2(\mathcal{A}) \xrightarrow{\iota} \Omega^1(\mathcal{A}) \otimes_{\mathcal{A}} \Omega^1(\mathcal{A}) \tag{3.1.40}$$

such that $\pi \circ \iota = 1$. This will allow us to identify $\Omega^2(\mathcal{A})$ as a submodule of $\Omega^1(\mathcal{A}) \otimes_{\mathcal{A}} \Omega^1(\mathcal{A})$ and consider π as a projection.

If $\mathcal{A}$ is a $*$-algebra the involution on $\Omega^1(\mathcal{A})$ can be uniquely extended to all of $\Omega^*(\mathcal{A})$ by extending the *reality condition* (3.1.32) on the differential. We set

$$(d\xi)^* = d\xi^*.$$

There is therefore in particular a unique involution $\imath_2$ of $\Omega^2(\mathcal{A})$ with

$$(\xi\eta)^* = \imath_2(\xi\eta).$$

We introduce also an involution $\jmath_2$ of $\Omega^1(\mathcal{A}) \otimes_{\mathcal{A}} \Omega^1(\mathcal{A})$ which is compatible with $\imath_2$ in the sense that

$$\pi \circ \jmath_2 = \imath_2 \circ \pi. \tag{3.1.41}$$

We can consistently define then

$$(\xi \otimes \eta)^* = \jmath_2(\xi \otimes \eta).$$

If the differential is based on real derivations then one sees immediately that

$$(df\,dg)^* = -dg^*\,df^*.$$

The product of two arbitrary hermitian 1-forms is not necessarily an hermitian 2-form.

If a frame θ^i exists then the construction can be made more explicit. With the identification (3.1.40) we have

$$\pi(\theta^i \otimes \theta^j) = P^{ij}{}_{kl}\theta^k \otimes \theta^l, \qquad P^{ij}{}_{kl} \in \mathcal{Z}(\mathcal{A}).$$

Since π is a projection we have

$$P^{ij}{}_{mn}P^{mn}{}_{kl} = P^{ij}{}_{kl} \qquad (3.1.42)$$

and the product $\theta^i\theta^j$ satisfies

$$\theta^i\theta^j = P^{ij}{}_{kl}\theta^k\theta^l. \qquad (3.1.43)$$

In some cases which we shall consider the θ^i anticommute. This corresponds to

$$P^{ij}{}_{kl} = \frac{1}{2}(\delta^i_k\delta^j_l - \delta^j_k\delta^i_l). \qquad (3.1.44)$$

Since the exterior derivative of θ^i is a 2-form it can necessarily be written as

$$d\theta^i = -\frac{1}{2}C^i{}_{jk}\theta^j\theta^k.$$

where, because of (3.1.43), the *structure elements* can be chosen to satisfy the constraints

$$C^i{}_{jk}P^{jk}{}_{lm} = C^i{}_{lm}.$$

Consider now Equation (3.1.38). From the identity $d^2 = 0$ one finds that

$$d(\theta f - f\theta) = [d\theta, f] + [\theta, [\theta, f]] = [d\theta + \theta^2, f] = 0.$$

It follows that if we write

$$d\theta + \theta^2 = -\frac{1}{2}K_{ij}\theta^i\theta^j \qquad (3.1.45)$$

we can deduce from (3.1.34) that the coefficients K_{ij} must lie in $\mathcal{Z}(\mathcal{A})$. Again from (3.1.43) they can be chosen to satisfy the constraints

$$K_{jk}P^{jk}{}_{lm} = K_{lm}.$$

Using the explicit form (3.1.37) for θ one finds immediately that (3.1.45) can be written as

$$(2\lambda_k\lambda_j + \lambda_i C^i{}_{jk} + K_{jk})\theta^j\theta^k = 0$$

from which we deduce that

$$2\lambda_m\lambda_l P^{lm}{}_{jk} + \lambda_i C^i{}_{jk} + K_{jk} = 0. \qquad (3.1.46)$$

More information can be obtained about the structure elements from the identity $d(f\theta^i - \theta^i f) = 0$. One deduces immediately that

$$[C^i{}_{jk} + 2\lambda_{(j}\delta^i_{k)}, f]\theta^j\theta^k = 0$$

from which one concludes that there exist elements $F^i{}_{jk} \in \mathcal{Z}(\mathcal{A})$ such that

$$(C^i{}_{jk} + 2\lambda_{(j}\delta^i_{k)} - F^i{}_{jk})\theta^j\theta^k = 0. \qquad (3.1.47)$$

Again because of (3.1.43) we can choose the $F^i{}_{jk}$ to satisfy the constraints

$$F^i{}_{jk}P^{jk}{}_{lm} = F^i{}_{lm}$$

and (3.1.47) can be written in the form

$$C^i{}_{jk} = F^i{}_{jk} - 2\lambda_l P^{(il)}{}_{jk}. \qquad (3.1.48)$$

Equation (3.1.47) can also be written in the form

$$d\theta^i = -[\theta, \theta^i] - \frac{1}{2}F^i{}_{jk}\theta^j\theta^k \qquad (3.1.49)$$

again with a graded commutator. From this we see that if $F^i{}_{jk} = 0$ then the differential d is given as a graded commutator on all elements of the differential calculus and the element θ plays the role of a generalized Dirac operator. Finally, from Equations (3.1.46) and (3.1.48) we deduce the equation

$$2\lambda_k\lambda_l P^{kl}{}_{ij} - \lambda_k F^k{}_{ij} - K_{ij} = 0. \qquad (3.1.50)$$

This is the basic equation from which many of the examples which we use will be derived. We have started from an integer n and a set of λ_i. The necessary conditions for the existence of the basis θ^i in (3.1.33) are expressed in the Equation (3.1.50). If $\Omega^2(\mathcal{A})$ is non-trivial there exists $P^{ij}{}_{kl} \neq 0$ in $\mathcal{Z}(\mathcal{A})$ such that (3.1.42) and (3.1.43) are satisfied. Conversely we could have started from elements $P^{ij}{}_{kl}$, $F^i{}_{jk}$, K_{ij} in $\mathcal{Z}(\mathcal{A})$ and looked for a solution λ_i to the Equation (3.1.50).

Define

$$C^{ij}{}_{kl} = \delta^i_k\delta^j_l - 2P^{ij}{}_{kl}. \qquad (3.1.51)$$

Then from (3.1.42) we find that

$$C^{ij}{}_{kl}C^{kl}{}_{mn} = \delta^i_m\delta^j_n. \qquad (3.1.52)$$

We can write then the first term

$$2\lambda_l\lambda_m P^{lm}{}_{jk} = \lambda_j\lambda_k - \lambda_l\lambda_m C^{lm}{}_{jk} \equiv [\lambda_j, \lambda_k]_C$$

as a sort of deformed bracket and Equation (3.1.50) can be rewritten in the form

$$[\lambda_j, \lambda_k]_C = F^i{}_{jk}\lambda_i + K_{jk}. \qquad (3.1.53)$$

If $P^{ij}{}_{kl}$ is given by (3.1.44) then we have

$$C^{ij}{}_{kl} = \delta^j_k\delta^i_l. \qquad (3.1.54)$$

Equation (3.1.53) defines a 'twisted' Lie algebra with a *central extension* and the $F^i{}_{jk}$ must satisfy a set of modified Jacobi identities.

Example 3.5 If we choose the λ_i so that the e_i are a basis of a Lie algebra of derivations then $P^{ij}{}_{kl}$ is given by (3.1.44), $K_{ij} = 0$ and the $F^i{}_{jk}$ are equal to the structure constants of the Lie algebra. An example with $K_{ij} \neq 0$ can be found. Examples with $F^i{}_{jk} = 0$ and $K_{ij} = 0$ have been also found. If $P^{ij}{}_{kl}$ is given by (3.1.44) with a plus instead of a minus sign, n is even and $F^i{}_{jk} = 0$ then a solution to (3.1.53) is given by Dirac matrices. If also $K_{ij} = 0$ then a solution is given by 'super-coordinates'. In these two cases the 1-forms θ^i commute. $\qquad\square$

If $\mathcal{A}$ is a $*$-algebra there are elements $I^{ij}{}_{kl}, J^{ij}{}_{kl} \in \mathcal{Z}(\mathcal{A})$ such that

$$(\theta^i \theta^j)^* = \imath_2(\theta^i \theta^j) = I^{ij}{}_{kl}\theta^k\theta^l, \qquad (\theta^i \otimes \theta^j)^* = \jmath_2(\theta^i \otimes \theta^j) = J^{ij}{}_{kl}\theta^k \otimes \theta^l.$$

We can suppose that

$$I^{ij}{}_{kl}P^{kl}{}_{mn} = I^{ij}{}_{mn}.$$

We have then

$$(I^{ij}{}_{kl})^* I^{kl}{}_{mn} = P^{ij}{}_{mn}, \qquad (J^{ij}{}_{kl})^* J^{kl}{}_{mn} = \delta^i_m \delta^j_n.$$

The compatibility condition (3.1.41) with the product becomes

$$(P^{ij}{}_{kl})^* J^{kl}{}_{mn} = I^{ij}{}_{kl}P^{kl}{}_{mn} = I^{ij}{}_{mn}.$$

But since the frame is hermitian and associated to derivations we have from (3.1.35)

$$(e_i f e_j g)^* I^{ij}{}_{kl}\theta^k\theta^l = (df\,dg)^* = -dg^*df^* = -e_j g^* e_i f^* \theta^j \theta^i = -(e_i f e_j g)^* \theta^j \theta^i$$

for arbitrary f and g and we must conclude that

$$I^{ij}{}_{kl} = -P^{ji}{}_{kl}. \tag{3.1.55}$$

The compatibility condition with the product implies then that

$$(P^{ij}{}_{kl})^* P^{lk}{}_{mn} = P^{ji}{}_{mn}.$$

For general $\xi, \eta \in \Omega^1(\mathcal{A})$ it follows from (3.1.55) that

$$(\xi\eta)^* = -\eta^*\xi^*.$$

In particular

$$(\theta^i\theta^j)^* = -\theta^j\theta^i.$$

The product of two frame elements is hermitian then if and only if they anticommute. Recall that the product of two hermitian elements f and g of the algebra is hermitian if and only if they commute. When the frame exists one has necessarily also the relations

$$(f\xi\eta)^* = (\xi\eta)^* f^*, \qquad (f\xi \otimes \eta)^* = (\xi \otimes \eta)^* f^*$$

for arbitrary $f \in \mathcal{A}$.

We have defined an involution on the algebra of forms using a reality condition on derivations, a procedure which is more or less a straightforward generalization of that which is used in the case of ordinary differential manifolds. On real manifolds one considers usually only real derivations and real forms. It is however often of interest on even-dimensional manifolds to introduce derivations which are not real and use them to define an involution on the module of 1-forms which is known as an almost-complex structure. We have given the general expression which the extension of the involution must have to $\Omega^2(\mathcal{A})$ as well as to $\Omega^1(\mathcal{A}) \otimes \Omega^1(\mathcal{A})$. This again parallels the procedure which is used in ordinary geometry. The involution, we have seen, depends on the form of the product projection π. In the following sections we shall introduce reality conditions also on calculi which are not based on derivations. The simplest examples of this are calculi over matrix algebras which are themselves matrix algebras; in this case the involution is the ordinary involution on matrices. In Example 3.10, for example, the matrix (3.2.9) is anti-hermitian as a matrix just as we saw above that θ is anti-hermitian with respect to the involution. These models are of interest also because they are simple examples of a general procedure which allows one to introduce involutions on differential calculi over algebras of operators.

Example 3.6 If in particular $P^{ij}{}_{kl}$ is given by (3.1.44) we can choose $\jmath_2$ to be the identity. In this case the $F^k{}_{ij}$ are hermitian and the K_{ij} anti-hermitian elements of $\mathcal{Z}(\mathcal{A})$. We have seen that this was the case of the differential calculi constructed over matrices which are based on derivations. In these cases $K_{ij} = 0$ and the derivations formed a Lie algebra.   □

Example 3.7 There is a general procedure which allows one to introduce involutions on differential calculi over algebras of operators, a procedure which is based on an operator used in the theory of von Neumann algebras known as the modular conjugation operator and which we shall mention in Example 4.57 of the next chapter. The discussion of the involution we have given is in principle valid for arbitrary associative algebras, not necessarily operator algebras, but if one is considering an algebra which is represented as an operator algebra and with a differential calculus defined by a generalized Dirac operator and with a frame as we have defined it then the involution induced by the modular conjugation operator on the frame would satisfy all of the conditions we have described for the involution $\imath$ and each representation would define a different, in general inequivalent, one. We recall that there is in general no unique way of defining an almost-complex structure on a manifold. We cannot discuss this in further detail since no concrete examples are known except for the trivial matrix algebras mentioned in the previous example. □

Example 3.8 Differential forms of odd degree have certain properties in common with fermions. They anticommute and they have operating on them a natural square root of the Laplace operator, the Kähler-Dirac operator. *Ghosts* have characteristics of fermions as well as of forms. With this analogy the map from the basis $d\lambda^a$ to θ^a is the equivalent of passing from *bare fermions* to *dressed fermions*. We shall mention also the dressing of operators in Section 4.3 and vacuum dressing in Section 7.3. □

3.2 Differential forms II

Having noticed that certain calculations which are performed when studying geometry on a manifold V can be expressed uniquely in terms of algebraic operations on the algebra $\mathcal{C}(V)$, we introduced a noncommutative geometry by replacing this algebra by the matrix algebra M_n. The manifold V was eliminated. The algebra of forms was constructed directly from M_n. We shall now pursue the same logic one step further and eliminate the distinction between the algebra M_n and the algebra of forms. We used derivations to define differential forms and the differential d but we saw in Section 3.1 that we could have defined d as acting on elements of M_n by the equation $df = -[\theta, f]$. The algebra of forms $\Omega^*(M_n)$ has a $\mathbb{Z}_2$ grading and θ is an odd element. There is here no explicit reference to the set of derivations. This construction can be generalized. We use the notation of Example 3.1. With the decomposition

$$M_n = M_n^+ \oplus M_n^-$$

it is possible to associate to M_n an *algebra of forms*. One can define a *graded derivation* $\hat{d}\alpha$ of $\alpha \in M_n$ by the formula

$$\hat{d}\alpha = -[\eta, \alpha], \tag{3.2.1}$$

where η is an arbitrary anti-hermitian element of M_n^- and the commutator is graded. Formula (3.2.1) is to be compared with (2.2.20). We find that $\hat{d}\eta = -2\eta^2$ and for any $\alpha \in M_n$,

$$\hat{d}^2\alpha = [\eta^2, \alpha]. \tag{3.2.2}$$

This is to be compared with (2.2.21).

If $n = 2m$ is even then it is possible to impose the condition

$$\eta^2 = -1. \tag{3.2.3}$$

The unit on the right-hand side of this equation is the unit in M_n. From (3.2.2) we see that $\hat{d}^2 = 0$. The map (3.2.1) is a *differential*. In this case we shall write

$$\hat{d} = d.$$

When (3.2.3) is satisfied the algebra M_n is a *differential algebra* with a $\mathbb{Z}_2$ grading and we can identify

$$M_n = \Omega^*(M_n^+).$$

Notice that (3.1.12) is not satisfied for η. We have instead

$$d\eta + \eta^2 = 1. \tag{3.2.4}$$

The reason for the difference is that initially a relation of the form (3.2.1) was valid only for elements of $\Omega^0(M_n)$. In defining the second differential d we have imposed it on all elements of M_n.

If n is not even or, in general, if η^2 is not proportional to the unit element of M_n then $\hat{d}^2$ given by (3.2.2) will not vanish and M_n will not be a differential algebra. Consider now an arbitrary grading of M_n and let η be an arbitrary odd element. It is possible to construct over M_n^+ an integer-graded *differential algebra* $\Omega_\eta^* = \Omega_\eta^*(M_n^+)$ based on Formula (3.2.1). Let $\Omega_\eta^0 = M_n^+$ and let $\Omega_\eta^1 \equiv \overline{d\Omega_\eta^0} \subset M_n^-$ be the M_n^+-bimodule generated by the image of Ω_η^0 in M_n^- under $\hat{d}$. Define

$$\Omega_\eta^0 \xrightarrow{\;d_\eta\;} \Omega_\eta^1 \tag{3.2.5}$$

using directly (3.2.1): $d_\eta = \hat{d}$. Let $\overline{d\Omega_\eta^1}$ be the M_n^+-module generated by the image of Ω_η^1 in M_n^+ under $\hat{d}$. It would be natural to try to set $\Omega_\eta^2 = \overline{d\Omega_\eta^1}$ and define

$$\Omega_\eta^1 \xrightarrow{\;d_\eta\;} \Omega_\eta^2 \tag{3.2.6}$$

using once again (3.2.1). Every element of Ω_η^1 can be written as a sum of elements of the form $f_0 \hat{d} f_1$. If we attempt to define an application (3.2.6) using again directly (3.2.1),

$$d_\eta(f_0 \hat{d} f_1) = \hat{d} f_0 \hat{d} f_1 + f_0 \hat{d}^2 f_1, \tag{3.2.7}$$

then we see that in general d_η^2 does not vanish. To remedy this problem we simply eliminate the unwanted terms. Let $\mathrm{Im}\,\hat{d}^2$ be the submodule of $\overline{d\Omega_\eta^1}$ consisting of those elements which contain a factor which is the image of $\hat{d}^2$ and define Ω_η^2 by

$$\Omega_\eta^2 = \overline{d\Omega_\eta^1}/\mathrm{Im}\,\hat{d}^2.$$

Then by construction the second term on the right-hand side of (3.2.7) vanishes as an element of Ω_η^2 and we have a well defined map (3.2.6) with $d_\eta^2 = 0$. This procedure can be continued to arbitrary order by iteration. For each $p \geq 2$ we let $\mathrm{Im}\,\hat{d}^2$ be the submodule of $\overline{d\Omega_\eta^{p-1}}$ defined as above and we define Ω_η^p by

$$\Omega_\eta^p = \overline{d\Omega_\eta^{p-1}}/\mathrm{Im}\,\hat{d}^2. \tag{3.2.8}$$

Since $\Omega_\eta^p \Omega_\eta^q \subset \Omega_\eta^{p+q}$ the complex Ω_η^* is a differential algebra. The Ω_η^p need not vanish for large values of p. In fact if $\eta^2 \propto 1$ we see that $\mathrm{Im}\,\hat{d}^2 = 0$. It follows that $\Omega_\eta^p \simeq M_n^+$ for p even and $\Omega_\eta^p \simeq M_n^-$ for p odd. We shall return to this construction in Section 6.1 after we have studied differential algebras from a more abstract point of view.

Example 3.9 Choose $n = 2$ and $m = 1$. The most general element of M_2 can be expanded using the 3 matrices λ_a with λ_1 and λ_2 odd and λ_3 and the identity even. The most general possible form for η is a linear combination of λ_1 and λ_2 and it can be normalized so that (3.2.3) is satisfied. Then M_2 is a differential algebra of dimension 4. The algebra of forms $\Omega^*(M_2)$ constructed using the differential of Section 3.1 is a $4 \times 8 = 32$ dimensional algebra. Over M_2 we have constructed then two differential calculi, one with an algebra of dimension 32 and a second of dimension 4, although more properly the latter should be considered as an algebra of forms over its even elements M_2^+. Both the $M_2^\pm$ are M_2^+-bimodules and one can write

$$M_2^- \otimes_{M_2^+} M_2^- = M_2^+.$$

The tensor product is of dimension 2 as a complex vector space whereas the tensor product over $\mathbb{C}$ would have been of dimension 4. By iteration one can see that, as bimodules, the tensor product of an even number of copies of M_2^- can be identified with M_2^+ and the tensor product of an odd number of copies with M_2^-. The tensor product coincides then with the product in the algebra of forms. Using M_2 one can construct a differential calculus over $C(\mathbb{R}^4) \times C(\mathbb{R}^4)$, the algebra of functions on a double-sheeted space-time. We shall return to this calculus in Example 8.5.   □

Example 3.10 Choose $n = 3$ and $m = 1$. The most general possible form for η is

$$\eta = \begin{pmatrix} 0 & 0 & a_1 \\ 0 & 0 & a_2 \\ -a_1^* & -a_2^* & 0 \end{pmatrix}. \tag{3.2.9}$$

For no values of the parameters can (3.2.3) be satisfied. The general construction yields $\Omega_\eta^0 = M_3^+ = M_2 \times M_1$ and $\Omega_\eta^1 = M_3^-$ as in the previous example but after that the quotient by elements of the form $\operatorname{Im} \hat{d}^2$ reduces the dimensions. One finds $\Omega_\eta^2 = M_1$ and $\Omega_\eta^p = 0$ for $p \geq 3$. As bimodules one can write as above

$$M_3^- \otimes_{M_3^+} M_3^- = M_3^+.$$

The tensor product is of dimension 5 as a complex vector space whereas the tensor product over $\mathbb{C}$ would have been of dimension 16. The tensor product does not coincide then with the product in the algebra of forms. We shall return often to this calculus.   □

Example 3.11 Set $N = n^2 2^{m^2-1}$. Then $\Omega_m^*(M_n)$ is a complex N-dimensional vector space with a natural $\mathbb{Z}_2$ grading. The algebra M_N has therefore also a natural $\mathbb{Z}_2$ grading. There is an embedding of $\Omega^*(M_n)$ in M_N by left multiplication. We designate as in Example 2.10 the image of a form α in M_N by $\hat{\alpha}$. The differential d

has an extension to all elements of M_N. Let L be an element of M_N with parity $|L|$ and α an arbitrary element of $\Omega^*(M_n)$. Define dL by

$$(dL)\alpha = d(L\alpha) - (-1)^{|L|} L d\alpha.$$

Using the graded commutator this can be written in the form

$$dL = [d, L]. \tag{3.2.10}$$

That is, d has a natural extension to all of M_N which is just the commutator of d considered as an element of M_N. In particular $d\hat{\alpha} = [d, \hat{\alpha}] = \widehat{d\alpha}$. We have seen that when acting on the elements of $\Omega^0(M_n) = M_n$ the exterior derivative d can be expressed as a commutator with the element θ. Although in general it is not possible to write the left-hand side of (3.2.10) as the image of a commutator in $\Omega^*(M_n)$, there is a sense in which this can almost be done. That is, for any $\alpha \in \Omega^*(M_n)$, we can write

$$d\hat{\alpha} = -[\eta, \hat{\alpha}] \tag{3.2.11}$$

where η is almost the image of an element of $\Omega^*(M_n)$.

Let $c^a = \hat{\theta}^a$ be the element of M_N which corresponds to left-multiplication by the element θ^a of $\Omega^1(M_n)$ and let $\bar{c}_a$ be the element of M_N which corresponds to the interior product by e_a. Both c^a and $\bar{c}_a$ are odd elements of M_N and their (graded) commutation rules are

$$[c^a, c^b] = 0, \qquad [\bar{c}_a, \bar{c}_b] = 0, \qquad [\bar{c}_a, c^b] = \delta_a^b.$$

They commute with the elements $\hat{\lambda}^a$. Let $\hat{\theta} = -\hat{\lambda}_a c^a$ be the image of θ in M_N and define η as

$$\eta = \hat{\theta} + \frac{1}{2} C^a{}_{bc} c^b c^c \bar{c}_a. \tag{3.2.12}$$

It is easy to verify that (3.2.11) is satisfied. It follows immediately from the definitions that

$$dc^a = -[\eta, c^a] = -\frac{1}{2} C^a{}_{bc} c^b c^c.$$

Let L_a be the Lie derivative with respect to e_a. It is interesting to note also that

$$d\bar{c}_a = L_a, \qquad dL_a = 0.$$

Because $\bar{c}_a$ is not in the image of $\Omega^*(M_n)$ in M_N the expression $d\bar{c}_a + [\eta, \bar{c}_a]$ does not vanish. It commutes however with all elements of the image of $\Omega^*(M_n)$ in M_N.

We have

$$\eta^2 = 0,$$

and therefore $d^2 = 0$. Consider the algebra Ω^*_η constructed from the η defined in (3.2.12). There is a natural embedding of differential algebras

$$\Omega^*(M_n) \xrightarrow{i} \Omega^*_\eta. \tag{3.2.13}$$

We construct the map i by iteration.   First of all there is the embedding of $\Omega^0(M_n) = M_n$ into $\Omega_\eta^0 = M_N^+$ given by left multiplication. We extend i to $\Omega^1(M_n)$ by setting

$$i(df) = -[\eta, \hat{f}] = d_\eta i(f).$$

Since both $\Omega^*(M_n)$ and Ω_η^* are graded differential algebras (3.2.13) is uniquely defined.

The notation in this example is adapted from the standard notation used in the application of 'ghosts' in the quantization of gauge fields. We shall return to an analogous construction in a series of examples in Section 6.1 which introduce ghosts and anti-ghosts in the study of constrained hamiltonian systems.   □

We are now in a position to give a quite general abstract definition of a differential calculus over an arbitrary associative algebra. Consider a graded algebra

$$\Omega^*(\mathcal{A}) = \bigoplus_{i \geq 0} \Omega^i(\mathcal{A})$$

which is the direct sum of a family of $\mathcal{A}$-bimodules. Suppose also that $\Omega^0(\mathcal{A}) = \mathcal{A}$. If the grading is a $\mathbb{Z}_2$ grading we write $\Omega^+(\mathcal{A}) = \mathcal{A}$. We have already encountered $\mathbb{Z}$-graded algebras and even a $\mathbb{Z}_2$-graded algebra in Example 3.9. A *differential* d is a graded derivation (of degree $+1$) of $\Omega^*(\mathcal{A})$ with $d^2 = 0$. If $\alpha \in \Omega^i(\mathcal{A})$ and $\beta \in \Omega^j(\mathcal{A})$ then $\alpha\beta \in \Omega^{i+j}(\mathcal{A})$ and $d(\alpha\beta) \in \Omega^{i+j+1}(\mathcal{A})$ is given by

$$d(\alpha\beta) = d\alpha\beta + (-1)^i \alpha d\beta.$$

A *differential algebra* is a graded algebra with a differential. We say also that $\Omega^*(\mathcal{A})$ is a *calculus* or a *differential calculus* over $\mathcal{A}$, in which case we refer to an element of $\Omega^p(\mathcal{A})$ as a *p-form*. In this and the previous section we have seen two different ways of constructing differential calculi over a matrix algebra. In Section 6.1 we shall construct calculi over more general algebras using the *universal calculus* $\Omega_u^*(\mathcal{A})$. We shall show in particular that there is one construction which yields a differential calculus which is uniquely defined by the structure of the space of 1-forms as a bimodule over the algebra. This differential calculus is in a sense the largest one which is consistent with the relations of the algebra $\mathcal{A}$ and the module structure of $\Omega^1(\mathcal{A})$. Until now we have considered only algebras of smooth functions and algebras of matrices. More general examples will be given in the next chapter.

The complete construction of the universal calculus will be given in Section 6.1. Here we construct the bimodule of 1-forms $\Omega_u^1(\mathcal{A})$. For an arbitrary algebra $\mathcal{A}$ consider the tensor product $\mathcal{A} \otimes \mathcal{A}$ over the complex numbers. Define the product map

$$\mathcal{A} \otimes \mathcal{A} \xrightarrow{\ m\ } \mathcal{A}$$

and the map

$$A \xrightarrow{d_u} A \otimes A$$

by

$$d_u f = 1 \otimes f - f \otimes 1. \tag{3.2.14}$$

We have $m \circ d_u = 0$ and the image of d_u is equal to the kernel of m. Define $\Omega_u^1(\mathcal{A}) \subset \mathcal{A} \otimes \mathcal{A}$ to be the bimodule generated by the image of d_u or by the kernel of m. Let $\Omega^1(\mathcal{A})$ be any other bimodule of 1-forms, with differential d. Then there exists a bimodule map

$$\Omega_u^1(\mathcal{A}) \xrightarrow{\phi_1} \Omega^1(\mathcal{A}) \tag{3.2.15}$$

given by

$$\phi_1(d_u f) = df.$$

Because $d1 = 0$ the map is well defined. The above can be written in the form of a diagram

$$
\begin{array}{ccc}
\mathcal{A} & \xrightarrow{d_u} & \Omega_u^1(\mathcal{A}) \\[4pt]
\| & & \phi_1 \downarrow \\[4pt]
\mathcal{A} & \xrightarrow{d} & \Omega^1(\mathcal{A})
\end{array}
\tag{3.2.16}
$$

and we can write

$$\Omega^1(\mathcal{A}) = \Omega_u^1(\mathcal{A})/\mathrm{Ker}\,\phi_1. \tag{3.2.17}$$

Every bimodule of 1-forms can be expressed as the quotient of the bimodule $\Omega_u^1(\mathcal{A})$ by some 2-sided ideal.

If $\mathcal{A}$ is a $*$-algebra we shall suppose that $\Omega^1(\mathcal{A})$ has also a $*$-involution which is such that the reality condition (3.1.32) on the differential is satisfied. The de Rham differential is real and if we extend the involution to the tensor product by

$$(f \otimes g)^* = f^* \otimes g^*$$

it follows that d_u is real. In general one cannot choose the involution so that $(f dg)^* = dg^* f^*$; this equality is not satisfied, for example, by d_u: in general $(f d_u g)^* \neq d_u g^* f^*$.

Example 3.12 Consider the case $\mathcal{A} = \mathcal{C}(V)$ of the algebra of functions on a smooth manifold V and let $\Omega^1(\mathcal{A}) \equiv \Omega^1(V)$ be the de Rham 1-forms. If $f \in \mathcal{A}$ then $d_u f$ defined by (3.2.14) is the function of 2 variables given by

$$d_u f(x, y) = f(y) - f(x).$$

The de Rham 1-form df on the other hand is given in local coordinates x^λ by

$$df = \partial_\lambda f dx^\lambda.$$

If we expand the function $f(y)$ about the point x,

$$f(y) = f(x) + (x^\lambda(y) - x^\lambda(x))\partial_\lambda f + \cdots,$$

we see that the map ϕ_1 of (3.2.15) is given by

$$\phi_1(x^\lambda(y) - x^\lambda(x)) = dx^\lambda$$

and that it annihilates any 1-form $f(x,y) \in \Omega_u^1(\mathcal{A})$ which is second order in $x - y$. One such form is $f d_u g - d_u g f$. It is given by

$$(f d_u g - d_u g f)(x,y) = -(f(y) - f(x))(g(y) - g(x)).$$

It does not vanish in $\Omega_u^1(\mathcal{A})$ but its image in $\Omega^1(\mathcal{A})$ under ϕ_1 is equal to zero. $\quad\square$

Example 3.13 Let $\mathcal{A} = M_n$ and let $\Omega^*(M_n)$ be the differential algebra defined in Section 3.1, with $m = n$. Let $f \in M_n$. Then the map ϕ_1 is defined by

$$\phi_1(1 \otimes f - f \otimes 1) = -[\theta, f]. \tag{3.2.18}$$

Recall the 1-forms θ^a defined in (3.1.8) and define

$$\theta_u^a = \lambda_b \lambda^a d_u \lambda^b = \lambda_b \lambda^a \otimes \lambda^b - \lambda_b \lambda^a \lambda^b \otimes 1.$$

If we anticipate the quantity ζ introduced in Example 3.40 then we can write

$$\theta_u^a = C^a{}_{bc} \lambda^b \otimes \lambda^c + n\lambda^a \zeta.$$

Using the associativity law $(\lambda_a \lambda_b)\lambda_c = \lambda_a(\lambda_b \lambda_c)$ and the relations (3.1.1) one can show that the θ_u^a satisfy also (3.1.34): $\theta_u^a \lambda^b = \lambda^b \theta_u^a$. Therefore the map

$$\theta_u^a \mapsto \theta^a$$

defines a bimodule isomorphism $\Omega_u^1(M_n) \simeq \Omega^1(M_n)$. Both are free left and right M_n-modules of rank $n^2 - 1$ and $\operatorname{Ker} \phi_1 = 0$. We can write $d_u f$ as a commutator:

$$d_u f = [1 \otimes 1, f].$$

However we cannot define (3.2.18) by $\phi(1 \otimes 1) = -\theta$ since $1 \otimes 1$ is not an element of Ω_u^1. Although $\Omega^p(M_n) = \Omega_u^p(M_n)$ for $p = 0, 1$ we shall see later in Example 6.4 that $\dim \Omega^p(M_n) < \dim \Omega_u^p(M_n)$ for larger values of p. $\quad\square$

Example 3.14 Let $\mathbb{C}^n = \mathbb{C}^m \oplus \mathbb{C}^m$ and consider the differential algebra $\Omega_\eta^*(M_n^+)$ constructed above with $\eta^2 = -1$. Then there is a map

$$\Omega_u^1(M_n^+) \xrightarrow{\phi} \Omega_\eta^1(M_n^+)$$

given by

$$\phi_1(1 \otimes f - f \otimes 1) = -[\eta, f].$$

In particular we shall see in Section 6.1 that $\Omega_u^*(M_2^+) \simeq \Omega_\eta^*(M_2^+)$. Again one is not allowed to define ϕ_1 by $\phi_1(1 \otimes 1) = -\eta$. In general $\dim \Omega_u^1(M_n^+) > \dim \Omega_\eta^1(M_n^+)$ but nevertheless, as in the universal calculus, we have

$$\Omega_\eta^1(M_n^+) \otimes_{M_n^+} \Omega_\eta^1(M_n^+) \simeq \Omega_\eta^2(M_n^+).$$

□

Example 3.15 Let $\mathcal{M}$ be an arbitrary $\mathcal{A}$-bimodule and Q an element of $\mathcal{M}$. Define a map d of $\mathcal{A}$ into $\mathcal{M}$ by $df = -[Q, f]$. The minus sign is a convention. Let $\Omega^1(\mathcal{A}) \subset \mathcal{M}$ be the bimodule generated by the image of d. Again one cannot define ϕ_1 by $\phi_1(1 \otimes 1) = -Q$ because $1 \otimes 1$ does not belong to $\Omega_u^1(\mathcal{A})$ and Q in general need not belong to $\Omega^1(\mathcal{A})$. Let now $\mathcal{A}$ be any (graded) algebra and d a (graded) derivation. It is possible to embed $\mathcal{A}$ into an algebra $\mathcal{B}$ such that the derivation on the image is given by a commutator. We construct the larger algebra by simply adding to the original one an (odd) element Q and defining the product Qa and aQ such that the (graded) commutator is given by $da = -[Q, a]$. In general d need not be defined on Q.   □

The universal calculus can be constructed over any algebra, for example the algebra of continuous functions on a compact topological space. There is therefore nothing about its existence which insures that the algebra over which it is constructed is in any way a noncommutative version of an algebra of 'smooth functions'. The Example 3.9 shows also that the existence of a generalized 'Dirac operator' is also not a sufficient condition. To construct calculi whose existence would impose a 'smoothness' condition on the algebra we have used the derivations of the algebra. The de Rham differential calculus is based on the $\mathcal{C}(V)$-module $\mathcal{X}(V)$ of all derivations. If we set $\mathcal{A} = \mathcal{C}(V)$ then the map $d : \mathcal{A} \to \Omega^1(\mathcal{A})$ is given by $df(X) = Xf$ where $X \in \mathcal{X}(V)$. This procedure defines a unique differential calculus over V because $\mathcal{X}(V)$ is a left $\mathcal{A}$-module. In the general case, when $\mathrm{Der}(\mathcal{A})$ is not a $\mathcal{A}$-module and X belongs to some subspace of $\mathrm{Der}(\mathcal{A})$ then a large number of different differential calculi can be so constructed. Since the algebras we shall consider are mostly formal algebras they would correspond to algebras of polynomials in the commutative case. We shall also consider them as given and fixed before the differential calculus is constructed. The 'smoothness' condition could therefore restrict only the way in which they are to be completed.

Example 3.16 We shall later encounter a situation in which an invariance property singles out two differential calculi over a given algebra which are in a sense 'complex conjugate' one to another. One can ask the question whether it is possible to construct from them in a 'natural' way a real differential calculus. Consider an algebra $\mathcal{A}$ with involution over which there are two differential calculi $(\Omega^*(\mathcal{A}), d)$ and $(\bar{\Omega}^*(\mathcal{A}), \bar{d})$ neither of which is necessarily real. Consider the product algebra $\tilde{\mathcal{A}} = \mathcal{A} \times \mathcal{A}$. This is the noncommutative analogue of the algebra of functions on the union of two disjoint copies of the same space. Consider over $\tilde{\mathcal{A}}$ the differential calculus

$$\tilde{\Omega}^*(\tilde{\mathcal{A}}) = \Omega^*(\mathcal{A}) \times \bar{\Omega}^*(\mathcal{A}). \tag{3.2.19}$$

It has a natural differential $\tilde{d} = (d, \bar{d})$. The embedding

$$\mathcal{A} \hookrightarrow \tilde{\mathcal{A}}$$

given by $f \mapsto (f, f)$ is well defined and compatible with the involution

$$(f, g)^* = (g^*, f^*) \tag{3.2.20}$$

on $\tilde{\mathcal{A}}$.

Let X and $\bar{X}$ be two derivations of $\mathcal{A}$. Then $\tilde{X} = (X, \bar{X})$ is a derivation of $\tilde{\mathcal{A}}$. It is real if

$$\tilde{X}(f, g)^* = (\tilde{X}(f, g))^*. \tag{3.2.21}$$

This can be written as the conditions

$$\bar{X}f^* = (Xf)^*, \qquad Xg^* = (\bar{X}g)^*.$$

The algebra $\mathcal{A}$ does not necessarily remain invariant under real derivations of $\tilde{\mathcal{A}}$.

Suppose that $\Omega^*(\mathcal{A})$ is defined in terms of a set of inner derivations $e_i = \mathrm{ad}\,\lambda_i$ and that $\bar{\Omega}^*(\mathcal{A})$ is defined in terms of a set of inner derivations $\bar{e}_i = \mathrm{ad}\,\bar{\lambda}_i$. Suppose also that the corresponding $\tilde{e}_i = (e_i, \bar{e}_i)$ are real derivations of $\tilde{\mathcal{A}}$. From (3.2.21) we see that this will be the case if and only if $\bar{\lambda}_i = -\lambda_i^*$. We define an involution on $\tilde{\Omega}^*(\tilde{\mathcal{A}})$ by the condition

$$(\tilde{d}(f, g))^*(\tilde{e}_i) = (\tilde{e}_i(f, g))^* = \tilde{e}_i(g^*, f^*).$$

From the argument which leads to (3.1.32) one sees that the differential $\tilde{d}$ is real.

Define $\mathcal{A}_r$ to be the smallest algebra which contains $\mathcal{A}$ and which is stable under the action of the derivations $\tilde{e}_i$. Then

$$\mathcal{A} \subseteq \mathcal{A}_r \subseteq \mathcal{A} \times \mathcal{A}.$$

Define e_{ri} to be the restriction of $\tilde{e}_i$ to $\mathcal{A}_r$ and d_r to be the restriction of $\tilde{d}$ to $\mathcal{A}_r$. We have then

$$d_r f(e_{ri}) = (e_i f, \bar{e}_i f) \tag{3.2.22}$$

and d_r is also real. We define

$$\Omega_r^1(\mathcal{A}_r) \subset \tilde{\Omega}^1(\tilde{\mathcal{A}}) \tag{3.2.23}$$

to be the $\mathcal{A}_r$-bimodule generated by the image of d_r. The module structure determines a differential calculus $(\Omega_r^*(\mathcal{A}_r), d_r)$.

If there exist a frame θ^i for $\Omega^*(\mathcal{A})$ and a frame $\bar{\theta}^i$ for $\bar{\Omega}^*(\mathcal{A})$ we can extend also the involution (3.2.20) to all of $\Omega_r^*(\mathcal{A}_r)$ by setting

$$(\theta^i)^* = \bar{\theta}^i$$

and we can define $\Omega_r^1(\mathcal{A})$ to be the $\mathcal{A}_r$-module generated by

$$\theta_r^i = (\theta^i, \bar{\theta}^i).$$

This is consistent with the previous definition since

$$d_r f = e_{ri} f \theta_r^i, \qquad e_{ri} f \in \mathcal{A}_r.$$

From the relations

$$\theta^i(e_j) = \delta_j^i, \quad \theta^i(\bar{e}_j) = 0,$$
$$\bar{\theta}^i(e_j) = 0, \quad \bar{\theta}^1(\bar{e}_j) = \delta_j^i$$

it follows that the frame dual to the derivation e_{ri} is indeed θ_r^i:

$$\theta_r^i(e_{rj}) = \delta_j^i.$$

In terms of the 1-form

$$\theta_r = -(\lambda_i, \bar{\lambda}_i)\theta_r^i = -(\lambda_i \theta^i, \bar{\lambda}_i \bar{\theta}^i)$$

introduced in (3.1.37) we find from Equation (3.2.22) that $d_r f = -[\theta_r, f]$ for all $f \in \mathcal{A}_r$. We have constructed a real calculus at the possible expense of increasing the size of the algebra from $\mathcal{A}$ to $\mathcal{A}_r$. We shall give examples of this construction in Section 4.1.   □

3.3   Tensor products

Let $\mathcal{A}$ and $\mathcal{A}'$ be two algebras with differential calculi $\Omega^*(\mathcal{A})$ and $\Omega^*(\mathcal{A}')$. Then there is a natural differential calculus over the *tensor product* $\mathcal{A} \otimes \mathcal{A}'$ given by

$$\Omega^*(\mathcal{A} \otimes \mathcal{A}') = \Omega^*(\mathcal{A}) \otimes \Omega^*(\mathcal{A}'). \tag{3.3.1}$$

If $\alpha \in \Omega^*(\mathcal{A})$, $\beta \in \Omega^p(\mathcal{A}')$, $\gamma \in \Omega^q(\mathcal{A})$ and $\delta \in \Omega^*(\mathcal{A}')$ then the product in $\Omega^*(\mathcal{A}) \otimes \Omega^*(\mathcal{A}')$ is given by

$$(\alpha \otimes \beta)(\gamma \otimes \delta) = (-1)^{pq} \alpha\gamma \otimes \beta\delta. \tag{3.3.2}$$

Equation (3.3.1) does not define the only choice of differential calculus over the product algebra. Consider the module of 1-forms

$$\Omega^1(\mathcal{A} \otimes \mathcal{A}') = \mathcal{A} \otimes \Omega^1(\mathcal{A}') \oplus \Omega^1(\mathcal{A}) \otimes \mathcal{A}'.$$

From the diagram (3.2.16) there is a map ϕ_1 of $\Omega^1_u(\mathcal{A} \otimes \mathcal{A}')$ onto $\Omega^1(\mathcal{A} \otimes \mathcal{A}')$ which can be used to construct a differential calculus $\Omega^*(\mathcal{A} \otimes \mathcal{A}')$ over the tensor product of the two algebras. It follows from the definition as well as from the conclusion of Example 6.1 below that this extension is in general larger than the tensor product $\Omega^*(\mathcal{A}) \otimes \Omega^*(\mathcal{A}')$. If θ^i is a frame for $\Omega^1(\mathcal{A})$ and θ'^j is a frame for $\Omega^1(\mathcal{A}')$ then

$$(\theta^i, \theta'^j) = (\theta^i \otimes 1, 1 \otimes \theta'^j)$$

is a frame for $\Omega^1(\mathcal{A} \otimes \mathcal{A}')$. The commutation relations (3.1.43) for each factor can be extended to the entire frame by the rule (3.3.2). In this case both constructions yield the same algebra of forms.

If $\mathcal{A} \otimes \mathcal{A}'$ is the tensor product of two algebras then with the natural identifications one has

$$\mathrm{Der}(\mathcal{A}) \oplus \mathrm{Der}(\mathcal{A}') \subset \mathrm{Der}(\mathcal{A} \otimes \mathcal{A}').$$

In fact if $X \in \mathrm{Der}(\mathcal{A})$ and $X' \in \mathrm{Der}(\mathcal{A}')$ then both have an extension to $\mathcal{A} \otimes \mathcal{A}'$ by

$$X(f \otimes 1) = Xf \otimes 1, \qquad X'(1 \otimes f') = 1 \otimes X'f'.$$

Example 3.17 Consider as example the tensor product $M = M' \otimes M''$ of two matrix algebras M' and M''. The product structure suggests that for the derivations we consider the algebra which is the direct sum of $\mathrm{Der}(M')$ and $\mathrm{Der}(M'')$. This is analogous to the case in ordinary geometry where the derivations of a product structure $V \times V'$ can be obtained from the derivations of each factor. In the noncommutative case there are additional derivations which have no analogue in ordinary geometry. If $M' = M_{n'}$ and $M'' = M_{n''}$ then

$$\dim \mathrm{Der}(M') = n'^2 - 1, \quad \dim \mathrm{Der}(M'') = n''^2 - 1, \quad \dim \mathrm{Der}(M) = (n'n'')^2 - 1$$

and one sees that

$$\dim \mathrm{Der}(M') + \dim \mathrm{Der}(M'') < \dim \mathrm{Der}(M).$$

The form algebra $\Omega^*(M', M'')$ based on the Lie algebra $\mathrm{Der}(M') \oplus \mathrm{Der}(M'')$ is given then by (3.3.1):

$$\Omega^p(M', M'') = \bigoplus_{i+j=p} \Omega^i(M) \otimes \Omega^j(M''). \tag{3.3.3}$$

It has a natural embedding

$$\Omega^*(M', M'') \hookrightarrow \Omega^*(M)$$

where by $\Omega^*(M)$ we understand the differential calculus based on the complete set of all derivations of M. We write d' (d'') for the exterior derivative in $\Omega^*(M')$ ($\Omega^*(M'')$) and we set $d = d' + d''$. Since $d'd'' + d''d' = 0$ we have $d^2 = 0$. Define

$$\theta = \theta' \otimes 1 + 1 \otimes \theta''.$$

Then θ satisfies Equation (3.1.12). This follows from the definition (3.3.2) of the exterior product of a tensor product. Since $(1 \otimes \theta'')(\theta' \otimes 1) = -\theta' \otimes \theta''$ the cross terms sum to zero and we have

$$\theta^2 = \theta'^2 \otimes 1 + 1 \otimes \theta''^2.$$

□

Example 3.18 Product structures using mixed types of differential forms can also be constructed. Consider

$$\Omega_\eta^* = \Omega^*(M') \otimes \Omega_{\eta''}^*,$$

and define

$$\eta = \theta' \otimes 1 + 1 \otimes \eta''.$$

Then η satisfies (3.2.4) if η'' does. A product structure using only the type of differential algebra defined in the previous section is given by

$$\Omega_\eta^* = \Omega_{\eta'}^* \otimes \Omega_{\eta''}^*.$$

□

More interesting structures, which we shall use in Chapter 8, can be obtained by taking the product of an algebra of functions by a matrix algebra. We shall now consider an extension of matrix geometry by considering the algebra $\mathcal{A}$ of matrix-valued functions on a manifold V. Let V be a parallelizable manifold which is not necessarily compact and M_n a matrix algebra. Then we can identify $\mathcal{A}$ with the tensor product over the real numbers of $\mathcal{C}(V)$ and M_n:

$$\mathcal{A} = \mathcal{C}(V) \otimes M_n. \tag{3.3.4}$$

Let $e_\alpha = e_\alpha^\mu \partial_\mu$ be a set of $\dim V$ generators of $\mathrm{Der}(\mathcal{C}(V))$ and let e_a be a basis of $\mathrm{Der}_m(M_n)$. Let $i = (\alpha, a)$. Then $1 \le i \le \dim V + m^2 - 1$. The set $e_i = (e_\alpha, e_a)$ is a set of generators of a Lie algebra of derivations $\mathrm{Der}(\mathcal{A})$ of $\mathcal{A}$.

We define differential forms over $\mathcal{A}$ just as we did in Section 2.1. For $f \in \mathcal{A}$ we define df by Equation (2.1.3) where now X is an element of $\mathrm{Der}(\mathcal{A})$. Choose the basis θ^α of the 1-forms $\Omega^1(V)$ on V dual to the e_α. Introduce $\theta^i = (\theta^\alpha, \theta^a)$ as generators of $\Omega^1(\mathcal{A})$ as an $\mathcal{A}$-module. Then if we define

$$\Omega_h^1 = \Omega^1(V) \otimes M_n, \qquad \Omega_v^1 = \mathcal{C}(V) \otimes \Omega^1(M_n),$$

we can write $\Omega^1(\mathcal{A})$ as a direct sum:

$$\Omega^1(\mathcal{A}) = \Omega_h^1 \oplus \Omega_v^1. \tag{3.3.5}$$

The *differential df* of an element f of $\mathcal{A}$ is given by

$$df = d_h f + d_v f. \tag{3.3.6}$$

We have written it as the sum of two terms, the horizontal and vertical parts, using notation from Kaluza-Klein theory, which we shall consider in more detail in Chapter 8. The horizontal component is the usual exterior derivative $d_h f = e_\alpha f \theta^\alpha$. The vertical component d_v, defined for example in Equation (3.1.13), is purely algebraic and it is what replaces the differential in the *internal manifold* of standard Kaluza-Klein theory. The algebra $\Omega^*(\mathcal{A})$ of differential forms is given in terms of the differential forms of each factor by a formula similar to (2.1.13):

$$\Omega^p(\mathcal{A}) = \bigoplus_{i+j=p} \Omega^i(V) \otimes \Omega^j(M_n).$$

It is again a *differential algebra* with a *differential d* which satisfies Equations (2.1.4) and (2.1.5).

The algebra $\mathcal{A}$ is an algebra of matrix-valued functions and it is convenient to consider it as such at times. However then one would not use the differential which has just been defined. If $df = 0$ then f is not only constant when considered as a matrix-valued function on V, but it is also a multiple of the identity.

If for some operator η we use the $\mathbb{Z}$-graded Ω_η^* instead of $\Omega^*(M_n)$ then the appropriate differential algebra which represents the product of the two structures is the tensor product of $\Omega^*(V)$ with Ω_η^*. It is the differential algebra $\Omega_\eta^*(V)$ defined by

$$\Omega_\eta^*(V) = \Omega^*(V) \otimes \Omega_\eta^*. \tag{3.3.7}$$

The differential is the sum of the differential d on V and d_η on Ω_η^*. The differential df of an element $f \in \Omega_\eta^*(V)$ is still given by (3.3.6) but now the vertical component is defined by (3.2.1) and not (3.1.13). If Ω_η^* is $\mathbb{Z}_2$-graded we define

$$\Omega_h^- = \Omega^-(V) \otimes \Omega_\eta^+, \quad \Omega_v^- = \Omega^+(V) \otimes \Omega_\eta^- \tag{3.3.8}$$

and the direct sum (3.3.5) is replaced by

$$\Omega_\eta^-(V) = \Omega_h^- \oplus \Omega_v^-. \tag{3.3.9}$$

3.4   Metrics

In ordinary differential geometry a metric is introduced to measure the distance between points. Since noncommutative geometry is essentially pointless the concept might seem at first sight rather useless in the noncommutative case. It is possible however to carry over one of the definitions of a metric which is used in commutative geometry and to give it a meaning in certain noncommutative cases as a measure of distance. Although all 'coordinates' cannot be measured simultaneously, a problem familiar from the quantized version of a phase space, it is possible to diagonalize one particular given 'coordinate' and at least in the case when it has a discrete spectrum it is desirable to be able to have a measure of the distance between spectral lines. Since in principle any two nearby 'points' can be joined by a 'coordinate' this implies that a general distance function can in principle be constructed. We shall return to this in Example 4.49.

Let $\Omega^*(\mathcal{A})$ be a differential calculus over an algebra $\mathcal{A}$. In complete analogy with the commutative case a *metric* g can be defined as an $\mathcal{A}$-bilinear, nondegenerate map

$$\Omega^1(\mathcal{A}) \otimes_{\mathcal{A}} \Omega^1(\mathcal{A}) \xrightarrow{\ g\ } \mathcal{A}. \tag{3.4.1}$$

It is important to notice here that the *bilinearity* is an alternative way of expressing *locality*. In ordinary differential geometry if ξ and η are 1-forms then the value of $g(\xi \otimes \eta)$ at a given point depends only on the values of ξ and η at that point. Bilinearity is an exact expression of this fact. In general the algebra introduces a certain amount of non-locality via the commutation relations and it is important to assure that all geometric quantities be just that nonlocal and not more. Without the bilinearity condition it is not possible to distinguish for example in ordinary space-time a metric which assigns a function to a vector field in such a way that the value at a given point depends only on the vector at that point from one which is some sort of convolution over the entire manifold.

In Section 3.1 we identified $\Omega^2(\mathcal{A})$ as the image of π in $\Omega^1(\mathcal{A}) \otimes_{\mathcal{A}} \Omega^1(\mathcal{A})$. As a *symmetry condition* on the metric we require that it vanish on the image of this projection; we require that

$$g \circ \pi = 0. \tag{3.4.2}$$

It would be possible also to introduce a *generalized flip*, a bilinear map

$$\Omega^1(\mathcal{A}) \otimes_{\mathcal{A}} \Omega^1(\mathcal{A}) \xrightarrow{\ \sigma\ } \Omega^1(\mathcal{A}) \otimes_{\mathcal{A}} \Omega^1(\mathcal{A}) \tag{3.4.3}$$

and to impose the condition $g \circ \sigma = g$. If $\sigma^2 = 1$ then $(1 + \sigma)/2$ is the projector onto the symmetric part of the tensor product in (3.4.3) and it is natural to impose the condition

$$\frac{1}{2}(1 + \sigma) + \pi = 1. \tag{3.4.4}$$

The two symmetry conditions are then equivalent. We shall refer to this as σ-*symmetry*. Since in general $\sigma^2 \neq 1$ we shall impose rather (3.4.2) and supplement it with the condition Equation (3.6.5) below which is weaker than (3.4.4).

If the algebra is a ∗-algebra then we shall require that the metric be real; if ξ and η are hermitian 1-forms then $g(\xi \otimes \eta)$ should be hermitian. The *reality condition* on the metric becomes therefore

$$g((\xi \otimes \eta)^*) = (g(\xi \otimes \eta))^*.$$

Using the involution $\jmath_2$ introduced in Section 3.1 we can write this condition as

$$g \circ \jmath_2(\xi \otimes \eta) = (g(\xi \otimes \eta))^*.$$

Anticipating the relation (3.6.8) which we shall find below between the symmetry condition and the reality condition we can write it also as

$$g \circ \sigma(\eta^* \otimes \xi^*) = (g(\xi \otimes \eta))^*. \tag{3.4.5}$$

If also $g \circ \sigma = g$ then we have

$$g(\eta^* \otimes \xi^*) = (g(\xi \otimes \eta))^*.$$

In the case where there exists a frame θ^i we define a metric by choosing a set of n^2 elements $g^{ij} \in \mathcal{A}$ and setting

$$g(\theta^i \otimes \theta^j) = g^{ij}. \tag{3.4.6}$$

Let f be an arbitrary element of $\mathcal{A}$. Then, using the bilinearity and the relation (3.1.34), by the sequence of identities

$$fg^{ij} = g(f\theta^i \otimes \theta^j) = g(\theta^i \otimes \theta^j f) = g^{ij} f$$

we conclude that g^{ij} must lie in $\mathcal{Z}(\mathcal{A})$. If the center is trivial then this is a very severe restriction and the metric is essentially unique. The σ-symmetry condition becomes

$$P^{ij}{}_{kl} g^{kl} = 0.$$

We shall write the metric as

$$g^{ab} = g_S^{ab} + g_A^{ab}, \tag{3.4.7}$$

the sum of a symmetric and an antisymmetric term (in the usual sense of the word). If the algebra is a ∗-algebra and if we express the action of σ on the basis elements by

$$\sigma(\theta^i \otimes \theta^j) = S^{ij}{}_{kl} \theta^k \otimes \theta^l$$

then the condition (3.4.5) becomes

$$S^{ij}{}_{kl} g^{kl} = (g^{ji})^*. \tag{3.4.8}$$

If $\sigma^2 = 1$ then g^{ij} is an hermitian matrix. If σ is the *ordinary flip* then $S^{ij}{}_{kl} = \delta^j_k \delta^i_l$ and g^{ij} is real and symmetric.

In ordinary geometry an equivalence class of moving frames determines a metric and all equivalence classes correspond to the same differential calculus, the ordinary de Rham differential calculus. The choice of differential calculus does not fix the metric. In the noncommutative case on the other hand, as we have defined it, each differential calculus determines a frame and thereby a metric. In the commutative limit all of the noncommutative differential calculi either are singular, if n is not equal to the classical dimension of the manifold, or have a common limit. The moving frame however and the associated metric remain as a 'shadow' of the noncommutative structure. This is an essential point and we shall return to it later.

Let $\mathcal{A}$ be a general $*$-algebra with a differential calculus such that the module of 1-forms is free of rank n and has a frame. Suppose also that the forms vanish when the degree is greater than n. We shall not use this condition explicitly but the construction is quite artificial when it is not satisfied. Consider the form

$$\eta = \sqrt{g}\theta^1 \cdots \theta^n, \qquad \sqrt{g} = \sqrt{|g_{ab}|}$$

in $\Omega^n(\mathcal{A})$. If $\mathcal{A}$ has a trace one can define the *integral* as

$$\int f\eta = \mathrm{Tr}(f). \tag{3.4.9}$$

If $\mathcal{A}$ is a noncommutative deformation of a commutative algebra $\tilde{\mathcal{A}}$ which can be represented as an algebra of functions on a differential manifold with a volume form $\tilde{\eta}$ then one can define the integral, and therefore the trace, as

$$\int f\eta = \int_V \tilde{f}\tilde{\eta}. \tag{3.4.10}$$

In Section 4.3 we shall see that under appropriate conditions the trace defines a representation of the algebra. The trace of the identity element exists only if the manifold V is compact. There is no simple generalization of Stokes' theorem.

Example 3.19 In the case where $\mathcal{A} = M_n$ and the differential calculus is that given in Section 3.1 we can choose σ to be the ordinary flip: $S^{ab}{}_{cd} = \delta^b_c \delta^a_d$. A metric can be defined by the condition that the moving frame θ^a be orthonormal with respect to the Killing metric:

$$g(\theta^a \otimes \theta^b) = g^{ab}.$$

One can define a duality map of $\Omega^p(M_n)$ into $\Omega^{m^2-1-p}(M_n)$ using the metric g exactly as in the commutative case. For each $p \leq m^2 - 1$, introduce the $(m^2 - 1 - p)$-form

$$*\theta_{a_1 \ldots a_p} = \frac{1}{(m^2 - 1 - p)!} \epsilon_{a_1 \ldots a_p a_{p+1} \ldots a_{m^2-1}} \theta^{a_{p+1}} \cdots \theta^{a_{m^2-1}}.$$

Then $*$ is defined as in (2.2.8) by the map

$$\theta^{a_1} \cdots \theta^{a_p} \longmapsto *\theta^{a_1 \ldots a_p} \tag{3.4.11}$$

where the indices on the right-hand side have been raised with g^{ab}. If $\alpha \in \Omega^p(M_n)$ then

$$* * \alpha = (-1)^{m^2 p}\alpha.$$

One can also define a map δ of $\Omega^p(M_n)$ into $\Omega^{p+1}(M_n)$ by

$$\delta\alpha = (-1)^{(m^2-1)p+m^2} * d * \alpha, \qquad \alpha \in \Omega^p(M_n).$$

For example, $\delta\theta^a = 0$ since C_{abc} is completely antisymmetric. If $\alpha = \alpha_a\theta^a$ is a 1-form then $\delta\alpha = -e_a\alpha^a$.

The *Laplace operator* Δ is defined as usual in terms of d and δ:

$$\Delta = d\delta + \delta d. \tag{3.4.12}$$

The Laplace equation,

$$\Delta\alpha = \omega^2\alpha,$$

is an ordinary matrix equation and it can be completely solved in principle. That is, the Laplace operator can be completely diagonalized.

An integral can be defined on M_n using the trace. Let α be an element of $\Omega^{m^2-1}(M_n)$. Then we have for some $f \in M_n$

$$\alpha = f\theta^1 \ldots \theta^{m^2-1}.$$

We define the integral of α as

$$\int \alpha = \operatorname{Tr}(f). \tag{3.4.13}$$

It depends in a weak sense on the metric because f is the dual of the $(m^2 - 1)$-form α. It satisfies the usual requirements of an integral on a compact manifold V. Let $\alpha \in \Omega^{m^2-2}(M_n)$, $\beta \in \Omega^p(M_n)$ and $\gamma \in \Omega^{m^2-1-p}(M_n)$. Then

$$\int d\alpha = 0, \qquad \int \beta\gamma = (-1)^{p(m^2-1-p)} \int \gamma\beta. \tag{3.4.14}$$

$\square$

Example 3.20 As in the ordinary case there is a natural inner product on the algebra of forms. If α is a p-form with coefficients $\alpha_{a_1\ldots a_p}$ define α^* to be the p-form with the adjoint coefficients $\alpha^*{}_{a_1\ldots a_p}$. Let β be another element of $\Omega^*(M_n)$. Then one can define the inner product (α, β) to be given by

$$(\alpha, \beta) = \int \alpha^* * \beta \tag{3.4.15}$$

if α and β have the same grading and to be zero otherwise. In particular we see from (3.4.11) that the norm $\|\alpha\|$ of α is given by

$$\|\alpha\|^2 = \frac{1}{p!}\operatorname{Tr}(\alpha^*{}_{a_1\ldots a_p}\alpha^{a_1\ldots a_p}).$$

As in the commutative case, δ is the adjoint of d with respect to (3.4.15). $\square$

Example 3.21 Although it was possible in Section 3.2 to define a d operator using only the $\mathbb{Z}_2$ grading, we have used the full structure of $\Omega^*(M_n)$ to define a metric. It would be necessary to introduce a duality operation on Ω_η^* to replace that of $\Omega^*(M_n)$ before one could define a metric on Ω_η^*. The norm however reduced to a simple trace. Since Ω_η^* is a matrix algebra then it is possible to define a norm on its elements by the same formula. For $\alpha \in \Omega_\eta^*$ we define

$$\|\alpha\|^2 = \mathrm{Tr}(\alpha^*\alpha). \tag{3.4.16}$$

$\square$

Consider the product algebra $\mathcal{A}$ defined in (3.3.4). The inner product on the algebra of forms is defined as in (2.2.11) and in (3.4.15). If α and β are elements of $\Omega^*(\mathcal{A})$ then one defines (α, β) to be given by

$$(\alpha, \beta) = \int \alpha^* * \beta \tag{3.4.17}$$

if α and β have the same grading and to be zero otherwise. The integration here includes the integral (3.4.13) as well as the integration over V. In particular, if α is a p-form we have

$$\|\alpha\|^2 = \frac{1}{p!}\mathrm{Tr}\int \alpha^*_{i_1\ldots i_p}\alpha^{i_1\ldots i_p}dx. \tag{3.4.18}$$

The integration here is over V only.

The Laplace operator on $\Omega^*(\mathcal{A})$ is defined as in (2.2.10) or (3.4.12) but using the exterior derivative given by (3.3.6). The duality operation which is used to define δ is the same as (2.2.8) or (3.4.11) but now using the basis θ^i.

Consider the algebra $\Omega_\eta^*(V)$ defined in (3.3.8). The norm of an element is given by the product of the norm defined by (2.2.11) for the $\Omega^*(V)$ factor and the norm (3.4.16) for the Ω_η^* factor. It has the same form as (3.4.18) if the α is considered as a p-form with values in Ω_η^*.

Example 3.22 If the differential calculus is based on a Lie algebra of derivations one can define an *interior product* i_X as is done in Chapter 2. We set $i_X\xi = \xi(X)$ on the 1-forms and extend it to the algebra of forms by the requirement that it be a graded derivation of degree -1. We can then define a *Lie derivative* L_X of the tensor algebra over the 1-forms to be the expression

$$L_X = i_X d + d i_X$$

the same that was derived (2.1.11) in Section 2.1. If there is a frame and we set $L_i = L_{e_i}$ we find in particular

$$L_i\theta^j = -C^j{}_{ik}\theta^k.$$

The Lie derivative acting on the form θ yields

$$L_i\theta = 0.$$

We find that $L_X\theta = 0$ for any derivation X and that to within multiplication by a complex number θ is the only such 1-form. The Lie derivation can be extended to the metric by the identity

$$(L_Xg)(\theta^i \otimes \theta^j) + g(L_X(\theta^i \otimes \theta^j)) = 0.$$

By formal analogy with the commutative case we can say that a derivation X is a *Killing derivation* if

$$L_Xg = 0. \qquad (3.4.19)$$

If the derivations do not close to form a Lie algebra then L_X can be defined as above but it will not in general be a derivation. If $g^{ij} = g(\theta^i \otimes \theta^j)$ is the Killing metric of the Lie algebra then (3.4.19) is satisfied. This will not be the case in Example 4.17 of the following chapter. □

Example 3.23 At the beginning of Chapter 2 we define a manifold V of dimension m by its embedding in a euclidean space $\mathbb{R}^n$ for some $n > m$. The embedding can be used to define an *induced metric* on V. If there are no topological complications then generically $n = m(m+1)/2$ since this is the number of independent components of a local metric on V. Let x^i be coordinates of the embedding space and e_i be a moving frame with respect to the flat metric. It can always be written in the form $e_i = \Lambda_i^j \partial_j$ with $\Lambda_i^j = \Lambda_i^j(x^k)$ a local rotation. In general a moving frame e_α exists only locally on V but, as we note in (2.1.19), it can always be completed by the $n - m$ basis elements of the normal bundle. Suppose that the resulting frame is a local rotation of a moving frame on the embedding space. Let θ^α be dual to the e_α. Then the induced metric on V is given by $g(\theta^\alpha \otimes \theta^\beta) = g^{\alpha\beta}$ where the $g^{\alpha\beta}$ are the components of the standard euclidean (or lorentzian) inner product on $\mathbb{R}^m$. We have already mentioned that noncommutative algebras are often explicitly constructed using generators and relations, a procedure which resembles the above definition of a manifold. It is to be expected then that a metric on a differential calculus over the algebra will bear resemblance to an induced metric. □

Example 3.24 The embedding ϵ of the complex numbers into $\Omega^0(M_n)$ by the diagonal matrices and the set of differentials d can be written out in the form of a sequence

$$0 \to \mathbb{C} \xrightarrow{\epsilon} \Omega^0(M_n) \xrightarrow{d} \Omega^1(M_n) \xrightarrow{d} \cdots$$
$$\xrightarrow{d} \Omega^{m^2-2}(M_n) \xrightarrow{d} \Omega^{m^2-1}(M_n) \xrightarrow{d} 0,$$

as in the commutative case. There is now however no notion of local exactness. As before if a p-form is in the kernel of d it is a p-*cocycle*; if it is in the image of d it is a p-*coboundary*. The p-th *cohomology group* $H^p(M_n; \mathbb{C})$ is the set of p-cocycles modulo the p-coboundaries and the groups $H^p(M_n; \mathbb{C})$ are a measure of the non-exactness of the sequence. Under the product induced from the product of forms they form a ring $H^*(M_n; \mathbb{C})$. The cohomology groups here are vector spaces over the complex numbers.

The operator Δ is a symmetric matrix and by definition an element of its kernel is an *harmonic form*. An arbitrary form α has then a decomposition as the sum $\alpha = \beta + \gamma$ of an harmonic form γ and a form β which lies in the image of Δ. From the definition of Δ follows then the Hodge decomposition

$$\Omega^*(M_n) = d\Omega^*(M_n) \oplus \delta\Omega^*(M_n) \oplus \Omega^*_\Delta \qquad (3.4.20)$$

of $\Omega^*(M_n)$ into three subspaces. As before $d\Omega^*(M_n)$ and $\delta\Omega^*(M_n)$ are respectively the exact and coexact forms and Ω^*_Δ are the harmonic forms. It follows also immediately from the identity (2.1.5) that the three spaces on the right-hand side of (3.4.20) are mutually orthogonal. A p-form α is harmonic if and only if $d\alpha = 0$ and $\delta\alpha = 0$. We have then a map

$$\Omega^p_\Delta \to H^p(M_n; \mathbb{C}). \qquad (3.4.21)$$

Let α be harmonic. If $\alpha = d\beta$ then $\delta d\beta = 0$ and therefore $\alpha = 0$. The kernel of (3.6.2) is therefore zero. Suppose on the other hand that α is closed. Then from the decomposition (3.4.20) we see that $\alpha - d\beta$ is harmonic for some $(p-1)$-form β. The image of (3.4.21) is therefore onto and the dimension of Ω^p_Δ is equal to the dimension of $H^p(M_n; \mathbb{C})$. This is the Hodge-de Rham theorem. $\square$

Example 3.25 For arbitrary $m \leq n$ the algebra $\Omega^*_m(M_n)$ can be identified with the tensor product of M_n and the exterior algebra of the dual of the vector space of derivations. The cohomology algebra $H^*_m(M_n)$ can be seen then to be equal to the tensor product of M_n and the cohomology algebra of the Lie algebra of $SL(m, \mathbb{C})$. It is a free graded-commutative algebra with unit, generated by elements $c_{2p-1} \in \Omega^{2p-1}_m(M_n)$ for $2 \leq p \leq m$. An harmonic 0-form therefore is necessarily a multiple of the unit matrix. There are no harmonic 1-forms or 2-forms. To within multiplication by a complex number there is only one harmonic 3-form. If $m = 2$ it can be chosen equal to the 3-form $\theta^1 \theta^2 \theta^3$. In this case the cohomology is equal to the cohomology of SU_2 which is a parallelizable U_1-bundle over S^2. $\square$

Example 3.26 It is only because of the particularly simple structure of the differential calculi over the matrix algebra that we have used that it was possible to introduce cohomology groups analogous to those used in describing the global structure of smooth manifolds. We shall discuss briefly in Section 6.2 a generalization which has been given which is applicable to a much larger class of differential calculi over more general algebras. There is no generalization of the Hodge-de Rham theory in these cases. □

Example 3.27 We have briefly hinted that to each noncommutative algebra with certain topological restrictions there is associated a space of pure states. We shall describe this is some more detail in the following chapter. On possible way of associating a metric or distance function to an algebra is then to define a metric on this space of pure states. It is to be expected that in the 'quasi-classical' approximation when the algebra is almost commutative the space of pure states is similar in some sense to the space which is present in the classical limit. In general however there are too many pure states. One has, for example, the impression that the algebra M_2 describes in some sense two points. Its space of pure states however is the entire 2-sphere. Also a space of pure states depends only on the algebra and is independent of the differential calculus. To use such a distance function one would have to restrict oneself to the algebras which we shall describe in Section 5.3 which are determined by the differential calculus. This would exclude many interesting algebras associated to quantum groups. □

Example 3.28 Consider the algebra and differential calculus of Example 3.10. Since $\eta^* = -\eta$ the differential is real. We write η in the form

$$\eta = \eta_1 - \eta_1^*$$

where

$$\eta_1 = \begin{pmatrix} 0 & 0 & a_1 \\ 0 & 0 & a_2 \\ 0 & 0 & 0 \end{pmatrix}. \tag{3.4.22}$$

Without loss of generality we can choose the euclidean 2-vector a_i of unit length: $\eta_{1i}^* \eta_{1i} = 1$. To form a basis for Ω_η^1 we must introduce a second matrix η_2. It is convenient to choose it of the same form (3.4.22) as η_1 and also with a 2-vector of unit length. We have then in Ω_η^2 the identity

$$\eta_i \eta_j^* = 0.$$

We shall further impose that

$$\eta_i^* \eta_j = \delta_{ij} e.$$

where e, the unit element in M_1, is a basis for the 2-forms. It follows that

$$d\eta_1 = e, \qquad d\eta_2 = 0.$$

There is a unitary element $u \in M_2 \subset M_3^+$ which exchanges η_1 and η_2: $\eta_2 = u\eta_1$. We have also $\eta_i u = 0$. The vector space of 1-forms is of dimension 4 over the complex numbers. The dimension of $\Omega^1_\eta \otimes \Omega^1_\eta$ is equal to 16 but the dimension of the tensor product $\Omega^1_\eta \otimes_{M^+_3} \Omega^1_\eta$ is equal to 5. One finds in fact over M^+_3 the relations

$$\eta_i \otimes \eta_j = 0, \quad \eta^*_i \otimes \eta^*_j = 0,$$
$$\eta^*_2 \otimes \eta_1 = 0, \quad \eta^*_1 \otimes \eta_2 = 0, \quad \eta^*_2 \otimes \eta_2 = \eta^*_1 \otimes \eta_1,$$

which leave

$$\eta_{ij} = \eta_i \otimes \eta^*_j, \qquad \zeta = \eta^*_1 \otimes \eta_1$$

as the 5 independent basis elements and we can make the identification

$$\Omega^1_\eta \otimes_{M^+_3} \Omega^1_\eta = M^+_3 \qquad (3.4.23)$$

using the map $\zeta \mapsto e$.

To define a symmetry condition we must first introduce the flip σ of (3.4.3). Because of the identification (3.4.23) it can be considered as a map from M^+_3 into itself and because of the bilinearity it is necessarily of the form

$$\sigma = \begin{pmatrix} \mu & 0 & 0 \\ 0 & \mu & 0 \\ 0 & 0 & -1 \end{pmatrix}, \qquad (3.4.24)$$

where $\mu \in \mathbb{C}$. The -1 in the lower right corner is imposed by the condition (3.6.5) below and which we have already encountered in Section 2.2. One finds that

$$\sigma(\eta_{ij}) = \mu \eta_{ij}, \qquad \sigma(\zeta) = -1.$$

The *Hecke relation* $(\sigma + 1)(\sigma - \mu) = 0$ is satisfied.

Suppose that $\mu \neq -1$ and define $\bigwedge^*$ and S^* to be the quotient of the tensor algebra by the ideal generated by the eigenvectors of respectively μ and -1. The images of the tensor products of order higher than 3 all vanish in both quotient spaces. For example consider the 3-tensor $\eta_i \otimes \zeta$. It vanishes in S^* because of the second factor and by the identity $\eta_i \otimes \zeta = \eta_{i1} \otimes \eta_1$ it vanishes in $\bigwedge^*$ because of the first factor. Therefore $\bigwedge^*$ is of dimension $5 + 4 + 1 = 10$ as a complex vector space and the map $\zeta \mapsto e$ induces an isomorphism of $\bigwedge^*$ with Ω^*_η. The S^* is of dimension $5 + 4 + 4 = 13$. It is an unusual fact that it is of finite dimension. If $\mu = -1$ then $\sigma = -1$. In this case it is natural to define $\bigwedge^*$ to be the entire tensor algebra. On the universal differential calculus the projection π of (3.6.5) is the identity and σ must be equal to -1.

Let g be a metric and set

$$h_{ij} = g(\eta_{ij}), \qquad h = g(\zeta).$$

Because of the bilinearity h_{ij} is given in terms of h_{11}. For example

$$h_{21} = u h_{11}.$$

If $\mu = 1$ then to within an overall scale the unique bilinear metric is given by

$$h_{ij} = \eta_i \eta_j^*, \qquad h = -e, \tag{3.4.25}$$

where the right-hand sides are considered as elements of M_3^+. Therefore h_{ij} takes its values in the M_2 factor of M_3^+ and h takes its values in the M_1 factor. From the form (3.4.24) of σ we see that the metric (3.4.25) is not symmetric; it is real on η_{ij} and imaginary on ζ. The unique metric on the Connes-Lott model is rather odd. $\square$

3.5   Yang-Mills connections

An important geometric problem is that of comparing vectors and forms defined at two different points of a manifold. The solution to this problem leads to the concepts of a *connection* and *covariant derivative*. There are two approaches, both of which were reviewed and compared within the context of classical geometry in some detail in Chapter 2. The traditional approach considers the connection as a primary object and the covariant derivative is defined in terms of it. But from the point of view of noncommutative geometry, which places primary importance on the algebra of functions, it is the second approach which is the more convenient and is the one which we shall consider here; the covariant derivative is defined as a linear map between modules which satisfies certain Leibniz rules. No attempt is made to define a noncommutative generalization of a connection as a 1-form on a principal fibre bundle.

We shall use here the expressions connection and covariant derivative synonymously. In fact we shall distinguish three different types of connections. A *left connection* or *Yang-Mills connection* is a connection on a left $\mathcal{A}$-module; it satisfies a left Leibniz rule. A *bimodule connection* is a connection on a general bimodule $\mathcal{M}$ which satisfies a left and a right Leibniz rule. In the particular case where $\mathcal{M}$ is the module of 1-forms we shall speak of a *linear connection*. The precise definitions are given below. A bimodule over an algebra $\mathcal{A}$ is also a left module over the enveloping algebra $\mathcal{A}^e$. So a bimodule can have a bimodule $\mathcal{A}$-connection as well as a left $\mathcal{A}^e$-connection.

Let $\mathcal{A}$ be an arbitrary algebra and $(\Omega^*(\mathcal{A}), d)$ a differential calculus over $\mathcal{A}$. One defines a Yang-Mills connection on a left $\mathcal{A}$-module $\mathcal{H}$ as a map

$$\mathcal{H} \xrightarrow{D} \Omega^1(\mathcal{A}) \otimes_{\mathcal{A}} \mathcal{H} \tag{3.5.1}$$

which satisfies the left Leibniz rule

$$D(f\psi) = df \otimes \psi + f D\psi \tag{3.5.2}$$

for arbitrary $f \in \mathcal{A}$ and $\psi \in \mathcal{H}$. This map has a natural extension

$$\Omega^*(\mathcal{A}) \otimes_{\mathcal{A}} \mathcal{H} \xrightarrow{D} \Omega^*(\mathcal{A}) \otimes_{\mathcal{A}} \mathcal{H} \tag{3.5.3}$$

given by
$$D(\alpha \otimes \psi) = d\alpha \otimes \psi + (-1)^p \alpha D\psi$$

for $\alpha \in \Omega^p(\mathcal{A})$. In particular, since

$$D(df \otimes \psi) = -df\, D\psi,$$

one easily verifies that the operator D^2 is left-linear:

$$D^2(f\psi) = fD^2\psi. \tag{3.5.4}$$

Since there is no ambiguity we shall occasionally omit the tensor product symbol:

$$D(\alpha\psi) = d\alpha\psi + (-1)^p \alpha D\psi.$$

Example 3.29 Suppose that $\mathcal{A}$ is a $*$-algebra and that $\mathcal{H}$ is a left $M_n(\mathcal{A})$-module and a right $\mathcal{A}$-module. One can introduce a metric g on $\mathcal{H}$ by the formula

$$g(\psi_1, \psi_2) = \psi_1^* \phi \psi_2, \qquad \phi \in M_n(\mathcal{A}). \tag{3.5.5}$$

We extend the metric to the tensor product image of D by

$$g(\psi_1, \xi \otimes \psi_2) = \xi g(\psi_1, \psi_2), \qquad \xi \in \Omega^1(\mathcal{A}).$$

One can impose a compatibility condition

$$d \circ g(\psi_1, \psi_2) = g(D\psi_1, \psi_2) + g(\psi_1, D\psi_2)$$

between the metric and the covariant derivative. We have used here the same symbol ϕ to describe the matrix which defines the metric as we shall use below to describe a component of the connections since it often happens in physical applications that they are one and the same. The metric (3.5.5) has been used to give a geometric description of the interaction of the pion field with the nucleon and the compatibility condition in this case is a relation between Yang-Mills potentials and the pion currents.   □

We are primarily interested here in linear connections but there are two examples of Yang-Mills connections which we mention since we shall have occasion to use them in Chapter 8. It is good notation to put a prime on the new potential as we do below but not on the D. The former depends on the gauge whereas the later is a map defined on all of $\mathcal{H}$.

Example 3.30 Consider the differential calculus $\Omega^*(M_n)$ constructed in Section 3.1. In Chapter 2 we showed that on the trivial U_1-bundle a connection is equivalent to a gauge potential, an anti-hermitian element of $\Omega^1(\mathcal{C}(V))$. Since it turns out to be possible in this case, as a pedagogical exercise we define by analogy a connection ω over the matrix algebra M_n to be an anti-hermitian element of $\Omega^1(M_n)$. To be strictly coherent with the notation of Chapter 2 we should designate this by A with an additional symbol to indicate that it is a noncommutative gauge potential. However, in the noncommutative case we shall have no occasion to refer to a connection as a 1-form on a principal bundle and so we can use the symbol ω to designate a general gauge potential.

The gauge transformations of the trivial U_1-bundle over V are the unitary elements of $\mathcal{A} = \mathcal{C}(V)$. We choose accordingly the unitary elements U_n of M_n to be the group of gauge transformations. An element g of U_n defines a map of $\Omega^1(M_n)$ into itself of the form

$$\omega' = g^{-1}\omega g + g^{-1}dg.$$

In particular because of the identity $dg = -[\theta, g]$ it can be readily seen that the 1-form θ we introduced in Section 3.1 is invariant under the action of U_n:

$$\theta' = \theta. \tag{3.5.6}$$

We use it then as a preferred origin for the connections and write

$$\omega = \theta + \phi. \tag{3.5.7}$$

It follows that ϕ transforms under the adjoint action of U_n:

$$\phi' = g^{-1}\phi g.$$

We define the *curvature form* Ω as in the commutative case by Equation (2.2.24) but where now ω is an anti-hermitian element of $\Omega^1(M_n)$ and d is defined from (3.1.9). By (3.1.12) we see that, as a connection, θ has vanishing curvature. The fact that it is invariant under a gauge transformation means in particular that it cannot be made to vanish by a choice of gauge. We have then a connection with vanishing curvature but which is not gauge-equivalent to zero. If we were considering an algebra of functions over a compact manifold, the existence of such a 1-form would be due to the non-trivial topology of the manifold. A short calculation yields

$$\Omega = \frac{1}{2}\Omega_{ab}\theta^a\theta^b \tag{3.5.8}$$

where

$$\Omega_{ab} = [\phi_a, \phi_b] - C^c{}_{ab}\,\phi_c. \tag{3.5.9}$$

Suppose we attempt to define a derivative $D\psi$ of $\psi \in \mathcal{H}$ as in (2.2.26):

$$D\psi = d\psi + \omega\psi. \tag{3.5.10}$$

Since $\mathcal{H}$ is an M_n-module it inherits a U_n-module structure, on which the definition of the product $\omega\psi = \omega \otimes \psi$ depends. If $\mathcal{H}$ is a left M_n-module, for example $\mathcal{H} = \mathbb{C}^n$, the action of U_n is left multiplication. If we impose the Leibniz rule

$$d(f\psi) = df \otimes \psi + f d\psi,$$

for d we see from (3.1.13) that we must define $d\psi$ as

$$d\psi = -\theta\psi.$$

This is possible because of the existence of the form θ. One of the problems with a decomposition of a derivative into the sum of two terms as in (3.5.10) is that in general it is not possible to give a meaning to the expression $d\psi$.

We can define a covariant derivative by the map

$$D_{(0)}\psi = -\theta\psi$$

since indeed from (3.1.13) $D_{(0)}$ satisfies the left Leibniz rule:

$$D_{(0)}(f\psi) = -\theta f\psi = [f,\theta]\psi - f\theta\psi = df\psi + fD_{(0)}\psi.$$

However, $D_{(0)}$ is not covariant under the left action of U_n since $\theta' \neq g^{-1}\theta g$. On the other hand from (3.5.7) we find

$$D\psi = \phi\psi$$

and so $D(g^{-1}\psi) = g^{-1}D\psi$. The derivative D is covariant under the left action of U_n but does not define a covariant derivative in the sense we have given the expression; the left Leibniz rule (3.5.2) is not respected. If one extends D to $\Omega^*(M_n) \otimes_{M_n} \mathcal{H}$ one finds that

$$D^2\psi = \Omega\psi \qquad\qquad\qquad (3.5.11)$$

but contrary to the commutative case (2.2.30) now the identity (3.5.4) is not satisfied.

If $\mathcal{H}$ is a bimodule, for example the free module $\mathcal{H} = M_n$, by the previous logic we must define $d\psi = -[\theta, \psi]$. There are now two possibilities for the action of U_n. We can choose $\mathcal{H}$ to be a bimodule with the adjoint action or a right module with right multiplication. If we decompose ω again we find in the first case

$$D\psi = d\psi + [\omega, \psi] = [\phi, \psi]. \qquad\qquad (3.5.12)$$

This is covariant under the adjoint action of U_n but again does not define a covariant derivative in the sense we have given the expression. In the second case we find

$$D\psi = d\psi - \psi\omega = -\theta\psi - \psi\phi = D_{(0)}\psi - \psi\phi. \qquad\qquad (3.5.13)$$

This is covariant only under the right action of U_n. By our conventions if we expand $\phi = \phi_a\theta^a$ then $\psi\phi \equiv \theta^a\psi\phi_a$.

In the U_1 case of Chapter 2 we identify the gauge transformations as the unitary elements of the algebra $\mathcal{A}$. A gauge transformation is an isomorphism of $\mathcal{H}$ as an $\mathcal{A}$-module; it commutes with the action of $\mathcal{A}$. In the noncommutative case if we wish a gauge transformation to be a morphism of an $\mathcal{A}$-module $\mathcal{H}$ then the order is important. If we choose the algebra to act on $\mathcal{H}$ from the left, for example, then the gauge transformations will have to act from the right. They must be completely independent of each other. If we try to maintain the analogy with the trivial U_1 case and identify the gauge transformations as the unitary elements of the algebra, they both must act on $\mathcal{H}$ in the same way and a gauge transformation can no longer be a morphism of $\mathcal{H}$ as an $\mathcal{A}$-module. A derivative must either satisfy (2.2.29) or be covariant under the left action of the unitary elements of the algebra. It cannot be both.

The only way to circumvent the incompatibility is to choose independent actions for the gauge group and the algebra. Since we have supposed that $\mathcal{H}$ is a left M_n-module we must choose it to be a right U_r-module where n and r are completely independent. We can then define a covariant derivative by (3.5.13) but with a ϕ which takes its values in the Lie algebra of U_r. The derivative $D_{(0)}$ is covariant under the right action of U_r:

$$D_{(0)}(\psi g) = (D_{(0)}\psi)g.$$

The most general covariant derivative D is the sum of $D_{(0)}$ and a left-module morphism $\mathcal{H} \to \Omega^1(M_n) \otimes_{\mathcal{A}} \mathcal{H}$ which, because of (3.1.34), can be written as a left-module morphism of $\mathcal{H}$. When $r = n$ this is the same as (3.5.13). If we extend D as above we find again (3.5.11). We must, however, correctly interpret the formula since the forms are supposed to be on the left of the module and the gauge action to be from the right:

$$D^2\psi = -\frac{1}{2}\theta^a\theta^b\psi\,\Omega_{ab}.$$

One verifies that the identity (3.5.4) is satisfied. Below in Section 3.7 we shall argue that this is a desirable feature.

In the commutative case, in order to identify a connection with a gauge potential one must be considering a trivial bundle, in which case the space of sections is a free module. When $\mathcal{H} = M_n$, the free module of rank 1, we can choose the unit element 1 as preferred basis and write $\psi = \psi \cdot 1$. If we define $D1 = -\omega$ then the covariant derivative of a general element $\psi \in \mathcal{H}$ is calculated using the left Leibniz rule (3.5.2):

$$D\psi = d\psi + \psi D1 = -[\theta, \psi] - \psi\omega = -\theta\psi - \psi\phi.$$

This again is Equation (3.5.13). In the case where the algebra acts on the right and the group on the left then we set $D1 = \omega$. We find

$$D\psi = d\psi + (D1)\psi = \psi\theta + \phi\psi \qquad (3.5.14)$$

and

$$D^2\psi = \frac{1}{2}\Omega_{ab}\psi\theta^a\theta^b.$$

□

Example 3.31 Using the geometry of M_n described in Example 3.19 one can write the Maxwell action for the gauge potential ω. From Example 2.16 it is given by

$$S[\phi] = -\frac{1}{4}\mathrm{Tr}(\Omega_{ab}\Omega^{ab}).$$

The minus sign is due to the euclidean signature and the trace is what replaces the integral of ordinary geometry. From (3.5.9) we see that $S[\phi]$ vanishes when $\phi = 0$ and also when it lies on the gauge orbit of λ_a : $\phi_a = g^{-1}\lambda_a g$. If one uses the differential calculus $\Omega_m^*(M_n)$ described in Section 3.1 one sees that $S[\phi]$ vanishes whenever ϕ_a is a representation of $\underline{su}_m$. In particular when $m = 2$ the number of zeros is given by the *partition function* $p(n)$, the number of ways one can partition the integer n into a set of non-increasing positive integers. The solution $\phi = 0$ corresponds to the partition $(1,\ldots,1)$ and the irreducible representation corresponds to the partition (n). If $n = 2$ there are no others. For large n the value of $p(n)$ is given by

$$p(n) \approx \frac{e^{\pi\sqrt{2n/3}}}{4n\sqrt{3}}.$$

We shall return to this action in Example 7.36 and Example 8.4.   □

Example 3.32 With slight modification the above formalism can be applied to the case of the matrix algebra M_n with a $\mathbb{Z}_2$ grading. A connection is defined to be an anti-hermitian element of Ω_η^-, the odd part of Ω_η^*, where $\eta^2 = -1$. The complex conjugate $f \mapsto f^*$ in M_n respects the grading. Define the group of gauge transformations to be the unitary elements of M_n^+. As in the previous example η is gauge invariant. We can use it as a preferred origin for the elements of Ω_η^- and write

$$\omega = \eta + \phi$$

where ϕ transforms as it did in Example 3.30. Instead of (3.5.8) we find that the *curvature form* Ω is given by

$$\Omega = 1 + \phi^2. \tag{3.5.15}$$

The constant comes from the right-hand side of (3.2.4). Any connection which is gauge invariant gives rise to a curvature which satisfies the equation

$$g^{-1}\Omega g = \Omega.$$

The curvature must therefore either vanish or be of the form (3.2.4) to within a normalization.

Suppose that $\mathcal{H}$ is a graded M_n-bimodule. A left connection is given by the map

$$D_{(0)}\psi = -\eta\psi.$$

The left Leibniz rule follows from (3.2.1). The most general covariant derivative can be written in the form

$$D\psi = -\eta\psi - \psi\phi.$$

If we extend D to $\Omega_\eta^* \otimes \mathcal{H}$ as a graded derivation we find

$$D^2\psi = -\psi\Omega.$$

$\square$

Example 3.33 Let $\mathcal{A}$ be an algebra and $\hat{\mathcal{A}}$ a deformation thereof. For the moment we have at our disposal only matrix algebras which are rather rigid, but in the following chapter we shall encounter more general infinite-dimensional algebras, like the algebras $\mathcal{A}_k$ of Example 4.7, which can be readily deformed. Sections 4.2 and 4.4 are both devoted to deformations of algebras. The algebra $\mathbb{C} \times \mathbb{C}$ of Example 3.32 is a singular contraction of the algebra M_2. Associated to the former are two gauge potentials, the photon γ and a massive neutral vector boson Z_0; the latter has also a massive charged W. The contraction can be implemented by letting the W mass tend to infinity. Another interesting example to which we shall return in Example 8.6, is the matrix algebra $\mathcal{A} = \mathbb{H} \times \mathbb{C}$, which is relevant in formulating noncommutative versions of the standard model. The $\mathbb{H}$ is the ring of quaternions. Let $\mathcal{G}$ be the group of unitary elements of $\mathcal{A}$. The group $\hat{\mathcal{G}}$ will be accordingly a deformation of $\mathcal{G}$. In both examples one has $\mathcal{G} = U_2$ but in the first $\hat{\mathcal{G}} = U_1 \times U_1$ and in the second $\hat{\mathcal{G}} \simeq SU_2 \times U_1$. One can think of the deformation parameter in this case as the Weinberg angle.

We assume an algebra homomorphism

$$\hat{\mathcal{A}} \xrightarrow{\rho} \mathcal{A} \qquad\qquad (3.5.16)$$

of $\hat{\mathcal{A}}$ onto $\mathcal{A}$ which in turn induces a group homomorphism of $\hat{\mathcal{G}}$ onto $\mathcal{G}$. Let $\mathcal{H}$ be a right $\mathcal{A}$-module and $\hat{\mathcal{H}}$ be a right $\hat{\mathcal{A}}$-module. We shall place a hat on an element of $\mathcal{H}$ whenever it is necessary to distinguish the $\hat{\mathcal{A}}$-module structure. For simplicity we shall suppose that both modules are free over their respective algebras and so the map ρ can be extended to a map

$$\hat{\mathcal{H}} \xrightarrow{\rho} \mathcal{H}$$

between the two of them. We shall simplify even further and suppose that the module is of rank one. It can be identified therefore with the respective algebra and each identification is equivalent to a choice of gauge. We choose $\psi_0 \in \mathcal{H}$ as basis of $\mathcal{H}$ as both $\mathcal{A}$-module and $\hat{\mathcal{A}}$-module and we write $\psi = \psi_0 * f$ and $\hat{\psi} = \psi_0 \hat{*} f$. This defines the map ρ in terms of the products. We shall suppose that the potential A lies in the Lie algebra $\underline{g}$ of $\mathcal{G}$ and likewise that $\hat{A}$ lies in the Lie algebra $\underline{\hat{g}}$ of $\hat{\mathcal{G}}$. We shall suppose that the gauge group acts on the left. The left action of $\mathcal{G}$ on $\mathcal{H}$ is compatible with the algebra action from the right. This condition is automatic in normal Yang-Mills theory where the two actions always commute. Since the derivative is covariant from the left one has also

$$D(g^{-1}\psi) = g^{-1}D\psi, \qquad g \in \mathcal{G}.$$

In ordinary geometry the case we are considering would be called an abelian gauge theory. This is in fact quite general enough since 'non-abelian' gauge theory can be incorporated simply by including a matrix factor, as discussed in Section 3.3. It is only important that the matrix factor be the same for both algebras since otherwise the map ρ in general would not be interesting.

If ρ is only a vector-space morphism then the induced map between the groups will not be a homomorphism and the corresponding map between the Lie algebras will not be compatible with the respective Lie brackets. Of interest are vector-space morphisms which are singular limits of algebra morphisms. The map between the Lie algebras will then be a *contraction*. The results of Section 2.3 are not applicable; even simple Lie algebras have contractions.

We suppose finally that there is a differential calculus $\Omega^*(\mathcal{A})$ over $\mathcal{A}$ and a differential calculus $\hat{\Omega}^*(\hat{\mathcal{A}})$ over $\hat{\mathcal{A}}$ and that the map ρ can be extended to an algebra morphism

$$\hat{\Omega}^*(\hat{\mathcal{A}}) \xrightarrow{\ \rho\ } \Omega^*(\mathcal{A})$$

of the latter onto the former. We can now come to the point of the example.

Let D and $\hat{D}$ be covariant derivatives defined on respectively $\mathcal{H}$ and $\hat{\mathcal{H}}$. We introduce the gauge potentials as usual by the conditions

$$D\psi_0 = \psi_0 * A, \qquad \hat{D}\psi_0 = \psi_0 \hat{*} \hat{A}.$$

These define D and $\hat{D}$ on all of $\mathcal{H}$ either by the Leibniz rule or by the gauge covariance. If $f \simeq 1 + h$ then to first order in h we can write

$$D\psi = \psi * (A + Dh), \qquad \hat{D}\psi = \psi \hat{*} (\hat{A} + \hat{D}h).$$

We have here introduced the covariant derivatives

$$Dh = dh + [A, h], \qquad \hat{D}h = \hat{d}h + [\hat{A}; h]$$

of an element $h \in \mathcal{A}\ (\hat{\mathcal{A}})$, with

$$[A, h] = A * h - h * A, \qquad [\hat{A}; h] = \hat{A} \hat{*} h - h \hat{*} \hat{A}.$$

Conversely, given A and $\hat{A}$ one can construct a map

$$\text{SW} : D \longrightarrow \hat{D}$$

between the two derivatives by assuring that the two Leibniz rules are satisfied. The map SW becomes then an equation because of integrability conditions; it must be well-defined on all of $\mathcal{H}$.

If ρ is an automorphism then $\hat{D} - D$ is a (right) module morphism. One can neglect the distinction between the two products and write

$$\hat{D}h = Dh + [\Gamma, h] \qquad (3.5.17)$$

with $\Gamma = \hat{A} - A$. If we define the variation

$$\delta_h \Gamma = \hat{D}h - Dh \qquad (3.5.18)$$

of Γ under multiplication by $f \simeq 1 + h$, we see that it is given by

$$\delta_h \Gamma = [\Gamma, h]. \qquad (3.5.19)$$

This is the well-known formula which expresses the gauge covariance of the difference between two connections. The map SW is a generalization of this formula to situations where the two connections in question are with respect to two different gauge groups.

In general, if ρ is not an automorphism, then Equation (3.5.19) will have no solution and we cannot define Γ as we have done. Since ρ is surjective we can introduce a function $\gamma(h)$ with values in $\mathcal{A}$ such that

$$\psi_0 \,\hat{*}\, (1 + h) = \psi_0 * (1 + h)(1 + \gamma).$$

This implies that $\psi_0 \,\hat{*}\, dh = \psi_0 * d(h + \gamma)$ and therefore that

$$\hat{D}\psi = \psi \,\hat{*}\, \hat{A} + \psi * \hat{D}(h + \gamma[h]).$$

Using the definition of $\delta_h \Gamma$ given above one obtains written as

$$\delta_h \Gamma = D\gamma + \hat{D}h - Dh = D\gamma + [\Gamma, h] + [\hat{A}; h] - [\hat{A}, h]. \qquad (3.5.20)$$

If ρ is not an automorphism then to compensate for the difference between ρ and an automorphism we have introduced an element $\gamma \in \underline{g}$, which is equivalent to interpreting the modification of the product as a change of gauge. This is one aspect of what is known as the *Seiberg-Witten map*. We shall return to it later at the end of Section 8.3.   □

Example 3.34 We saw in Equation (3.5.3) that both D and $\hat{D}$ can be extended to the entire differential calculus. In the special cases we are considering here both of the differential calculi can be written in the form (6.1.9) which we shall mention in Chapter 6:

$$\Omega^*(\mathcal{A}) = \mathcal{A} \otimes \bigwedge{}^*_C$$

where the second factor is the deformed exterior algebra over the vector space spanned by the frame. If

$$\bigwedge{}^*_C = \hat{\bigwedge}{}^*_{\hat{C}}$$

then both ρ and SW can be extended also. We can write

$$D\psi = \theta^a D_a \psi, \qquad \hat{D}\psi = \hat{\theta}^a \hat{D}_a \psi.$$

We shall restrict our attention here to the important special case with the projector $P^{ab}{}_{cd}$ given by the expression (3.1.44). Anticipating (4.2.7) below, one finds to lowest order, the expression

$$\hat{e}_a f = e_a f + i \hbar \theta^{bc} [\lambda_b, \lambda_a] * e_c f$$

for the 'partial derivatives'. The frame is gauge invariant: $\delta_h \theta^a = 0$. The solution to Equation 3.5.20 is difficult to find in general but if the deformation matrix θ^{ab} which defines the algebra $\hat{\mathcal{A}}$ in terms of $\mathcal{A}$ is small a formal Taylor-series expansion can be given. In general the Seiberg-Witten map must be extended to the case where the covariant derivative includes a gravitational contribution. We have changed the structure of the algebra without changing that of the differential calculus and this is not always possible.   □

Example 3.35 Let again $\mathcal{A}$ be an algebra with $\hat{\mathcal{A}}$ a deformation and suppose that over both algebras are differential calculi with frames, respectively θ^i and $\hat{\theta}^i$, We argued in the previous section that these frames are to a certain extent unique and they considerably restrict the class of metrics g and $\hat{g}$ on the respective algebras. We would like to require that each metric be symmetric as defined in (3.4.2). There is however a certain ambiguity in the choice of 'symmetry'. There is a natural operation given by $S^{ij}{}_{kl} = C^{ij}{}_{kl}$, as we saw in Section 3.1, but there are other possibilities, as discussed in Example 3.41. In particular, ordinary differential geometry can have metrics in the present formalism which are symmetric with respect to exotic symmetry operations. We shall give an example of this in Section 4.1. We shall write the inverse matrix g_{ij} to the matrix g^{ij} of the frame components of such a metric in the form

$$g_{ij} = \eta_{ij} + B_{ij} \tag{3.5.21}$$

with

$$\eta_{ij} = \frac{1}{2} g_{(ij)}, \qquad B_{ij} = \frac{1}{2} g_{[ij]}$$

respectively the symmetric and antisymmetric parts of the metric matrix, with respect to the ordinary flip. We shall normally choose η_{ij} to be the ordinary Minkowski or euclidean metric matrix.   □

3.6   Linear connections

A covariant derivative on the module $\Omega^1(\mathcal{A})$ must satisfy the left Leibniz rule (3.5.2). But $\Omega^1(\mathcal{A})$ has also a natural structure as a right $\mathcal{A}$-module and one must be able to write a corresponding right Leibniz rule in order to hope to construct a bilinear curvature. Quite generally let $\mathcal{M}$ be an arbitrary bimodule. We wish to construct a covariant derivative

$$\mathcal{M} \xrightarrow{D} \Omega^1(\mathcal{A}) \otimes_{\mathcal{A}} \mathcal{M} \tag{3.6.1}$$

which satisfies both a left and a right Leibniz rule. We propose to introduce a generalized flip

$$\mathcal{M} \otimes_{\mathcal{A}} \Omega^1(\mathcal{A}) \xrightarrow{\sigma} \Omega^1(\mathcal{A}) \otimes_{\mathcal{A}} \mathcal{M} \tag{3.6.2}$$

and define the right Leibniz rule as

$$D(\xi f) = \sigma(\xi \otimes df) + (D\xi)f \tag{3.6.3}$$

for arbitrary $f \in \mathcal{A}$ and $\xi \in \mathcal{M}$. The purpose of the map σ is to bring the differential to the left while respecting the order of the factors. In general $\sigma^2 \neq 1$. In the case of ordinary differential geometry it follows immediately that σ is necessarily of the form

$$\sigma(\xi \otimes \eta) = \eta \otimes \xi.$$

Taking the covariant derivative of the identity $(\xi f)g = \xi(fg)$ yields the identity

$$\sigma(\xi \otimes dfg) = D(\xi f)g - (D\xi)fg.$$

If one expands $D(\xi f)$ and uses once more (3.6.3) then one finds that

$$\sigma(\xi \otimes dfg) = \sigma(\xi \otimes df)g.$$

The map σ must be right $\mathcal{A}$-linear. Taking the covariant derivative of the identity $(f\xi)g = f(\xi g)$ yields the identity

$$\sigma(f\xi \otimes dg) = f(D\xi g) - f(D\xi)g.$$

If one expands $D(\xi g)$ and uses once more (3.6.3) then one finds that

$$\sigma(f\xi \otimes dg) = f\sigma(\xi \otimes dg).$$

The map σ must be left $\mathcal{A}$-linear.

Example 3.36 The bilinearity of σ is related to a *cocycle condition*. In Section 6.2 we introduce the Hochschild homology of an algebra with values in an $\mathcal{A}$-bimodule $\mathcal{M}$. Dual to this one can define Hochschild cohomology. A p-cochain is a multilinear map ϕ of the tensor product of p copies of $\mathcal{A}$ into $\mathcal{M}$. In Section 6.2 we mention the particular case $\mathcal{M} = \mathbb{C}$. In the general case the first two terms in the sequence of cochains

$$C^1(\mathcal{A}; \mathcal{M}) \xrightarrow{\delta} C^2(\mathcal{A}; \mathcal{M}) \xrightarrow{\delta} C^3(\mathcal{A}; \mathcal{M})$$

are given by

$$\delta\phi_1(f, g) = f\phi_1(g) - \phi_1(fg) - \phi_1(f)g,$$
$$\delta\phi_2(f, g, h) = f\phi_2(g, h) - \phi_1(fg, h) + \phi_1(f, gh) - \phi_1(f, g)h,$$

where $\phi_p \in C^p(\mathcal{A}; \mathcal{M})$. Consider in particular the cochains defined by

$$\phi_1(f) = D(df), \qquad \phi_2(f, g) = -(1 + \sigma)(df \otimes dg).$$

A short calculation yields the identity $\phi_2 = \delta\phi_1$. The map ϕ_2 must satisfy then the cocycle condition $\delta\phi_2 = 0$. From the identity

$$\delta\phi_2(f, g, h) = \sigma(f dg \otimes dh) - f\sigma(dg \otimes dh) - \sigma(df \otimes dgh) + \sigma(df \otimes dg)h$$

we see then that the cocycle condition connects the left-linearity of σ with the right-linearity.   □

We define a bimodule $\mathcal{A}$-connection to be the couple (D, σ). If, in particular, $\mathcal{M} = \Omega^1(\mathcal{A})$ then we shall refer to the bimodule $\mathcal{A}$-connection as a *linear connection*. Although we shall here be concerned exclusively with this case we shall sometimes use the more general notation to be able to distinguish the two copies of $\Omega^1(\mathcal{A})$ on the right-hand side of (3.6.1).

We define the *torsion map*

$$\Theta : \Omega^1(\mathcal{A}) \to \Omega^2(\mathcal{A}) \tag{3.6.4}$$

by $\Theta = d - \pi \circ D$. It is left-linear. A short calculation yields

$$\Theta(\xi)f - \Theta(\xi f) = \pi \circ (1 + \sigma)(\xi \otimes df).$$

We shall impose the condition

$$\pi \circ (\sigma + 1) = 0 \tag{3.6.5}$$

on σ. It could also be considered as a condition on π. In fact in ordinary geometry it is the definition of π; a 2-form can be considered as an antisymmetric tensor. Because of this condition the torsion is a bilinear map.

Using σ one can also construct an extension

$$\mathcal{M} \otimes_{\mathcal{A}} \mathcal{M} \xrightarrow{D_2} \Omega^1(\mathcal{A}) \otimes_{\mathcal{A}} \mathcal{M} \otimes_{\mathcal{A}} \mathcal{M}$$

by

$$D_2(\xi \otimes \eta) = D\xi \otimes \eta + \sigma_{12} \circ (\xi \otimes D\eta). \tag{3.6.6}$$

We have here used again the notation $\sigma_{12} = \sigma \otimes 1$ which we shall use often in Section 4.4. The operator $D_2 \circ D$ is not in general left-linear. However, from the condition (3.6.5) follows the relation, with $\pi_{12} = \pi \otimes 1$,

$$D(\xi \otimes \eta) = \pi_{12} \circ D_2(\xi \otimes \eta) + \Theta(\xi) \otimes \eta$$

between the D given in the extension (3.5.3) and D_2. It follows that

$$D^2 = \pi_{12} \circ D_2 \circ D + \Theta.$$

The left-hand side of this equation is defined for a left $\mathcal{A}$-connection whereas the right-hand side is defined only in the case of a linear connection. In particular one can conclude that $\pi_{12} \circ D_2 \circ D$ is left-linear.

We introduce the notion of metric-compatibility exactly as in the commutative case as expressed by Formula (2.2.7). Let g be a metric (3.4.1) and (D, σ) a linear connection. Both $g_{23} \circ D_2(\xi \otimes \eta)$ and $dg(\xi \otimes \eta)$ are elements of $\Omega^1(\mathcal{A})$. The linear connection is *metric-compatible* if

$$g_{23} \circ D_2 = d \circ g. \tag{3.6.7}$$

Example 3.37 Return to the Example 3.10 on which we constructed a metric in Example 3.28. A covariant derivative is given by

$$D_{(0)}\xi = -\eta \otimes \xi + \sigma(\xi \otimes \eta)$$

Because of the identification (3.4.23) any other D must differ from this one by a bimodule morphism ϕ of Ω^1_η into M_3^+. Consider the images of the basis η_i of Ω^1_η. By the bilinearity we must have $\phi(\eta_i)u = 0$. But u is unitary in M_2. Therefore $\phi = 0$ and $D \equiv D_{(0)}$ defines the unique linear connection. The torsion vanishes because of the identity

$$\Theta_{(0)}(\xi) = d\xi + [\eta, \xi] = 0$$

which follows from the definition (3.2.6) of the extension of the differential. The condition (3.6.7) that the connection be compatible with the (unique) metric (3.4.25) is expressed by the equations

$$dh_{11} = -\mu\eta_1 h + \eta_1^* h_{11},$$

$$dh = -\eta_1 h + \mu\eta_1^* h_{11}.$$

They have no solution unless $\mu^2 = 1$. □

Example 3.38 In Section 6.1 we shall construct the universal calculus over an algebra $\mathcal{A}$ and we shall see that the projection π of (3.1.39) is the identity map. The ordinary differential d_u is clearly a covariant derivative. From (3.6.5) it follows that $\sigma = -1$ so D_2 is also given by d_u. The torsion vanishes and, anticipating the results of Section 3.7, so does the curvature. Conversely let D define an arbitrary linear connection on the 1-forms of the universal calculus. If we require the torsion to vanish then $D = d_u$. The only torsion-free linear connection is the trivial one. $\square$

If the algebra $\mathcal{A}$ is a $*$-algebra we shall require the *reality condition*

$$D\xi^* = (D\xi)^*$$

on the connection. This must be consistent with the Leibniz rules. There is little one can conclude in general but if the differential is based on real derivations then from the equalities

$$(D(f\xi))^* = D((f\xi)^*) = D(\xi^* f^*)$$

one finds the conditions

$$(df \otimes \xi)^* + (f D\xi)^* = \sigma(\xi^* \otimes df^*) + (D\xi^*)f^*.$$

Since this must be true for arbitrary f and ξ we conclude that

$$(df \otimes \xi)^* = \sigma(\xi^* \otimes df^*)$$

and

$$(f D\xi)^* = (D\xi^*)f^*.$$

We shall suppose that the involution is such that in general

$$(\xi \otimes \eta)^* = \sigma(\eta^* \otimes \xi^*). \tag{3.6.8}$$

A change in σ therefore implies a change in the definition of an hermitian tensor. From the compatibility conditions (3.1.41) and (3.6.5) one can deduce (3.1.55). The condition that the star operation be in fact an involution places a constraint on the map σ:

$$(\sigma(\eta^* \otimes \xi^*))^* = (\xi \otimes \eta).$$

It is clear that there is an intimate connection between the reality condition and the right Leibniz rule.

Let $\mathcal{A}^e$ be the enveloping algebra of $\mathcal{A}$. A bimodule $\mathcal{M}$ can also be considered then as a left $\mathcal{A}^e$-module. The differential calculus $\Omega^*(\mathcal{A})$ has a natural extension to a differential calculus $\Omega^*(\mathcal{A}^e)$ given by

$$\Omega^*(\mathcal{A}^e) = \Omega^*(\mathcal{A}) \otimes \Omega^*(\mathcal{A}^{op}) = (\Omega^*(\mathcal{A}))^e$$

with $d(a \otimes b) = da \otimes b + a \otimes db$. We noticed however in Section 3.3 that this is not the only choice. Suppose that $\mathcal{M}$ has a left $\mathcal{A}^{\mathrm{e}}$-connection

$$\mathcal{M} \xrightarrow{D^{\mathrm{e}}} \Omega^1(\mathcal{A}^{\mathrm{e}}) \otimes_{\mathcal{A}^{\mathrm{e}}} \mathcal{M}. \qquad (3.6.9)$$

From the equality

$$\Omega^1(\mathcal{A}^{\mathrm{e}}) = (\Omega^1(\mathcal{A}) \otimes \mathcal{A}^{\mathrm{op}}) \oplus (\mathcal{A} \otimes \Omega^1(\mathcal{A}^{\mathrm{op}}))$$

and using the identification

$$(\mathcal{A} \otimes \Omega^1(\mathcal{A}^{\mathrm{op}})) \otimes_{\mathcal{A}^{\mathrm{e}}} \mathcal{M} \simeq \mathcal{M} \otimes_{\mathcal{A}} \Omega^1(\mathcal{A}) \qquad (3.6.10)$$

given by

$$(1 \otimes \xi) \otimes \eta \mapsto \eta \otimes \xi$$

we find that we have

$$\Omega^1(\mathcal{A}^{\mathrm{e}}) \otimes_{\mathcal{A}^{\mathrm{e}}} \mathcal{M} = (\Omega^1(\mathcal{A}) \otimes_{\mathcal{A}} \mathcal{M}) \oplus (\mathcal{M} \otimes_{\mathcal{A}} \Omega^1(\mathcal{A})). \qquad (3.6.11)$$

The covariant derivative D^{e} splits then as the sum of two terms

$$D^{\mathrm{e}} = D_L + D_R. \qquad (3.6.12)$$

From the identifications it is obvious that D_L (D_R) satisfies a left (right) Leibniz rule and is right (left) $\mathcal{A}$-linear.

Example 3.39 In the case of ordinary geometry with $\mathcal{A}$ equal to the algebra $\mathcal{C}(V)$ of smooth functions on a smooth manifold V the algebra $\mathcal{A}^{\mathrm{e}}$ is the algebra of smooth functions in two variables. So D^{e} cannot exist if $\Omega^*(\mathcal{A})$ is the algebra of de Rham differential forms. In fact D_L would satisfy a left Leibniz rule and be left linear since left and right multiplication are equal. In general let $\mathcal{M}$ be the $\mathcal{A}$-module of smooth sections of a vector bundle over V. Then $\mathcal{M}$ is an $\mathcal{A}^{\mathrm{e}}$-module. Although it is projective as an $\mathcal{A}$-module it is never projective as an $\mathcal{A}^{\mathrm{e}}$-module since a free $\mathcal{A}^{\mathrm{e}}$-module is formed using 2-point functions.   □

Example 3.40 The free M_n-bimodule of rank 1 is of dimension n^4 (over $\mathbb{C}$) and the dimension of $\Omega^1(M_n)$ is equal to $(n^2 - 1)n^2 < n^4$. The element

$$\zeta = \frac{1}{n^2} 1 \otimes 1 - \frac{1}{n} \lambda_a \otimes \lambda^a$$

is a projector in $M_n^{\mathrm{e}} - M_n \otimes M_n^{\mathrm{op}}$ which commutes with the elements of M_n. We have thus the direct-sum decomposition

$$M_n^{\mathrm{e}} = \Omega^1(M_n) \oplus M_n \zeta$$

and $\Omega^1(M_n)$ is projective as a bimodule.   □

One can write a (noncommutative) triangular diagram

$$\mathcal{M}$$

$$D_L \nearrow \qquad \searrow D_R$$

$$\Omega^1(\mathcal{A}) \otimes_{\mathcal{A}} \mathcal{M} \xleftarrow{\sigma} \mathcal{M} \otimes_{\mathcal{A}} \Omega^1(\mathcal{A})$$

from which one sees that given an arbitrary bimodule homomorphism (3.6.2) and a covariant derivative (3.6.9) one can construct a covariant derivative (3.6.1) by the formula

$$D = D_L + \sigma \circ D_R \qquad\qquad (3.6.13)$$

which satisfies both Leibniz rules (3.5.2) and (3.6.3).

Suppose further that the differential calculus is such that the differential d of an element $f \in \mathcal{A}$ is of the form (3.1.13): $df = -[\theta, f]$ for some element $\theta \in \Omega^1(\mathcal{A})$. Then obviously particular choices for D_L and D_R are the expressions

$$D_{L(0)}\xi = -\theta \otimes \xi, \qquad D_{R(0)}\xi = \xi \otimes \theta. \qquad\qquad (3.6.14)$$

Let τ be a bimodule homomorphism from $\mathcal{M}$ into $\Omega^1(\mathcal{A}^e) \otimes_{\mathcal{A}^e} \mathcal{M}$ and decompose

$$\tau = \tau_L + \tau_R$$

according to the decomposition (3.6.11). The most general D_L and D_R are of the form

$$D_L\xi = -\theta \otimes \xi + \tau_L(\xi), \qquad D_R\xi = \xi \otimes \theta + \tau_R(\xi). \qquad\qquad (3.6.15)$$

Using (3.6.13) we can construct a covariant derivative

$$D_{(0)}\xi = -\theta \otimes \xi + \sigma(\xi \otimes \theta) \qquad\qquad (3.6.16)$$

from (3.6.14). In Example 3.37 we studied a differential calculus for which this is the unique covariant derivative. If $\mathcal{A}$ is a $*$-algebra then from (3.6.8) it follows that $D_{(0)}$ is real.

In the diagram (3.2.16) we constructed a projection ϕ_1 of a submodule of $\mathcal{A} \otimes \mathcal{A}$ onto $\Omega^1(\mathcal{A})$. If the differential is of the form (3.1.13) then ϕ_1 has an extension to a bimodule map of $\mathcal{A} \otimes \mathcal{A}$ onto $\Omega^1(\mathcal{A})$ given by $\phi_1(1 \otimes 1) = -\theta$. The map ϕ_1 can be also considered as a projection from the free left $\mathcal{A}^e$-module of rank 1 onto $\Omega^1(\mathcal{A})$ considered as a left $\mathcal{A}^e$-module. In these terms we write $(a \otimes b) \mapsto -a\theta b$. That is we have

$$\Omega^1(\mathcal{A}) = \mathcal{A}^e \theta.$$

If we suppose that $\Omega^1(\mathcal{A})$ is a projective bimodule we can identify $\Omega^1(\mathcal{A})$ as a submodule of the free $\mathcal{A}^e$-module of rank 1 and we can write

$$\mathcal{A}^e = \Omega^1(\mathcal{A}) \oplus \mathcal{N}. \qquad\qquad (3.6.17)$$

We have already seen this decomposition in Example 3.40.

A left $\mathcal{A}^e$-connection on $\mathcal{A}^e$ as a left $\mathcal{A}^e$-module is a map of the form (3.6.1) with $\mathcal{M} = \mathcal{A}^e$. The ordinary differential d^e on $\mathcal{A}^e$,

$$\mathcal{A}^e \xrightarrow{\ d^e\ } \Omega^1(\mathcal{A}^e),$$

is clearly a covariant derivative in this sense for the same reasons as given in Example 3.38. The right-hand side can be written using (3.6.11) as

$$\Omega^1(\mathcal{A}^e) = (\Omega^1(\mathcal{A}) \otimes_{\mathcal{A}} \mathcal{A}^e) \oplus (\mathcal{A}^e \otimes_{\mathcal{A}} \Omega^1(\mathcal{A}))$$

and so we can split d^e as the sum of two terms d_L and d_R. Let $f \otimes g$ be an element of $\mathcal{A}^e$. Then we have

$$d_L(f \otimes g) = -[\theta, f] \otimes g = -\theta \otimes (f \otimes g) + (f \otimes g)(\theta \otimes 1).$$

In the first term on the right-hand side the first tensor product is over the algebra and the second is over the complex numbers; in the second term the first tensor product is over the complex numbers and the second is over the algebra. To write the second equality we have used the identifications

$$\Omega^1(\mathcal{A}) \otimes \mathcal{A}^{\mathrm{op}} = \Omega^1(\mathcal{A}) \otimes_{\mathcal{A}} \mathcal{A}^e.$$

If we restrict d_L to an element ξ in the first term in the direct-sum decomposition (3.6.17) we obtain

$$d_L \xi = -\theta \otimes \xi + \xi(\theta \otimes 1).$$

The product in the second term is defined by considering ξ as an element of $\mathcal{A}^e$. Similarly we have

$$d_R(f \otimes g) = -f \otimes [\theta, g] = (f \otimes g) \otimes \theta - (f \otimes g)(1 \otimes \theta).$$

In both terms in the last expression the first tensor product is over the complex numbers and the second is over the algebra. To write the second equality we have used the identifications

$$\mathcal{A} \otimes \Omega^1(\mathcal{A}^{\mathrm{op}}) = \mathcal{A}^e \otimes_{\mathcal{A}^{\mathrm{op}}} \Omega^1(\mathcal{A}^{\mathrm{op}}) = \mathcal{A}^e \otimes_{\mathcal{A}} \Omega^1(\mathcal{A}).$$

The second equality here is a particular case of (3.6.10). If we restrict d_R to an element ξ in the first term in the direct-sum decomposition (3.6.17) we obtain

$$d_R \xi = \xi \otimes \theta - \xi(1 \otimes \theta).$$

The product in the second term is defined by considering ξ as an element of $\mathcal{A}^e$.

Because of the relation

$$\Omega^1(\mathcal{A}) \otimes_{\mathcal{A}} \Omega^1(\mathcal{A}) = (\Omega^1(\mathcal{A}) \otimes_{\mathcal{A}} \mathcal{A}^e)(1 \otimes \theta)$$

we can define a left-covariant derivative D_L as

$$D_L \xi = -(d_L \xi)(1 \otimes \theta).$$

But we can write $\xi = -\xi\theta$ if on the right-hand side we consider ξ as an element of $\Omega^1(\mathcal{A})$ and on the left as an element of $\mathcal{A}^e$. We obtain then the first of equations (3.6.15) with

$$\tau_L(\xi) = -\xi(\theta \otimes \theta).$$

Here, on the right-hand side, the tensor product is over the algebra and the left multiplication by ξ is defined by considering it as an element of $\mathcal{A}^e$. Similarly one can construct a D_R as

$$D_R \xi = -(d_R \xi)(\theta \otimes 1).$$

Decompose $\xi^e \in \Omega^1(\mathcal{A}^e)$ as $\xi^e = \xi_L + \xi_R$ and define the projector P from $\Omega^1(\mathcal{A}^e)$ onto $\Omega^1(\mathcal{A}) \otimes_{\mathcal{A}} \Omega^1(\mathcal{A})$ given by

$$P\xi^e = -\xi_L(1 \otimes \theta) - \xi_R(\theta \otimes 1).$$

Then we can define a D^e by equation (3.6.12):

$$D^e \xi = P d^e \xi. \tag{3.6.18}$$

Example 3.41 Using the fact that π is a projection one sees that the most general solution to the constraint (3.6.5) is given by

$$1 + \sigma = (1 - \pi) \circ \tau$$

where τ is an arbitrary bilinear map

$$\Omega^1(\mathcal{A}) \otimes_{\mathcal{A}} \Omega^1(\mathcal{A}) \xrightarrow{\tau} \Omega^1(\mathcal{A}) \otimes_{\mathcal{A}} \Omega^1(\mathcal{A}).$$

If we choose $\tau = 2$ then we find $\sigma = 1 - 2\pi$ and $\sigma^2 = 1$. The eigenvalues of σ are then equal to ± 1. In this case one has the identity (3.4.4) and with the identification ι of Equation (3.1.40) an element of $\Omega^1(\mathcal{A}) \otimes_{\mathcal{A}} \Omega^1(\mathcal{A})$ can be written uniquely as the sum of a symmetric element and an element of $\Omega^2(\mathcal{A})$. In general this will not be the case. If τ is proportional to the identity matrix but not necessarily equal to 2 then (3.4.2) is equivalent to the condition

$$g \circ \sigma \propto g.$$

More generally, if τ is invertible then one can write

$$(1 + \sigma) \circ \tau^{-1} + \pi = 1$$

instead of (3.4.4) and identify the first term on the left-hand side as the projection of the tensor product onto the symmetric part. □

The general formalism can be applied in particular to a differential calculus which has a frame θ^i. The map σ can be defined by its action on the basis elements:

$$\sigma(\theta^i \otimes \theta^j) = S^{ij}{}_{kl}\theta^k \otimes \theta^l. \tag{3.6.19}$$

By the sequence of identities

$$f S^{ij}{}_{kl}\theta^k \otimes \theta^l = \sigma(f\theta^i \otimes \theta^j) = \sigma(\theta^i \otimes \theta^j f) = S^{ij}{}_{kl}f\theta^k \otimes \theta^l \tag{3.6.20}$$

we conclude that the coefficients $S^{ij}{}_{kl}$ must lie in $\mathcal{Z}(\mathcal{A})$.

Since $\Omega^1(\mathcal{A})$ is a free module a covariant derivative can be defined by its action on the basis elements:

$$D\theta^i = -\omega^i{}_k \otimes \theta^k, \qquad \omega^i{}_k = \omega^i{}_{jk}\theta^j. \tag{3.6.21}$$

The coefficients $\omega^i{}_{jk}$ here are elements of the algebra. The *torsion form* is defined as usual as

$$\Theta^i = d\theta^i - \pi \circ D\theta^i.$$

Using the $F^i{}_{jk}$ defined in (3.1.47) one can define a covariant derivative $D_{(0)}$ by

$$D_{(0)}\theta^i = -\theta \otimes \theta^i + \sigma(\theta^i \otimes \theta) - \frac{1}{2}F^i{}_{jk}\theta^j \otimes \theta^k, \tag{3.6.22}$$

which is torsion-free by (3.1.49). When $F^i{}_{jk} = 0$ this coincides with (3.6.16). The corresponding coefficients are given by

$$\omega_{(0)}{}^i{}_{jk} = \lambda_l(S^{il}{}_{jk} - \delta^l_j\delta^i_k) + \frac{1}{2}F^i{}_{jk}. \tag{3.6.23}$$

The most general D is of the form

$$D = D_{(0)} + \chi \tag{3.6.24}$$

where χ is an arbitrary bimodule morphism

$$\Omega^1(\mathcal{A}) \xrightarrow{\chi} \Omega^1(\mathcal{A}) \otimes_{\mathcal{A}} \Omega^1(\mathcal{A}).$$

If we write

$$\chi(\theta^i) = -\chi^i{}_{jk}\theta^j \otimes \theta^k$$

then by a sequence of identities similar to (3.6.20) we conclude that $\chi^i{}_{jk} \in \mathcal{Z}(\mathcal{A})$. In general a covariant derivative is torsion-free provided the condition

$$\omega^i{}_{lm}P^{lm}{}_{jk} = \frac{1}{2}C^i{}_{jk} \tag{3.6.25}$$

is satisfied. The covariant derivative (3.6.24) is torsion free if and only if

$$\pi \circ \chi = 0.$$

If the frame θ^i exists then one sees that the condition (3.6.7) that the connection be *metric-compatible* becomes the condition

$$\omega^i{}_{kl}g^{lj} + \omega^j{}_{ln}S^{il}{}_{km}g^{mn} = 0. \tag{3.6.26}$$

One can understand this odd condition by introducing a 'covariant derivative' $D_i X^j$ of a constant 'vector' by the formula

$$D_i X^j = \omega^j{}_{ik} X^k.$$

The covariant derivative $D_i(X^j Y^k)$ of the product of two such 'vectors' must be defined as

$$D_i(X^j Y^k) = D_i X^j Y^k + S^{jl}{}_{im} X^m D_l Y^k$$

since there is a 'flip' as the index on the derivation crosses the index on the first 'vector'. The condition (3.6.27) becomes then simply

$$D_i g^{jk} = 0.$$

If $F^i{}_{jk} = 0$ then the condition that the connection defined by (3.6.23) be metric-compatible becomes the equation

$$S^{im}{}_{ln}g^{np}S^{jk}{}_{mp} = g^{ij}\delta^k_l. \tag{3.6.27}$$

This must be solved with (3.4.2) and (3.6.5) as equations for g and σ. If the algebra is a $*$-algebra then one must look for a real solution to the compatibility condition (3.6.26). The map D_2 is given by

$$D_2(\theta^i \otimes \theta^j) = -(\omega^i{}_{pq}\delta^j_r + S^{ik}{}_{pq}\omega^j{}_{kr})\theta^p \otimes \theta^q \otimes \theta^r.$$

If the algebra is a $*$-algebra then the expression (3.6.8) for the involution on tensor products becomes the identity

$$J^{ij}{}_{kl} = S^{ji}{}_{kl}. \tag{3.6.28}$$

This is consistent with (3.1.55) because of (3.6.5). It forces also the constraint

$$(S^{ji}{}_{kl})^* S^{lk}{}_{mn} = \delta^i_m \delta^j_n \tag{3.6.29}$$

on σ. Equation (3.6.28) can be also read from right to left as a definition of the right Leibniz rule in terms of the hermitian structure.

The condition that the connection (3.6.21) be real can be written as

$$(\omega^i{}_{jk})^* = \omega^i{}_{lm}(J^{lm}{}_{jk})^*. \tag{3.6.30}$$

One verifies immediately that the connection (3.6.23) with $F^i{}_{jk} = 0$ is real.

Example 3.42 Return to the generic matrix example of Section 3.1. Introduce a metric by setting $g(\theta^a \otimes \theta^b) = g^{ab}$, the components of the SU_n Killing metric. The linear connection defined by

$$D\theta^a = -\omega^a{}_b \otimes \theta^b, \qquad \omega^a{}_b = -\frac{1}{2}C^a{}_{bc}\theta^c \tag{3.6.31}$$

has vanishing torsion and is compatible with the metric. With this connection the geometry of M_n looks like the invariant geometry of the group SU_n. Since the elements of the algebra commute with the frame θ^a, we can define D on all of $\Omega^*(M_n)$ using the left Leibniz rule. The map σ is necessarily given by

$$\sigma(\theta^a \otimes \theta^b) = \theta^b \otimes \theta^a. \tag{3.6.32}$$

It follows that D satisfies also the right Leibniz rule (3.6.3) and the metric satisfies the symmetry condition (3.4.2). With the same σ the covariant derivative (3.6.16) is given by $D_{(0)}\theta^a = 0$ and so the curvature vanishes. The connection has a non-vanishing torsion form given by $\Theta^a = d\theta^a$. We can suppose a general linear connection to be of the form

$$D\theta^a = -\omega^a{}_{bc}\,\theta^b \otimes \theta^c$$

with $\omega^a{}_{bc}$ arbitrary elements of M_n. Suppose that σ is given by (3.6.32). From the Leibniz rules we find that

$$0 = D([f, \theta^a]) = [f, D\theta^a]$$

and so the $\omega^a{}_{bc}$ must be all in the center of M_n. They are complex numbers. If we require that the torsion vanish then we have by (3.6.25) the condition

$$\omega^a{}_{[bc]} = C^a{}_{bc}.$$

If we impose the condition (3.6.7) that the connection be metric-compatible we find that

$$\omega^a{}_{(bc)} = 0.$$

The linear connection (3.6.31) is the unique torsion-free metric connection on $\Omega^1(M_n)$. If $\sigma^2 = 1$ then σ is necessarily given by (3.6.32). If the coefficients $S^{ab}{}_{cd}$ are real then we have seen that the expression (3.6.8) forces this condition on σ. If we allow the $S^{ab}{}_{cd}$ to be complex numbers then we cannot prove that (3.6.31) is the unique torsion-free metric connection. □

Example 3.43 It is often difficult to find general solutions to Equation (3.1.50) and associated metrics and linear connections. It is therefore of interest also to examine how one calculates first-order perturbations around a given solution. If we perturb

σ we necessarily perturb $\jmath_2$ and the definition of an hermitian element in the tensor product of 1-forms changes accordingly. We set $\hat{e}_a = \mathrm{ad}\,\hat{\lambda}_a$ where

$$\hat{\lambda}_a = \lambda_a + h_a$$

and where h_a is a set of n elements of the algebra $\mathcal{A}$ with $|h_a| \ll 1$. If $\mathcal{A}$ is a *-algebra we suppose that the h_a are anti-hermitian so that the $\hat{e}_a$ are real. If we write the perturbed version of (3.1.50) as

$$2\hat{\lambda}_c\hat{\lambda}_d\hat{P}^{cd}{}_{ab} - \hat{\lambda}_c\hat{F}^k{}_{ab} - \hat{K}_{ab} = 0$$

and expand the coefficients as

$$\hat{P}^{cd}{}_{ab} = P^{cd}{}_{ab} + \pi^{cd}{}_{ab}, \qquad \hat{F}^c{}_{ab} = F^c{}_{ab} + \phi^c{}_{ab}, \qquad \hat{K}_{ab} = K_{ab} + \kappa_{ab}$$

we find the equation

$$[\lambda_a, h_b] - [\lambda_b, h_a] + 2\lambda_c\lambda_d\pi^{cd}{}_{ab} - h_cF^c{}_{ab} - \lambda_c\phi^c{}_{ab} - \kappa_{ab} = 0 \qquad (3.6.33)$$

for the general perturbation. The $\pi^{ab}{}_{cd}$ satisfy constraints

$$P^{ab}{}_{ef}\pi^{ef}{}_{cd} + \pi^{ab}{}_{ef}P^{ef}{}_{cd} + \pi^{ab}{}_{ef}\pi^{ef}{}_{cd} = \pi^{ab}{}_{cd}$$

which come from the fact that $P^{ab}{}_{cd}$ and $\hat{P}^{ab}{}_{cd}$ are both projectors. The components $\pi^{[cd]}{}_{(ab)}$ and $\pi^{(cd)}{}_{[ab]}$ are of first order and to within terms of third order the constraints can be written in the form

$$\pi^{(cd)}{}_{(ab)} = \frac{1}{4}\pi^{(cd)}{}_{[ef]}\pi^{[ef]}{}_{(ab)}, \qquad \pi^{[cd]}{}_{[ab]} = \frac{1}{4}\pi^{[cd]}{}_{(ef)}\pi^{(ef)}{}_{[ab]}. \qquad (3.6.34)$$

The $\pi^{cd}{}_{ab}$ satisfy also constraints which follow from the braid relation (6.1.8) we shall derive below. It is obvious that an inner automorphism $\hat{\lambda}_a = u^{-1}\lambda_a u$ leads to a new solution of Equation (3.1.50) with the same coefficients. Since, as one sees for example in (3.6.23), the geometry depends on the coefficients these transformations are not particularly interesting. If $F^c{}_{ab} = 0$ and $P^{ab}{}_{cd}$ is of the form (3.1.44) then to first order the Equations (3.6.33) become

$$e_{[a}h_{b]} - \frac{1}{2}\lambda_c\phi^c{}_{[ab]} - \frac{1}{2}\kappa_{[ab]}.$$

If we write a power-series expression

$$h_a(x^i) = f_a + f_{ai}x^i + \frac{1}{2}f_{aij}x^ix^j + \cdots$$

for h_a with the coefficients completely symmetric an all but the first index then there is an obvious solution $h_a = e_ah$ with $\phi^c{}_{ab} = 0$ and $\kappa_{ab} = 0$ which corresponds to the inner automorphisms mentioned above. A more interesting solution is given by

$$f_{[ab]} = \kappa_{[ab]}, \qquad f_{[ab]c} = -K_{cd}\phi^d{}_{ab}$$

and the remaining coefficients equal to zero. For this solution to exist when $d \geq 3$ the integrability condition

$$K_{ad}\phi^d{}_{bc} + K_{bd}\phi^d{}_{cd} + K_{cd}\phi^d{}_{ab} = 0 \tag{3.6.35}$$

must be satisfied. If we introduce the function

$$h = g_i x^i + \frac{1}{2} g_{ij} x^i x^j + \frac{1}{3!} g_{ijk} x^i x^j x^k + \cdots$$

where without loss of generality we can suppose that the coefficients are all completely symmetric in all indices, then we find that

$$e_a h = g_a + g_{ai} x^i + \frac{1}{3} g_{ajk} x^j x^k + \cdots.$$

We can use a contribution of this form to set $f_a = 0$, $f_{(ab)} = 0$. We can also use it to set

$$f_{abc} = \frac{1}{3} K_{(bd}\phi^d{}_{c)a}.$$

The first-order solution is given then by

$$h_a = \frac{1}{2} \kappa_{[ai]} x^i + \frac{1}{3!} K_{(id}\phi^d{}_{j)a} x^i x^j + \cdots.$$

$\square$

3.7 Curvature

One of the more important technical points of noncommutative geometry which has not been clarified in a satisfactory manner is the correct definition of the curvature of a linear connection. It is not obvious that the ordinary definition of curvature taken directly from differential geometry is the quantity which is most useful in the noncommutative theory. The main interest of curvature in the case of a smooth manifold is the fact that it is a local quantity which entirely characterizes the linear connection. On a smooth manifold V the curvature can be defined as a $\mathcal{C}(V)$-bilinear, symmetric map

$$\mathrm{Curv} : \Omega^1(V) \to \Omega^2(V) \otimes_{\mathcal{C}(V)} \Omega^1(V).$$

If $\xi \in \Omega^1(\mathcal{C}(V))$ then $\mathrm{Curv}(\xi)$ at a given point depends only on the value of ξ at that point. This can be expressed as a bilinearity condition; the above map is a $\mathcal{C}(V)$-bimodule map. If $f \in \mathcal{C}(V)$ then

$$f\,\mathrm{Curv}(\xi) = \mathrm{Curv}(f\xi), \qquad \mathrm{Curv}(\xi f) = \mathrm{Curv}(\xi)f.$$

It is important to notice here that the bilinearity is an alternative way of expressing *locality*: if ξ is a 1-form then the value of $\mathrm{Curv}(\xi)$ at a given point depends only on the value of ξ at that point. Bilinearity is an exact expression of this fact.

Consider a general algebra $\mathcal{A}$ and a differential calculus $\Omega^1(\mathcal{A})$ over $\mathcal{A}$. We shall require that curvature be a $\mathcal{A}$-bilinear map

$$\mathrm{Curv} : \Omega^1(\mathcal{A}) \rightarrow \Omega^2(\mathcal{A}) \otimes_{\mathcal{A}} \Omega^1(\mathcal{A}).$$

In ordinary differential geometry the curvature is given in terms of the covariant derivative by the expression

$$\mathrm{Curv} = -D^2.$$

We saw in Equation (3.5.4) that D^2 is left-linear. It is however not in general right-linear. Consider the covariant derivative (3.6.1). We can define a right-linear curvature by factoring out in the image of D^2 all those elements ($\mathcal{J} =$ 'junk') which do not satisfy the desired condition. Define $\mathcal{J}$ as the vector space

$$\mathcal{J} = \{ \sum D^2(\xi f) - D^2(\xi)f \; : \; \xi \in \mathcal{M}, f \in \mathcal{A} \}.$$

In fact $\mathcal{J}$ is a sub-bimodule of $\Omega^2(\mathcal{A}) \otimes_{\mathcal{A}} \mathcal{M}$. It is obviously a left submodule. Consider the element $\alpha = D^2(\xi g) - D^2(\xi)g \in \mathcal{J}$ and let $f \in \mathcal{A}$. We can write

$$\alpha f = (D^2(\xi gf) - D^2(\xi)gf) - (D^2(\xi gf) - D^2(\xi g)f).$$

Therefore $\alpha f \in \mathcal{J}$ and $\mathcal{J}$ is also a right submodule. Let p be the projection

$$\Omega^2(\mathcal{A}) \otimes_{\mathcal{A}} \mathcal{M} \xrightarrow{p} (\Omega^2(\mathcal{A}) \otimes_{\mathcal{A}} \mathcal{M})/\mathcal{J}.$$

A possible definition of the curvature of D is the combined map

$$\mathrm{Curv}' = -p \circ D^2. \qquad (3.7.1)$$

By construction Curv' is left- and right-linear:

$$\mathrm{Curv}'(f\xi) = f\mathrm{Curv}'(\xi), \qquad \mathrm{Curv}'(\xi f) = \mathrm{Curv}'(\xi)f.$$

We have here used explicitly only the left Leibniz rule. In later examples we shall illustrate the role which the right Leibniz rule plays in this construction.

Example 3.44 If there exists a frame θ^i then one can write

$$\mathrm{Curv}(\theta^i) = \Omega^i{}_j \otimes \theta^j, \qquad \Omega^i{}_j = \frac{1}{2} R^i{}_{jkl} \theta^k \theta^l. \qquad (3.7.2)$$

If Curv is bilinear then one concludes immediately from (3.1.34) that the coefficients $R^i{}_{jkl}$ must belong to the center of the algebra. This is an unsatisfactory situation to which we have no satisfactory remedy.   □

Example 3.45 When $F^i{}_{jk} = 0$ the curvature of the covariant derivative $D_{(0)}$ defined in (3.6.16) can be readily calculated. One finds the expression

$$\frac{1}{2}R^i_{(0)jkl} = S^{im}{}_{rn}S^{np}{}_{sj}P^{rs}{}_{kl}\lambda_m\lambda_p - \frac{1}{2}\delta^i_j K_{kl}.$$

This can also be written in the form

$$\mathrm{Curv}_{(0)}(\theta^i) = -\pi_{12}\sigma_{12}\sigma_{23}\sigma_{12}(\theta^i \otimes \theta \otimes \theta) - \theta^2 \otimes \theta^i.$$

The indices have the same meaning as they had in (3.6.6). If $\xi = \xi_i\theta^i$ is a general 1-form then since Curv is left-linear one can write

$$\mathrm{Curv}_{(0)}(\xi) = -\pi_{12}\sigma_{12}\sigma_{23}\sigma_{12}(\xi \otimes \theta \otimes \theta) - \theta^2 \otimes \xi.$$

We have used here the fact that θ^2 commutes with the elements of the algebra. The left linearity here follows immediately from Equation (3.1.50). Therefore the definition (3.7.1) yields

$$\mathrm{Curv}'_{(0)}(\xi) = -\theta^2 \otimes \xi = (d\theta + \theta^2) \otimes \xi.$$

$\square$

Example 3.46 Consider the covariant derivative (3.6.9) and its decomposition given in (3.6.12). One can define a map

$$\mathrm{Curv}_L = -D^2_L$$

from $\mathcal{M}$ into $\Omega^2(\mathcal{A}) \otimes_{\mathcal{A}} \mathcal{M}$. It is bilinear because by construction it is trivially right-linear. In the case where the differential d is given by (3.1.38) and $D_L = D_{L(0)}$ we find that $\mathrm{Curv}_{L(0)}$ is given by the formula

$$\mathrm{Curv}_{L(0)}(\xi) = (d\theta + \theta^2) \otimes \xi.$$

The bilinearity here follows from the identity (3.1.45). We have seen however that D_L has no commutative limit. $\square$

Example 3.47 The covariant derivative (3.6.9) has also a bilinear curvature 2-form

$$\mathrm{Curv}^e = -D^{e2}.$$

It takes its values in a space which can be naturally identified with $\Omega^2(\mathcal{A}) \otimes_{\mathcal{A}} \Omega^1(\mathcal{A})$. The curvature of the particular connection (3.6.18) is given by

$$\mathrm{Curv}^e(\xi) = -D^{e2}\xi = -d^e P d^e P P \xi.$$

There is a similar construction in Section 5.1. $\square$

Example 3.48 From the definition we find that in the Example 3.42 the map Curv is given by (3.7.2) with

$$R^a{}_{bcd} = \frac{1}{4} C'^a{}_{be} C'^e{}_{cd}.$$

□

Example 3.49 Return again to the Example 3.10 on which we constructed a metric in Example 3.28 and a linear connection in Example 3.37. Using D_2 defined by (3.6.6) we find

$$D_2 \eta_{11} = \zeta \otimes \eta_1^* - \mu \eta_{11} \otimes \eta_1, \quad D_2 \eta_{12} = \zeta \otimes \eta_2^*,$$

$$D_2 \eta_{21} = -\mu \eta_{21} \otimes \eta_1, \qquad\qquad D_2 \zeta = \mu \zeta \otimes \eta_1^* - \eta_{11} \otimes \eta_1,$$

from which we conclude that

$$D_2 \circ D\eta_1 = (\mu^2 - 1)\eta_{11} \otimes \eta_1, \qquad\qquad D_2 \circ D\eta_2 = \mu^2 \eta_{21} \otimes \eta_1,$$

$$D_2 \circ D\eta_1^* = (\mu + 1)(\eta_{11} \otimes \eta_1 - \zeta \otimes \eta_1^*), \quad D_2 \circ D\eta_2^* = -\zeta \otimes \eta_2^*$$

and therefore

$$D^2 \eta_1 = 0, \qquad\qquad D^2 \eta_2 = 0,$$

$$D^2 \eta_1^* = -(\mu + 1)e \otimes \eta_1^*, \quad D^2 \eta_2^* = -e \otimes \eta_2^*.$$

Since $\eta_1 = \eta e$ and $\eta_1^* = -e\eta$ one could choose η as a sort of 'frame'. We have then

$$D^2 \eta = -(\mu + 1)e \otimes \eta,$$

We have already noticed that $\mu = -1$ is a degenerate value. It is easy to see that with the definition (3.7.1) Curv $\equiv 0$. Since there is only one metric connection it is difficult to understand what this condition implies.   □

Example 3.50 The holonomy group of classical geometry is a group in the Lie algebra of which the curvature form takes its values. If one neglects the fact that it is a 2-form the map Curv can be considered as a map of the module of 1-forms into itself. Using (3.1.36) we can thus identify Curv as an element of $M_n(\Omega^1(\mathcal{A}))$. A representation of $\mathcal{A}$ in a space $\mathcal{H}$ yields a representation of $M_n(\Omega^1(\mathcal{A}))$ in $\mathcal{H} \otimes \mathbb{C}^n$. If $\mathcal{A}$ is the commutative algebra of smooth functions on a smooth manifold then one can localize at a point and define the resulting image in $M_n(\mathbb{C}^n)$ to be an element of the Lie algebra of the holonomy group. In more general situations one would have to consider the holonomy 'group' or rather its 'Lie algebra' as an object such as $\mathcal{U}_q$ described in Section 4.4.   □

In order for the curvature to be real we must require that the extension of the involution to the tensor product of three elements of $\Omega^1(\mathcal{A})$ be such that

$$\pi_{12} \circ D_2(\xi \otimes \eta)^* = (\pi_{12} \circ D_2(\xi \otimes \eta))^*.$$

We shall impose a stronger condition. We shall require that D_2 be real:

$$D_2(\xi \otimes \eta)^* = (D_2(\xi \otimes \eta))^*.$$

If we introduce an involution $\jmath_3$ of $\Omega^1(\mathcal{A}) \otimes_{\mathcal{A}} \Omega^1(\mathcal{A}) \otimes_{\mathcal{A}} \Omega^1(\mathcal{A})$ then the reality condition for D_2 can be written

$$D_2 \circ \jmath_2 = \jmath_3 \circ D_2. \tag{3.7.3}$$

The form of $\jmath_3$ can be made more explicit when a frame exists.

To solve the reality condition (3.7.3) we introduce elements $J^{ijk}{}_{lmn} \in \mathcal{Z}(\mathcal{A})$ such that

$$(\theta^i \otimes \theta^j \otimes \theta^k)^* = \jmath_3(\theta^i \otimes \theta^j \otimes \theta^k)^* = J^{ijk}{}_{lmn}\theta^l \otimes \theta^m \otimes \theta^n.$$

Using (3.6.28) one finds then that the condition can be written in the form

$$J^{ij}{}_{pq}(\omega^p{}_{lm}\delta^q_n + J^{rp}{}_{lm}\omega^q{}_{rn}) = ((\omega^i{}_{pq})^*\delta^j_r + (J^{si}{}_{pq})^*(\omega^j{}_{sr})^*)J^{pqr}{}_{lmn}. \tag{3.7.4}$$

This equation must be solved for $J^{ijk}{}_{lmn}$ as a function of $J^{ij}{}_{kl}$. One cannot simply cancel the factor $\omega^i{}_{jk}$ since it satisfies constraints. As a test case we choose (3.6.22) with $F^i{}_{jk} = 0$. We find that (3.7.4) is satisfied provided

$$J^{ijk}{}_{lmn} = J^{ij}{}_{pq}J^{pk}{}_{lr}J^{qr}{}_{mn} = J^{jk}{}_{pq}J^{iq}{}_{rn}J^{rp}{}_{lm}. \tag{3.7.5}$$

The second equality is the *Yang-Baxter equation* (4.4.12) written out with indices. Using this equation it follows that $\jmath_3$ is indeed an involution:

$$(J^{ijk}{}_{lmn})^* J^{lmn}{}_{rst} = \delta^i_r \delta^j_s \delta^k_t.$$

Using (3.7.5) Equation (3.7.4) can be written in the form

$$J^{ij}{}_{pn}\omega^p{}_{lm} - J^{ip}{}_{mn}\omega^j{}_{lp} + J^{ij}{}_{pq}J^{rp}{}_{lm}\omega^q{}_{rn} - J^{pj}{}_{lr}J^{qr}{}_{mn}\omega^i{}_{pq} = 0. \tag{3.7.6}$$

This can be rewritten more concisely in the form

$$D_2 \circ \sigma = \sigma_{23} \circ D_2.$$

The connection then must satisfy two reality conditions, (3.6.30) as well as (3.7.6).

It is reasonable to suppose that even in the absence of a frame the involution constraint on σ and the Yang-Baxter condition still hold. The latter can be written in the form

$$(\xi \otimes \eta \otimes \zeta)^* = \sigma_{12}\sigma_{23}\sigma_{12}(\zeta^* \otimes \eta^* \otimes \xi^*)$$

with ξ, η, ζ arbitrary 1-forms. Because of (3.6.28) the Yang-Baxter condition for $\jmath_2$ becomes the *braid relation* (4.4.13) for σ:

$$\sigma_{12}\sigma_{23}\sigma_{12} = \sigma_{23}\sigma_{12}\sigma_{23}.$$

Since we are not sure of how one should form a curvature 'tensor' we are even less sure of curvature invariants, in particular the 'Ricci scalar'. A Ricci map

$$\mathrm{Ric} : \Omega^1(V) \to \Omega^1(V)$$

can be defined by

$$\mathrm{Ric} = g_{23} \circ \mathrm{Curv}.$$

The appropriate definition of the Ricci scalar is not evident. What is missing is a module structure for the derivations. If a frame exists one can define a map from the derivations $\mathrm{Der}(\mathcal{A})$ into the 1-forms $\Omega^1(\mathcal{A})$ by $e_i \mapsto \theta^i$ but, contrary to the commutative case, this map is not onto.

If the geometry of a manifold V is described (locally) by a moving frame θ^α then a change of moving frame is a map

$$\theta^\alpha \mapsto \hat\theta^\alpha = \Lambda^\alpha_\beta \theta^\beta$$

with Λ^α_β smooth functions of V. If the metric is to remain invariant then there are restrictions on the Λ^α_β:

$$\Lambda^\alpha_\gamma \Lambda^\beta_\delta g^{\gamma\delta} = g^{\alpha\beta}.$$

In the noncommutative case one could define a change of frame θ^i using the same formula

$$\theta^i \mapsto \hat\theta^i = \Lambda^i_j \theta^j$$

but with Λ^i_j arbitrary invertible elements of the algebra. If $\Lambda^i_j \in \mathcal{Z}(\mathcal{A})$ then the new frame has the same status as the old; it is dual to a set of derivations. Otherwise the change is purely formal. From the left-linearity of D^2 we conclude nevertheless that

$$\mathrm{Curv}(\Lambda^i_j \theta^j) = \Lambda^i_j \mathrm{Curv}(\theta^j).$$

The noncommutative equivalent of a diffeomorphism ϕ of a manifold V is an automorphism of the algebra $\mathcal{A}$. Consider the *inner automorphism*

$$\phi^* f = u^{-1} f u.$$

Let X_f be an inner derivation. Then from the equalities

$$(\phi_* X_f)g = \phi^{*-1} X_f(\phi^* g) = \phi^{*-1}[f, u^{-1}gu] = [ufu^{-1}, g] = X_{\phi^{*-1}f}g$$

we conclude that

$$\phi_* X_f = X_{\phi^{*-1}f}. \tag{3.7.7}$$

In particular, if the differential calculus is derivation-based as defined in (3.1.31) then we have

$$\hat\lambda_i = \phi^* \lambda_i = u^{-1} \lambda_i u$$

and a solution to Equation (3.1.50) is transformed into another solution with the same values of $P^{ij}{}_{kl}$, $F^i{}_{jk}$ and K_{jk}. If the algebra is a $*$-algebra and the automorphism is to respect a reality condition then the elements u must be chosen unitary. The derivation e_i is transformed into $\hat{e}_i$ given by

$$\hat{e}_i = \phi_* e_i = \mathrm{ad}\,(\phi^{*-1} f).$$

We see then that the automorphism leaves invariant the differential calculus. It commutes with the differential. We chose the relation (3.7.7) so that this would be the case.

A more interesting morphism is one which changes the differential calculus but leaves the algebra invariant. These morphisms have no analogue in ordinary differential geometry. Within the set of differential calculi which are based on a family of derivations then the possibility of such morphisms is due to the fact that the derivations do not form a module over the algebra. Let ϕ^* be a map

$$\lambda_i \mapsto \hat{\lambda}_i = \phi^* \lambda_i \qquad (3.7.8)$$

from the vector space spanned by a set of λ_i to that spanned by another set $\hat{\lambda}_i$. Then $e_i \mapsto \hat{e}_i$ and $d \mapsto \hat{d}$. If we suppose that ϕ^* has the form of an inner automorphism then it will leave invariant the solutions of Equation (3.1.50). Although this is a very particular case it is of some interest in that it is possible to express explicitly $\hat{d}$ in terms of d:

$$\hat{d}f(\hat{e}_i) = df(e_i) + [u^{-1}e_i u, f].$$

Example 3.51 The equations we gave at the beginning of Section 3.1 are with respect to an arbitrary basis λ^a but they are all tensorial in character with respect to a change of basis

$$\lambda^a \mapsto \hat{\lambda}^a = \Lambda^a_b \lambda^b, \qquad (\Lambda^a_b) \in SO_{m^2-1}.$$

We have written SO_{m^2-1} instead of GL_{m^2-1} to conserve the form of the Killing metric. We find

$$\hat{C}^a{}_{bc} = \Lambda^a_d \Lambda^{-1e}_b \Lambda^{-1f}_c C^d{}_{ef},$$

and therefore g_{ab} transforms as

$$\hat{g}_{ab} = \Lambda^c_a \Lambda^d_b g_{cd} = g_{ab}.$$

From the expression for θ^a we see that it transforms as

$$\theta^a \mapsto \hat{\theta}^a = \Lambda^a_b \theta^b.$$

The equivalent of a global map of the manifold V onto itself is an automorphism of M_n, given by

$$\lambda^a \mapsto \hat{\lambda}^a = g^{-1} \lambda^a g, \qquad g \in U_n. \qquad (3.7.9)$$

We have written U_n instead of GL_n because of the condition that the matrices λ^a be anti-hermitian. The set $\{\lambda^a\}$ remains invariant only if $m = n$. The Equation (3.7.9) defines then a map of U_n into SO_{n^2-1} and so a change of basis.   □

Notes

The formalism of Section 3.1 is based on Madore (1989b) and Dubois-Violette *et al.* (1989b, 1989a, 1990b, 1990a, 1991). The identities used in the proof of (3.1.8) are taken from the article by Macfarlane *et al.* (1968). For a more detailed discussion of the symplectic form (3.1.23) we refer to Dubois-Violette (1991). The relation between the matrix algebras and the Heisenberg algebra is taken from Holstein & Primakoff (1940). The Weyl algebra was introduced by Weyl (1950). The finite-dimensional version discussed in Section 3.1 was introduced by Sylvester (1884) and independently by Weyl to describe a finite-dimensional version of quantum mechanics. It was developed later for this purpose by Schwinger (1960). The idea of using a set of derivations to define a differential calculus is due to Dimakis and is described in Dimakis & Madore (1996) and in Madore & Mourad (1998). The particular case in which the set forms a Lie subalgebra of all derivations was first used by Connes & Rieffel (1987). The even more particular case of a differential calculus based on the complete set of all derivations has been studied by Dubois-Violette (1988). For something similar to Example 3.8 on Poisson manifolds we refer to Kosmann-Schwarzbach (1997).

The differential algebra defined in Section 3.2 was adapted from Connes & Lott (1991) and Coquereaux *et al.* (1991). More details of the differential calculi associated to $\mathbb{Z}_2$-graded algebras are to be found in the book by Kastler (1988). The idea of extending the notion of an exterior derivative to noncommutative geometry and using it to define noncommutative 'differential' geometry is due to Connes (1986). The Examples 3.9 and 3.10 are due to Connes & Lott (1991). More details of Example 3.11 are to be found in Kastler *et al.* (1997).

The rule (3.3.2) is not the only way to define the product of two differential algebras. We refer, for example, to Majid (1995) for a discussion of other possibilities. The definition of a metric as a distance on the space of states associated to the algebra has been proposed by Connes (1994). Example 3.28 is taken from Madore *et al.* (1995).

The algebraic approach to classical differential geometry, which places prime importance on the covariant derivative as a map between modules over the algebra of functions of a manifold, is reviewed in the Tata lecture notes by Koszul (1960). It was generalized to noncommutative geometry by Karoubi (1982) and Connes (1986). Compare also the earlier work of Segal (1968) and others (Rideau 1981). An extension to certain noncommutative cases of the definition of a linear connection as a differential form on a principal fibre bundle has been proposed by Hajac (1996), Majid (1999) and Durdević (1998). See Example 4.71 in the following chapter. More details of the physical aspects of Example 3.29 can be found, for example, in Madore (1981). Example 3.30 is taken from Dubois-Violette *et al.* (1989b, 1990a) and Example 3.32 is taken from Connes & Lott (1991); in a particular case it was shown by Madore *et al.* (1997) that the second is a singular contraction of the first. Example 3.33 is based on the work of Schomerus (1999), Seiberg and Witten (1999) and others (Ho & Wu 1997; Chu et al. 1999; Madore et al. 2000b; Madore et al. 2000a).

The Sections 3.6 and 3.7 are adapted from Dubois-Violette *et al.* (1996) and Fiore & Madore (1998). The idea of using a generalized flip to complete the definition of a connection on a bimodule is due to Mourad (1995) and Dubois-Violette & Michor (1996). The bilinearity was proven by Dubois-Violette *et al.* (1995). Example 3.36 is an unpublished Appendix to this paper. See also Kastler *et al.* (1997). If one requires a right Leibniz rule only on the 'center' of the bimodule $\mathcal{M}$, the elements which commute with the algebra, then one obtains another possible definition of a linear connection (Dabrowski et al. 1996). More details of Example 3.43 can be found in Madore (1997). The left- and right-covariant derivatives were introduced by Cuntz & Quillen (1995). See also Bresser *et al.* (1996). A detailed discussion of curvature has been given by Dubois-Violette *et al.* (1996). For another variation on the theme of Example 3.50 we refer to Cotta-Ramusino & Rinaldi (1992); for a discussion of holonomy within the context of quantum gravity we refer to Gambini *et al.* (1996).

4 Noncommutative Geometry

It is currently believed that when expressed in terms of classical fields fundamental physics involves only scalar fields, Yang-Mills potentials and spinors, as well as the gravitational field. The actions which determine the dynamics of these fields can be written in terms of exterior derivatives, covariant derivatives and Dirac operators, as well as the space-time metric. In Chapter 5 we shall see that in a rather large class of algebras a generalization of the Dirac operator can be introduced. It would seem then that field theory could be studied on a large class of noncommutative geometries. If the associative algebra which underlies the geometry is too abstract or too different from an ordinary algebra of functions on space-time, it is impossible to interpret its elements in terms of known classical observables but in principle every associative algebra $\mathcal{A}$ can be used to define a noncommutative geometry. In Section 6.1 we shall show in fact how one can construct a differential calculus over arbitrary $\mathcal{A}$. This construction contains however absolutely no information about the structure of $\mathcal{A}$ and even for finite algebras the corresponding algebra of forms is of infinite dimension.

The purpose of the present chapter is to introduce a large class of associative algebras of infinite dimension which can be used in constructing noncommutative geometries which might be of interest in physics. In the first section we make a few general remarks about formal algebras and we present some examples which have been used in the construction of noncommutative geometries. In Section 4.2 we discuss the commutative limit and make a few remarks about Poisson structures. In Section 4.3 we introduce topological algebras and take the occasion to recall some elementary results from quantum field theory as far as they help in understanding noncommutative geometry. Quantum groups are introduced and briefly studied in the last section.

4.1 General algebras

ΑΓΕΩΜΕΤΡΗΤΟΣ ΜΗΔΕΙΣ ΕΙΣΙΤΩ *

Just as in the case of ordinary differential geometry and matrix geometry derivations will play an important role in the noncommutative geometries which we describe in this section. The complete set of all derivations of an algebra $\mathcal{A}$ is a natural analogue of the space of all smooth vector fields on a manifold V. A derivation X_f defined in terms of the commutator by $X_f g = [f, g]$ is said to be an *inner derivation*; otherwise it is said to be an *outer derivation*. The Leibniz rule for an inner derivation is the same as the *Jacobi identity* for the commutator. All noncommutative algebras have

---

*Inscribed over the entrance to Plato's Academy in Athens

derivations. If the algebra has an involution then we define a real derivation as in the commutative case, using (2.1.33). A real derivation takes an hermitian element of the algebra into another hermitian element; it is not to be confused with an hermitian operator on $\mathcal{A}$ considered as a Hilbert space.

Let $\mathcal{T}$ be the *free algebra* generated by a finite set of n elements x^i called *generators*. It is a $\mathbb{Z}$-*graded* algebra

$$\mathcal{T} = \bigoplus_0^\infty \mathcal{T}_i$$

which is the direct sum of the vector spaces $\mathcal{T}_i$ of all homogeneous polynomials of degree i. Since we wish the algebra to have a unit we have included the polynomials $\mathcal{T}_0 = \mathbb{C}$ of degree zero. The algebra has the obvious property that the product of an element $f \in \mathcal{T}_i$ and an element $g \in \mathcal{T}_j$ is an element $fg \in \mathcal{T}_{i+j}$. One can consider the product as the tensor product in which case $\mathcal{T} = \mathcal{T}(\mathcal{H})$ is the *tensor algebra* over $\mathcal{H} = \mathcal{T}_1$. Consider the two-sided ideal $\mathcal{I} \subset \mathcal{T}$ generated by a finite set of polynomials $R_p(x^i)$. The quotient algebra $\mathcal{A} = \mathcal{T}/\mathcal{I}$ is defined by a set of generators and a set of *relations*

$$R_p(x^i) = 0.$$

Most of the algebras which we shall use can be considered as being defined in terms of generators and relations.

One can think of the x^i as 'coordinates' of an 'embedding space' and the algebra $\mathcal{A}$ as the algebra of all 'polynomials' on the 'subspace' defined by $\mathcal{I}$. However, the 'dimension' of the 'subspace' will bear no relation to the difference between the number of generators and the number of relations. In fact in the commutative limit, when it exists, some of the relations might become vacuous. We proposed in Section 3.1 a general definition of dimension and it will become obvious from the examples that this dimension has nothing to do with the number of generators and relations, except of course in the commutative limit. An outer derivation ∂_i of $\mathcal{T}$ can be defined by its action $\partial_i x^j = \delta_i^j$ on the generators but in general a derivation X of $\mathcal{T}$ can determine a derivation of $\mathcal{A}$ only if it is compatible with the relations: $X(\mathcal{I}) \subset \mathcal{I}$. In the case of an actual embedding of a smooth manifold this would become the condition that the corresponding vector field be tangent to the manifold.

Example 4.1 The *Heisenberg algebra* is the formal $*$-algebra algebra with a single generator a and its adjoint a^* which satisfy the commutation relations $[a, a^*] = k$. For later convenience we have here introduced a parameter k with the units of $(\text{length})^2$. If one writes $a = (x + iy)/\sqrt{2}$ then the relation can be written in the form $[x, y] = ik$. The Heisenberg algebra has a representation on *Fock space*. We choose units with $k = 1$. Let $\mathcal{F}$ be an infinite-dimensional Hilbert space with basis $\{|i\rangle\}$ for integer $i \geq 0$ and inner product $\langle i|j \rangle = \delta_{ij}$. Define the (unbounded) operators a and a^* by

$$a|i\rangle = \sqrt{i}\,|i-1\rangle, \qquad a^*|i\rangle = \sqrt{i+1}\,|i+1\rangle.$$

Then one verifies that in fact $[a, a^*] = 1$. The vector $|i\rangle$ is the i-particle state and the operator N defined by $N|i\rangle = i|i\rangle$ is called the *number operator*. It is given by $N = a^*a$.   □

Example 4.2 The *Weyl algebra* is the formal *-algebra generated by two unitary elements which satisfy the relation $uv = qvu$ of (3.1.25) but not necessarily the constraints involving the integer n. The number $q \in \mathbb{C}$ is of unit modulus. The Weyl algebra has two outer derivations δ_1, δ_2 given by

$$\begin{aligned} \delta_1 u = ir^{-1}u, && \delta_1 v = 0, \\ \delta_2 u = 0, && \delta_2 v = ir^{-1}v. \end{aligned} \tag{4.1.1}$$

They form a Lie algebra with commutation relation $[\delta_1, \delta_2] = 0$. They are to be compared with the ordinary derivations defined by (2.1.26) and with the inner derivations defined by (3.1.29), both of which satisfy the same commutation relations. There is of course also an infinite-dimensional vector space of inner derivations. If $q^n = 1$ then u^n and v^n commute and we saw in Example 3.4 that the algebra can be identified with a matrix algebra over a commutative algebra.   □

If the ideal $\mathcal{I}$ which defines the relations is a subset of $\mathcal{T}_2$ then the algebra is called a *quadratic algebra*. These will play an important role in the construction of examples. It is often convenient to fix a lexicographical ordering of the n generators x^i of $\mathcal{T}$. We shall choose a decreasing order and we shall say that $x^i < x^j$ if $i > j$. Suppose the relations are of the form

$$x^i x^j = q^{a(i,j)} x^j x^i + p^{ij} \tag{4.1.2}$$

where p^{ij} is a quadratic polynomial which is strictly less than $x^i x^j$, q is an arbitrary complex number and $a(i, j)$ is an exponent which depends in general on the indices i and j. Then one says that the quotient algebra is a *polynomial algebra*. The relations have been so chosen that it is possible to write every element of $\mathcal{T}$ in a given order. If $n = 2$ there are essentially two possibilities.

Consider two copies of the projective space $\mathbb{P}^1$ with coordinates (x, y) and (x', y') respectively and consider an automorphism

$$(x, y) \mapsto (x', y') = (ax + by, cx + dy)$$

of $\mathbb{P}^1$ and the corresponding graph $\Gamma \subset \mathbb{P}^1 \times \mathbb{P}^1$. This is the set of points $(x, y) \times (x', y')$ with $x' = ax + by$ and $y' = cx + dy$. Write the relation $x'y' = y'x'$ as $(ax + by)y' = (cx + dy)x'$ ordered so that the primed coordinates are to the right. Then the equation $(x, y) = (x', y')$ of Γ becomes

$$(ax + by)y = (cx + dy)x.$$

Up to conjugation one can show that there are two canonical forms for the automorphism. A conjugation between the automorphisms gives rise to an inner automorphism between the corresponding quadratic algebras.

Example 4.3 Consider the formal algebra $\mathcal{A}$ generated by elements x and y which
satisfy the relation

$$xy = qyx.$$

This is (4.1.2) with $a(1,2) = 1$ and $p^{12} = 0$. The graph Γ is given by

$$\begin{pmatrix} a & b \\ c & d \end{pmatrix} = \begin{pmatrix} 1 & 0 \\ 0 & q \end{pmatrix}.$$

We shall add two extra elements x^{-1} and y^{-1} with the standard relations of inverses.
If $\mathcal{A}$ is a $*$-algebra and $x = u$ and $y = v$ are unitary and $|q| = 1$ then the algebra is
the Weyl algebra of the previous example. $\square$

Example 4.4 Consider the $*$-algebra $\mathcal{A}$ generated by hermitian elements x and y
which satisfy the relation

$$[x, y] = hy^2 \tag{4.1.3}$$

where $h \in i\mathbb{R}$. This is (4.1.2) with $a(1,2) = 0$ and $p^{12} = hy^2$. The graph Γ is given
by

$$\begin{pmatrix} a & b \\ c & d \end{pmatrix} = \begin{pmatrix} 1 & -h \\ 0 & 1 \end{pmatrix}.$$

If one adds the element y^{-1} with the standard relation of inverses then Equation (4.1.3) can be written as $[x, y^{-1}] = -h$. In this form we see that the algebra has
the same formal structure as the Heisenberg algebra of Example 4.1. $\square$

As a particular case of a quadratic algebra we can choose $\mathcal{I}$ to be generated by
all elements of $T_2(\mathcal{H})$ of the form $\alpha \otimes \beta + \beta \otimes \alpha$. Then $\mathcal{A}$ is the *exterior algebra* over
$\mathcal{H}$; it is *graded commutative*:

$$\mathcal{A} = \bigoplus_0^\infty \mathcal{A}_i \tag{4.1.4}$$

with the product of an element $\alpha \in \mathcal{A}_i$ and an element $\beta \in \mathcal{A}_j$ satisfying
$\alpha\beta = (-1)^{ij}\beta\alpha$. As a result $\mathcal{A}_i = 0$ for all $i \geq n + 1$. An algebra can also be
$\mathbb{Z}_2$-*graded* with only two components, in which case we write $\mathcal{A} = \mathcal{A}^+ \oplus \mathcal{A}^-$. A
superalgebra $\mathcal{A}$ is an algebra of the form $\mathcal{A} = \mathcal{A}^+ \oplus \mathcal{A}^-$ with a product which is
graded commutative. The graded commutative algebra $\mathcal{A}$ above is a superalgebra
with

$$\mathcal{A}^+ = \bigoplus_0^\infty \mathcal{A}_{2i}, \qquad \mathcal{A}^- = \bigoplus_0^\infty \mathcal{A}_{2i+1}.$$

If the algebra $\mathcal{A}$ is graded then the derivations can be also graded. A *graded derivation* (of degree 0) of $\mathcal{A}$ is a $\mathbb{C}$-linear map X of $\mathcal{A}$ into itself which satisfies the *graded Leibniz rule*: if $f \in \mathcal{A}_i$ and $g \in \mathcal{A}_j$ then $fg \in \mathcal{A}_{i+j}$ and $X(fg) \in \mathcal{A}_{i+j}$ is given by $X(fg) = Xfg + (-1)^i fXg$. In Section 2.1 we defined a graded derivation d of degree $+1$ and a graded derivation i_X of degree -1. A graded derivation which is inner is defined using a graded commutator.

Example 4.5 Let E be a vector space with basis θ^α and dual basis e_α. Then the associated exterior algebra $\bigwedge^* E$ is a superalgebra. The interior product i_{e_α} is a graded derivation and the most general graded derivation is of the form $X = X^\alpha i_{e_\alpha}$ with $X^\alpha \in \bigwedge^* E$. If there is an exterior derivative d then (2.1.7) is satisfied for some $C^\alpha{}_{\beta\gamma}$. This defines a commutator $[e_\alpha, e_\beta] = C^\gamma{}_{\alpha\beta} e_\gamma$ on the dual basis. The condition $d^2 = 0$ is equivalent to the Jacobi identity and E is a Lie algebra. Conversely if E is a Lie algebra then one can define an exterior derivative by (2.1.7).   □

Example 4.6 As an important example consider the algebra of polynomials $\mathcal{P}(\mathbb{R}^4)$ generated by the coordinates $\tilde{x}^\mu$ of Minkowski space. Define $\mathcal{A}$ to be the tensor product of $\mathcal{P}(\mathbb{R}^4)$ and the exterior algebra over 4 variables, a Weyl spinor θ^a and its conjugate spinor $\bar{\theta}^{\dot{a}}$ ($1 \leq a, \dot{a} \leq 2$). The algebra $\mathcal{A}$ has then 4 bosonic generators $\tilde{x}^\mu$ which commute and 4 fermionic generators $(\theta^a, \bar{\theta}^{\dot{a}})$ which anticommute. The Lie algebra of the Poincaré group which acts on $\mathcal{A}$ has a natural extension to a graded Lie algebra. One adds to the bosonic generators P_μ and $M_{\mu\nu}$ the 4 fermionic generators $Q_a, \bar{Q}_{\dot{a}}$, which commute with P_μ and which satisfy the (graded) commutation rules

$$[Q_a, Q_b] = 0, \qquad [\bar{Q}_{\dot{a}}, \bar{Q}_{\dot{b}}] = 0, \qquad [Q_a, \bar{Q}_{\dot{a}}] = 2\sigma^\mu{}_{a\dot{a}} P_\mu.$$

The structure constants $2\sigma^\mu{}_{a\dot{a}}$ are determined by Lorentz invariance. The generators P_μ and $M_{\mu\nu}$ as well as the four generators $Q_a, \bar{Q}_{\dot{a}}$ can be realized as derivations of $\mathcal{A}$. The P_μ are the standard ones but the $M_{\mu\nu}$ have extra terms which contain derivatives with respect to the fermionic coordinates. The $Q_a, \bar{Q}_{\dot{a}}$ are given by

$$Q_a = \frac{\partial}{\partial \theta^a} - i\sigma^\mu{}_{a\dot{a}} \bar{\theta}^{\dot{a}} \partial_\mu, \qquad \bar{Q}_{\dot{a}} = \frac{\partial}{\partial \bar{\theta}^{\dot{a}}} - i\sigma^\mu{}_{a\dot{a}} \theta^a \partial_\mu.$$

They are also known as super-translations. The most general graded derivation X of $\mathcal{A}$ is of the form

$$X = X^\mu P_\mu + X^a Q_a + X^{\dot{a}} \bar{Q}_{\dot{a}},$$

with $X^\mu, X^a, X^{\dot{a}} \in \mathcal{A}$.   □

Example 4.7 For us an important family of algebras are the algebras $\mathcal{A}_k$ with four generators x^λ which satisfy the commutation relations

$$[x^\mu, x^\nu] = ik J^{\mu\nu}.$$

We shall study these algebras in more detail in Chapter 7. The parameter $\hat{k}$ is a fundamental area scale which we shall suppose to be of the order of the Planck area:

$$\hat{k} \simeq \mu_P^{-2} = G\hbar. \tag{4.1.5}$$

Further details of the structure of $\mathcal{A}_{\hat{k}}$ will be contained, for example, in the commutation relations $[x^\lambda, J^{\mu\nu}]$. The $J^{\mu\nu}$ can be also considered as extra generators and the equations which define them as extra relations. In this case the $J^{\mu\nu}$ cannot be chosen arbitrarily. They must satisfy the four Jacobi identities

$$[x^\lambda, J^{\mu\nu}] + [x^\mu, J^{\nu\lambda}] + [x^\nu, J^{\lambda\mu}] = 0.$$

One can define in fact recursively an infinite sequence of elements by setting, for $p \geq 1$,

$$[x^\lambda, J^{\mu_1 \cdots \mu_p}] = i\hat{k}J^{\lambda\mu_1 \cdots \mu_p}. \tag{4.1.6}$$

With our choice of normalization $J^{\mu_1 \cdots \mu_p}$ has units of mass to the power $p - 2$. Even this infinite sequence cannot determine the algebra. □

Some basic definitions of the algebraic theory of a noncommutative *ring* can be briefly summarized since we shall have little occasion to use them. Consider for simplicity of notation only left modules with respect to a given ring $\mathcal{A}$. A module is said to be *semisimple* if every submodule has a complement; it is said to be *simple* if it has no submodules. Trivially then a simple module is semisimple. A ring $\mathcal{A}$ is said to be semisimple if every $\mathcal{A}$-module is semisimple. Every left ideal of $\mathcal{A}$ is an $\mathcal{A}$-module. The intersection of all maximal left ideals is known as the (Jacobson) *radical* Rad($\mathcal{A}$) of $\mathcal{A}$. If a left ideal $\mathcal{I}$ is such that $\mathcal{I}^n = 0$ for some integer n then $\mathcal{I}$ is said to be *nilpotent*. If it exists, the largest nilpotent ideal is the Wedderburn radical. Under certain 'finiteness' conditions the Wedderburn radical can be shown to exist and in these cases the two radicals coincide. One can prove then the theorem that all modules are semisimple if and only if the radical vanish. In these cases one can prove also the *Wedderburn-Artin theorem*: the ring is the direct sum of a finite set of rings of matrices. A derivation of a simple ring is necessarily an inner derivation.

Example 4.8 If the vector space $\mathcal{H}$ considered at the beginning of Section 3.1 is the space $\mathbb{C}^\infty$ then the algebra $\mathcal{L}(\mathcal{H})$ contains as subalgebra the algebra M_∞ of *sparse matrices*, matrices with zero in all but a finite number of entries. Considered as a ring it does not satisfy the necessary conditions for the Wedderburn-Artin theorem to be true; it obviously cannot be written as the direct sum of a finite set of finite matrix rings. □

Example 4.9 Consider again the exterior algebra (4.1.4). The ideal

$$\mathcal{R} = \bigoplus_1^\infty \mathcal{A}_i$$

is the unique maximal (left/right) ideal of the algebra. It is nilpotent with $\mathcal{R}^{n+1} = 0$ and so $\text{Rad}(\mathcal{A}) = \mathcal{R}$. Although as a vector space the dimension of $\mathcal{A}$ is 2^n one might like to think of it as describing a 'space' with n points. This is to be compared with the interpretation of Example 2.9   □

One convenient way to construct a noncommutative algebra is to consider the algebra $\mathcal{D}$ of differential operators on a manifold. These are discussed in some detail in Section 5.1. The algebra $\mathcal{D}$ is a graded algebra with the grading given by the order of the operator. It is often possible to represent a given abstract algebra as an algebra of differential operators on a manifold. Since the latter act on a vector space of smooth functions the representation gives a concrete realization of the abstract algebra. The vector space can be completed and the differential operators extended to operators on the resulting Hilbert space. This is a convenient way of representing the abstract algebras as algebras of operators on Hilbert spaces. In general the representation is not unique.

Example 4.10 The elements x and y of the Heisenberg algebra in Example 4.1 can be identified as unbounded operators on the space $L^2(\mathbb{R})$ of square-integrable functions on the line by the action

$$(xf)(x) = xf(x), \qquad (yf)(x) = -i\hbar\frac{df}{dx}(x)$$

on smooth functions.   □

Example 4.11 Consider the associative algebra $\mathcal{A}_s$ generated by the differential operators

$$J_- = -\partial_x, \qquad J_+ = x^2\partial_x + 2sx, \qquad J_3 = x\partial_x + s$$

on the real line and designate by $\underline{sl}(s)$ the Lie algebra over the complex numbers by taking the commutator as Lie bracket. The commutation relations are the same as those given in Example 3.3 and so $\underline{sl}(s)$ is a representation of the Lie algebra $\underline{sl}_2(\mathbb{C})$ of the special linear group $SL_2(\mathbb{C})$. If one imposes the condition that $2s+1 = n \in \mathbb{Z}$ then the space of n-th order polynomials is invariant under the action of the operators and we have $\mathcal{A}_s = M_n$. Otherwise $\mathcal{A}_s$ is of infinite dimension. Another associative algebra one can form from the Lie algebra $\underline{sl}_2(\mathbb{C})$ is its (universal) *enveloping algebra* $\mathcal{U}(\underline{sl}_2(\mathbb{C}))$ (not to be confused with the enveloping algebra $\mathcal{A}^e$ of an associative algebra $\mathcal{A}$) which is the quotient of the tensor algebra $T(\underline{sl}_2(\mathbb{C}))$ over $\underline{sl}_2(\mathbb{C})$ by the ideal generated by the elements

$$R(X,Y) = [X,Y] - X \otimes Y + Y \otimes X, \qquad X,Y \in \underline{sl}_2(\mathbb{C}).$$

Denote by $\mathcal{I}(s)$ the 2-sided ideal of $\mathcal{U}(\underline{sl}_2(\mathbb{C}))$ generated by the polynomial

$$R(J_a) = J_+J_- + J_3(J_3 - 1) - s(s + 1). \tag{4.1.7}$$

Then one has the equality
$$\mathcal{A}_s = \mathcal{U}(\underline{sl}_2(\mathbb{C}))/\mathcal{I}(s).$$
When $2s+1$ is an integer this states that the algebra M_n is the part of the enveloping
algebra which corresponds to the value $s(s+1)$ of the Casimir operator $C(J_a)$.   □

It is of interest to note that the construction we gave in Section 3.1 of a differential
calculus based on derivations has a natural generalization to any arbitrary algebra
which possesses an action of a Lie group.

Example 4.12 Let $\mathcal{A}$ be an arbitrary associative algebra with unit element and let
$\mathrm{Der}_G(\mathcal{A})$ be a Lie subalgebra of derivations. If $\mathcal{A}$ is noncommutative there will at
least be the inner derivations. One can define an algebra of forms over $\mathcal{A}$ just as
in ordinary differential geometry. First define $\Omega_D^0(\mathcal{A}) = \mathcal{A}$. Then for each element
$a \in \mathcal{A}$ define the differential da by $da(X) \equiv Xa$ for an arbitrary derivation X.
Define $\Omega_D^1(\mathcal{A})$ to be the smallest $\mathcal{A}$-bimodule which contains all the da. For $p \geq 1$
define $\Omega_D^p(\mathcal{A})$ to be the $\mathcal{A}$-bimodule generated by exterior products of p elements
of $\Omega_D^1(\mathcal{A})$, as in Section 2.1. The exterior algebra $\Omega^*(V)$ of Section 2.1 and the
differential algebras $\Omega_m^*(M_n)$ defined in Section 3.1 are particular examples of the
above construction. If G is a group of automorphisms of $\mathcal{A}$ and $\mathrm{Der}_G(\mathcal{A})$ the Lie
algebra of derivations generated by the action of the Lie algebra of G, then the above
construction yields a differential algebra $\Omega_G^*(\mathcal{A})$ which depends on G.   □

The general formalism developed in Chapter 3 can be carried over to many
infinite-dimensional algebras. The algebra and calculi in the following examples
have been chosen so as to be invariant under the *coaction* of a quantum group. One
uses the word 'coaction' since the 'space' is in fact the noncommutative equivalent of
an algebra of functions on a space. As we saw in Equation (2.1.8) the action on the
algebra is in the opposite direction from the action on the space. The coaction can
be extended as in Chapter 2 to any algebra of forms. We shall introduce quantum
groups later in Section 4.4. Here we are more interested in the various geometries
associated with the algebras and the invariance properties play a secondary role.

Example 4.13 The q-deformed plane or *quantum plane* is the formal algebra $\mathcal{A}$
mentioned in Example 4.3, with two generators x and y and one relation

$$xy = qyx,$$

and with an associated covariant differential calculus. Introduce the differentials
$\xi = dx$ and $\eta = dy$. Then the structure of the algebra of forms is given by the
commutation relations

$$x\xi = q^2\xi x, \quad x\eta = q\eta x + (q^2 - 1)\xi y,$$
$$y\xi = q\xi y, \quad y\eta = q^2\eta y$$

$$(4.1.8)$$

and

$$\xi^2 = 0, \qquad \eta^2 = 0, \qquad \eta\xi + q\xi\eta = 0. \tag{4.1.9}$$

Here q is a complex number, which we shall suppose in general not to be a root of unity. Consider the 1-form

$$\kappa = x\eta - qy\xi.$$

It is easily seen that

$$\kappa^2 = 0. \tag{4.1.10}$$

It follows from the results of Example 4.72 that κ is covariant under the coaction of the quantum group $SL_q(2, \mathbb{C})$:

$$\kappa' = 1 \otimes \kappa.$$

It is in fact, to within multiplication by a complex number, the only invariant element of $\Omega^1(\mathcal{A})$. We shall discuss $SL_q(2, \mathbb{C})$ later in Example 4.63. From the structure of the algebra of forms we deduce the commutation relations

$$x^a\kappa = q\kappa x^a, \qquad \xi^a\kappa = -q^{-3}\kappa\xi^a.$$

To fix the definition of a covariant derivative we must first introduce the operator σ used to define the right Leibniz rule (3.6.3). If we take the covariant derivative on both sides of the Equation (4.1.8), which expresses how x and y commute with ξ and η, we find that σ must be given by

$$\sigma(\xi \otimes \xi) = q^{-2}\xi \otimes \xi, \qquad\qquad \sigma(\xi \otimes \eta) = q^{-1}\eta \otimes \xi,$$
$$\sigma(\eta \otimes \xi) = q^{-1}\xi \otimes \eta - (1 - q^{-2})\eta \otimes \xi, \quad \sigma(\eta \otimes \eta) = q^{-2}\eta \otimes \eta.$$

We shall see later in Example 4.72 that this means that σ is equal to the inverse of the matrix $q\hat{R}$. The extension to $\Omega^1(\mathcal{A}) \otimes_{\mathcal{A}} \Omega^1(\mathcal{A})$ is given by $\mathcal{A}$-linearity. One verifies immediately that the important consistency condition (3.6.5) is satisfied. As a result of the linearity one finds

$$\sigma(\xi \otimes \kappa) = q^{-3}\kappa \otimes \xi, \quad \sigma(\kappa \otimes \xi) = q\xi \otimes \kappa - (1 - q^{-2})\kappa \otimes \xi,$$
$$\sigma(\eta \otimes \kappa) = q^{-3}\kappa \otimes \eta, \quad \sigma(\kappa \otimes \eta) = q\eta \otimes \kappa - (1 - q^{-2})\kappa \otimes \eta,$$

as well as

$$\sigma(\kappa \otimes \kappa) = q^{-2}\kappa \otimes \kappa.$$

Although $\sigma^2 \neq 1$, one finds that σ satisfies the *Hecke relation*

$$(\sigma + 1)(\sigma - q^{-2}) = 0.$$

Suppose that $q^2 \neq -1$. The exterior algebra is obtained by dividing the tensor algebra over $\Omega^1(\mathcal{A})$ by the ideal generated by the three eigenvectors $\xi \otimes \xi$, $\eta \otimes \eta$ and $\eta \otimes \xi + q\xi \otimes \eta$ corresponding to the eigenvalue q^{-2}. The exterior algebra can be then identified with the algebra of forms $\Omega^*(\mathcal{A})$. The symmetric algebra is obtained by dividing the tensor algebra by the ideal generated by the eigenvector $\xi \otimes \eta - q\eta \otimes \xi$ corresponding to the eigenvalue -1. If $q^2 = -1$ then $\sigma = -1$. We shall see later that the curvature must vanish when $q^2 = -1$.

There is a unique one-parameter family of covariant derivatives compatible with the algebraic structure of $\Omega^*(\mathcal{A})$. It is given by

$$D\xi^a = \mu^4 x^a \kappa \otimes \kappa. \tag{4.1.11}$$

The parameter μ must have the dimensions of inverse length. From the invariance of κ it follows that D is covariant under the coaction of $SL_q(2, \mathbb{C})$. From Equation (4.1.10) one sees that the torsion vanishes.

Using the extension D_2 of D defined in (3.6.6) one finds the equality

$$D^2\xi^a = \Omega^a \otimes \kappa \tag{4.1.12}$$

where the 2-form Ω^i is given by

$$\Omega^a = \mu^4 q^{-2}(q^2 + 1)(q^4 + 1)x^a \xi \eta.$$

It vanishes for $q = \pm i$ and $q^2 = \pm i$ but it does not vanish when $q = 1$. There is a preferred family of non-trivial linear connections on the ordinary complex 2-plane which are stable under the quantum deformation. Equation (4.1.12) can also be written in the form

$$D^2\xi^a = -\Omega^a{}_b \otimes \xi^b,$$

with the 'curvature' 2-form given by

$$\Omega^a{}_b = \mu^4(1 + q^{-2})(1 + q^{-4}) \begin{pmatrix} q^2 xy & -qx^2 \\ q^2 y^2 & -xy \end{pmatrix} \xi\eta. \tag{4.1.13}$$

The operator D^2 is a left-module morphism by construction. Since Ω^3 vanishes the Bianchi identities are trivially satisfied. The form κ satisfies the equation

$$D^2\kappa = 0.$$

The metric is a $\mathcal{A}$-linear map (3.4.1) which satisfies the symmetry condition (3.4.2). There exist metrics but the connection (4.1.11) is not compatible with any of them. This could be inferred from the absence of any symmetry in the matrix on the right-hand side of (4.1.13). $\square$

The *extended quantum plane* is the algebra of the previous example with the addition the two extra generators x^{-1} and y^{-1} and the usual relations between inverses. We shall introduce two differential calculi over it, one of which is covariant under the coaction of the quantum group $SL_q(2, \mathbb{C})$ and one which is not.

Example 4.14 The extended quantum plane has two outer derivations defined by

$$e_1^{(0)}x = x, \quad e_1^{(0)}y = 0,$$
$$e_2^{(0)}x = 0, \quad e_2^{(0)}y = y.$$

These are the same as Equation (4.1.1) written with no regard to reality. The corresponding θ^a are given by

$$\theta^1 = x^{-1}dx, \qquad \theta^2 = y^{-1}dy$$

and our construction yields then the ordinary flat metric. To obtain a metric which is almost flat one can add a 'small' inner derivation but using λ_a which are 'small' of the order of some expansion parameter as was done in Example 3.43.   □

Example 4.15 Let now $\mathcal{A}$ be a *-algebra, q of unit modulus and x and y hermitian elements. Define with $n = 2$, for $q^4 \neq 1$, the elements

$$\lambda_1 = -\frac{q^4}{q^4 - 1}x^{-2}y^2, \qquad \lambda_2 = \frac{q^2}{q^4 - 1}x^{-2}.$$

The normalization has been chosen so that the λ_a are singular in the limit $q \to 1$ and that they are anti-hermitian if q is of unit modulus. We find for $q^2 \neq -1$

$$e_1 x = \frac{q^2}{(q^2 + 1)}x^{-1}y^2, \quad e_1 y = \frac{q^4}{q^2 + 1}x^{-2}y^3,$$
$$e_2 x = 0, \qquad\qquad e_2 y = -\frac{q^2}{q^2 + 1}x^{-2}y.$$

(4.1.14)

These derivations are extended to arbitrary polynomials in the generators by the Leibniz rule. Using them we find

$$\xi = \frac{q^2}{(q^2 + 1)}x^{-1}y^2\theta^1, \qquad \eta = \frac{q^2}{q^2 + 1}x^{-2}y(q^2y^2\theta^1 - \theta^2)$$

and solving for the θ^a we obtain

$$\theta^1 = (q^2 + 1)xy^{-2}\xi, \qquad \theta^2 = -(q^2 + 1)x(xy^{-1}\eta - \xi).$$

The module structure which follows from the condition (3.1.34) that the θ^a commute with the elements of the algebra is given by

$$x\xi = q^2\xi x, \quad x\eta = q\eta x + (q^2 - 1)\xi y,$$
$$y\xi = q\xi y, \quad y\eta = q^2\eta y.$$

These are again Equations (4.1.8). We shall see in Example 4.72 that they are invariant under the coaction of the quantum group $SL_q(2,\mathbb{C})$. This invariance was encoded in the choice of λ_a.

Since we can write

$$\theta^1 = \frac{q^4}{q^4 - 1}(\lambda_2^{-1}\lambda_1)^{-1}d\lambda_2^{-1}, \qquad \theta^2 = -\frac{1}{q^4 - 1}(\lambda_2^{-1}\lambda_1)d\lambda_1^{-1}$$

for the θ defined by Equation (3.1.37) we find the expression

$$\theta = \frac{1}{1 - q^4}(q^4\lambda_1^{-1}d\lambda_1 - \lambda_2^{-1}d\lambda_2).$$

It has the same general structure as the analogous 1-form defined in Section 3.1 using the derivations $\text{Der}_2(M_n)$.

The structure of the differential algebra is given by the relations

$$(\theta^1)^2 = 0, \qquad (\theta^2)^2 = 0, \qquad q^4\theta^1\theta^2 + \theta^2\theta^1 = 0.$$

This can be written in the form (3.1.43). If we reorder the indices $(11, 12, 21, 22) = (1, 2, 3, 4)$ then the $C^{ab}{}_{cd}$ introduced in (3.1.51) is given by the expression

$$C = \begin{pmatrix} 1 & 0 & 0 & 0 \\ 0 & 0 & q^{-4} & 0 \\ 0 & q^4 & 0 & 0 \\ 0 & 0 & 0 & 1 \end{pmatrix}.$$

That is, $C^{12}{}_{21} = q^{-4}$ and $C^{21}{}_{12} = q^4$.

The structure elements are given by

$$C^1{}_{12} = (q^4 - 1)\lambda_2, \qquad C^2{}_{21} = (q^4 - 1)\lambda_1.$$

The coefficients $F^a{}_{bc}$ and K_{ab} of Equation (3.1.50) vanish and Equation (3.1.48) becomes

$$C^a{}_{bc} = -2\lambda_d P^{(ad)}{}_{cb}.$$

From (3.6.23) the most general linear connection is given then by

$$\omega^a{}_{bc} = \lambda_d(S^{ad}{}_{bc} - \delta^d_b\delta^a_c). \tag{4.1.15}$$

The condition that this be metric-compatible can be written as (3.6.27). With our index conventions the metric is written as $g^{ab} = (g^1, g^2, g^3, g^4)$ and so the condition can be written in the matrix form

$$\begin{pmatrix} S^1{}_1 & S^1{}_2 & S^1{}_3 & S^1{}_4 \\ S^2{}_1 & S^2{}_2 & S^2{}_3 & S^2{}_4 \\ S^3{}_1 & S^3{}_2 & S^3{}_3 & S^3{}_4 \\ S^4{}_1 & S^4{}_2 & S^4{}_3 & S^4{}_4 \end{pmatrix} \times (S_{(g)}) = \begin{pmatrix} g^1 & 0 & g^3 & 0 \\ 0 & g^1 & 0 & g^3 \\ g^2 & 0 & g^4 & 0 \\ 0 & g^2 & 0 & g^4 \end{pmatrix}. \tag{4.1.16}$$

where we have introduced the matrix $S_{(g)}$ defined by

$$S_{(g)} = \begin{pmatrix} S^1{}_1 g^1 + S^1{}_2 g^3 & S^1{}_3 g^1 + S^1{}_4 g^3 & S^3{}_1 g^1 + S^3{}_2 g^3 & S^3{}_3 g^1 + S^3{}_4 g^3 \\ S^1{}_1 g^2 + S^1{}_2 g^4 & S^1{}_3 g^2 + S^1{}_4 g^4 & S^3{}_1 g^2 + S^3{}_2 g^4 & S^3{}_3 g^2 + S^3{}_4 g^4 \\ S^2{}_1 g^1 + S^2{}_2 g^3 & S^2{}_3 g^1 + S^2{}_4 g^3 & S^4{}_1 g^1 + S^4{}_2 g^3 & S^4{}_3 g^1 + S^4{}_4 g^3 \\ S^2{}_1 g^2 + S^2{}_2 g^4 & S^2{}_3 g^2 + S^2{}_4 g^4 & S^4{}_1 g^2 + S^4{}_2 g^4 & S^4{}_3 g^2 + S^4{}_4 g^4 \end{pmatrix}.$$

$$(4.1.17)$$

The consistency condition (3.6.5) is equivalent to the conditions

$$S^1{}_3 = 0, \qquad q^4 S^2{}_3 = S^2{}_2 + 1, \qquad q^4 S^3{}_3 = S^3{}_2 - q^4, \qquad q^4 S^4{}_3 = S^4{}_2. \qquad (4.1.18)$$

A solution to (4.1.16), (4.1.18) is given by

$$S = \begin{pmatrix} q^{-4} & 0 & 0 & 0 \\ 0 & 0 & q^{-4} & 0 \\ 0 & q^4 & 0 & 0 \\ 0 & 0 & 0 & q^4 \end{pmatrix}. \qquad (4.1.19)$$

It tends to the ordinary flip as $q \to 1$. The σ and π are related as in Example 3.41 with the matrix $T^{ab}{}_{cd}$ which defines τ given by

$$T = \begin{pmatrix} 1 + q^{-4} & 0 & 0 & 0 \\ 0 & 2 & 0 & 0 \\ 0 & 0 & 2 & 0 \\ 0 & 0 & 0 & 1 + q^4 \end{pmatrix}. \qquad (4.1.20)$$

There is a unique σ-symmetric metric given by

$$g^{ab} = \begin{pmatrix} 0 & q^{-2} \\ q^2 & 0 \end{pmatrix}. \qquad (4.1.21)$$

It is of indefinite signature and in 'light-cone' coordinates. The linear connection (4.1.15) is given by

$$\omega^a{}_b = (q^4 - 1) \begin{pmatrix} q^{-4} & 0 \\ 0 & -1 \end{pmatrix} \theta.$$

The curvature vanishes because of the identities $d\theta = 0$, $\theta^2 = 0$; with the choice (4.1.19) of flip the quantum plane is flat. The reality condition (3.4.8) on the metric and the constraint (3.6.29) both force q to be of unit modulus. The solution (4.1.19) is not unique. The other possibilities are of less interest.

We shall give below in Example 4.54 the form of the connection of a simple line bundle over the algebra mentioned in Example 3.4. If one neglects the signature and the reality conditions then in the form of $\omega^a{}_b$ one can see the expression of the fact that the tangent bundle is the direct sum of two such line bundles.   □

In principle differential calculi can be constructed using arbitrary values of n over any given algebra. If $n > 2$ then there is an essential difference with the previous two cases in that relations of the form (4.1.8) which allow one to pass from one side of the differential to the other no longer necessarily exist; the difference is given in terms of the extra elements of the frame. What one does in fact is extend the definition of ξ and η to extra derivations and the extension satisfies quite naturally fewer relations. The left (or right) module $\Omega^1(\mathcal{A})$ is now of rank n instead of 2 as in the two previous examples.

The next two examples satisfy also a reality condition.

Example 4.16 The h-deformed plane or *jordanian deformation* is the formal algebra $\mathcal{A}$ mentioned in Example 4.4, with two generators x and y and one relation

$$[x, y] = hy^2$$

and with an associated covariant differential calculus. Introduce the differentials $\xi = dx$ and $\eta = dy$. Then the structure of the algebra of forms is determined by the commutation relations

$$x\xi = \xi x - h\xi y + h\eta x + h^2\eta y, \quad x\eta = \eta x + h\eta y,$$
$$y\xi = \xi y - h\eta y, \quad\quad\quad\quad\quad y\eta = \eta y, \quad\quad\quad\quad (4.1.22)$$

and

$$\xi^2 = h\xi\eta, \quad\quad \xi\eta = -\eta\xi, \quad\quad \eta^2 = 0.$$

Here h is a complex number which we shall later assume to be pure imaginary so that the x and y can be chosen hermitian. Consider the 1-form

$$\kappa = x\eta - y\xi - hy\eta.$$

It is easily seen that

$$\kappa^2 = 0 \quad\quad\quad\quad\quad\quad\quad (4.1.23)$$

and that κ is covariant under the coaction of the quantum group $SL_h(2, \mathbb{C})$:

$$\kappa' = 1 \otimes \kappa.$$

It is in fact, to within multiplication by a complex number, the only invariant element of $\Omega^1(\mathcal{A})$. We shall define $SL_h(2, \mathbb{C})$ later in Example 4.65. From the structure of the algebra of forms we deduce the commutation relations

$$x^a\kappa = \kappa x^a, \quad\quad \xi^a\kappa = -\kappa\xi^a. \quad\quad\quad (4.1.24)$$

Now, as in the Example 4.13, we take the covariant derivative of both sides of Equation (4.1.22). In terms of the braid matrix $\hat{R}$ introduced below in (4.4.18) we find the relations

$$\sigma(\xi^a \otimes \xi^b) = \hat{R}^{ab}{}_{cd}\xi^c \otimes \xi^d, \quad\quad x^a D\xi^b = \hat{R}^{ab}{}_{cd}(D\xi^c)x^d.$$

This yields the form of σ:

$$\sigma(\xi \otimes \xi) = \xi \otimes \xi - h\xi \otimes \eta + h\eta \otimes \xi + h^2 \eta \otimes \eta,$$
$$\sigma(\xi \otimes \eta) = \eta \otimes \xi + h\eta \otimes \eta,$$
$$\sigma(\eta \otimes \xi) = \xi \otimes \eta - h\eta \otimes \eta,$$
$$\sigma(\eta \otimes \eta) = \eta \otimes \eta.$$

The extension to $\Omega^1(\mathcal{A}) \otimes_{\mathcal{A}} \Omega^1(\mathcal{A})$ is given by $\mathcal{A}$-linearity. As a result of the linearity one finds

$$\sigma(\xi \otimes \kappa) = \kappa \otimes \xi, \quad \sigma(\kappa \otimes \xi) = \xi \otimes \kappa,$$
$$\sigma(\eta \otimes \kappa) = \kappa \otimes \eta, \quad \sigma(\kappa \otimes \eta) = \eta \otimes \kappa,$$
$$\sigma(\kappa \otimes \kappa) = \kappa \otimes \kappa.$$

The σ is an involution in this case:

$$\sigma^2 = 1.$$

There is a two-parameter family of covariant derivatives compatible with the algebraic structure of $\Omega^*(\mathcal{A})$. It is given by

$$D\xi^a = \mu^4 x^a \kappa \otimes \kappa + \nu^2 (\xi^a \otimes \kappa + \kappa \otimes \xi^a).$$

The parameters μ and ν must have the dimensions of inverse length. From the invariance of κ it follows that D is covariant under the coaction of $SL_h(2, \mathbb{C})$. From Equations (4.1.23) and (4.1.24) one sees that the torsion vanishes.

For simplicity suppose that $\nu = 0$. Using the extension D_2 of D defined in (3.6.6) one finds the equality

$$D^2 \xi^a = -\Omega^a{}_b \otimes \xi^b,$$

with the curvature 2-form given by

$$\Omega^a{}_b = 4\mu^4 \begin{pmatrix} xy & -x^2 + hxy \\ y^2 & -yx + hy^2 \end{pmatrix} \xi\eta.$$

The 1-form κ satisfies the equation

$$D^2 \kappa = 0.$$

The general covariant derivative is compatible with no metric. $\square$

Example 4.17 The *extended h-deformed plane* is the algebra of the previous example with the addition of the two extra generators x^{-1} and y^{-1} and the usual relations between inverses. Using the inverse elements a frame θ^a can be constructed. The extended h-deformed plane is more interesting than the extended q-deformed one in that the metric and linear connection it supports have an interesting commutative limit. We shall here suppose that $\mathcal{A}$ is a $*$-algebra. We require that $h \in i\mathbb{R}$ and that the generators x and y be hermitian.

If we introduce

$$u = xy^{-1} + \frac{1}{2}h, \qquad v = y^{-2}$$

then the commutation relation becomes

$$[u, v] = -2hv. \tag{4.1.25}$$

This choice of generators is useful in studying the commutative limit. If x and y are hermitian then so are u and v.

A real frame is given by

$$\theta^1 = y\xi - (x - hy)\eta = v^{-1}du,$$
$$\theta^2 = 2y^{-1}\eta = -v^{-1}dv. \tag{4.1.26}$$

The module structure (4.1.22) is equivalent to the condition (3.1.34) that this frame commute with the elements of the algebra. The structure of the algebra of forms is given by the relations

$$(\theta^1)^2 = 0, \qquad (\theta^2)^2 = 0, \qquad \theta^1\theta^2 + \theta^2\theta^1 = 0.$$

These can be written as

$$\theta^a\theta^b = P^{ab}{}_{cd}\theta^c\theta^d,$$

where $P^{ab}{}_{cd}$ is given by (3.1.44). Therefore, from Equation (3.1.48) we have $C^a{}_{bc} = F^a{}_{bc}$. In particular the structure elements $C^a{}_{bc}$ are real numbers:

$$C^1{}_{12} = -C^1{}_{21} = 1, \qquad C^2{}_{ab} = 0.$$

If we introduce the derivations $e_a = \text{ad } \lambda_a$ with

$$\lambda_1 = \frac{1}{2h}y^{-2} = \frac{1}{2h}v, \qquad \lambda_2 = \frac{1}{2h}xy^{-1} + \frac{1}{4} = \frac{1}{2h}u$$

we see that Equation (3.1.33) is satisfied. From Equation (4.1.25) we find

$$[\lambda_1, \lambda_2] = \lambda_1.$$

The λ_a form a solvable Lie algebra and satisfy Equation (3.1.50) with $K_{ab} = 0$. From the form of the expression of the λ_a in terms of u and v it is obvious that to within a factor it is they which have been used as the new generators, just as in Example 4.15. Written in terms of them the Dirac operator becomes

$$\theta = \lambda_2(\lambda_1^{-1}d\lambda_1 - \lambda_2^{-1}d\lambda_2).$$

It is an anti-hermitian form which satisfies (3.1.45) with $K_{ab} = 0$.

Since a frame exists a metric is given by Equation (3.4.6):

$$g(\theta^a \otimes \theta^b) = g^{ab}$$

and a linear connection is given by Equations (3.6.19), (3.6.21):

$$\sigma(\theta^a \otimes \theta^b) = S^{ab}{}_{cd}\theta^c \otimes \theta^d, \qquad D\theta^a = -\omega^a{}_{bc}\theta^b\theta^c.$$

We shall suppose that $S^{ab}{}_{cd} = C^{ab}{}_{cd}$ and that $g^{ab} = \delta^{ab}$. We have then in terms of the generators x and y

$$g(\xi \otimes \xi) = y^{-2} + x^2/4, \quad g(\xi \otimes \eta) = xy/4,$$
$$g(\eta \otimes \xi) = yx/4, \qquad g(\eta \otimes \eta) = y^2/4$$

and in terms of the generators u and v

$$g(du \otimes du) = v^2, \quad g(du \otimes dv) = 0,$$
$$g(dv \otimes du) = 0, \quad g(dv \otimes dv) = v^2.$$

The unique torsion-free, metric-compatible linear connection is given by

$$D\theta^1 = -\theta^1 \otimes \theta^2, \qquad D\theta^2 = \theta^1 \otimes \theta^1.$$

The curvature map defined by Equation (3.7.2) becomes

$$D^2\theta^1 = \theta^1\theta^2 \otimes \theta^2, \qquad D^2\theta^2 = -\theta^1\theta^2 \otimes \theta^1.$$

If one sets as usual $R_{abcd} = g_{ae}R^e{}_{bcd}$ then one finds that the Gaussian curvature is given by

$$R_{1212} = -1.$$

The geometry is a noncommutative version of the Poincaré half-plane.

If we introduce a third hermitian element

$$w = -\frac{1}{2}(u^2 - 2hu + 1 + 2h^2)v^{-1}$$

and add a third derivation defined in terms of

$$\lambda_3 = \frac{1}{2h}w$$

then the resulting derivations e_i form a representation of the Lie algebra $\underline{sl}_2(\mathbb{R})$ of $SL(2,\mathbb{R})$. We have found a frame with 2 generators since the Poincaré half-plane is a parallelizable manifold and the module of 1-forms is a free (left or right) module. This is not so in the case of the 2-sphere; the module of 1-forms in this case is a nontrivial submodule of a free module of rank 3. The Lie algebra of Killing vector fields of the Poincaré half-plane and the sphere are different real realizations of $\underline{sl}_2(\mathbb{C})$.

A differential calculus can be defined using the three 1-forms θ^i dual to the derivations e_i. From the identity $df = [\lambda_i, f]\theta^i$ we can conclude that

$$du = v\theta^1 - w\theta^3, \qquad dv = -v\theta^2 + u\theta^3, \qquad dw = -u\theta^1 + w\theta^2.$$

Provided that $h \neq 0$ these can be inverted to obtain equations for the θ^i in terms of du, dv and dw:

$$\theta^1 = \frac{1}{2h}u^{-1}[w, du], \qquad \theta^2 = \frac{1}{2h}v^{-1}[u, dv], \qquad \theta^3 = \frac{1}{2h}u^{-1}[v, du].$$

This differential calculus has fewer relations than the one defined above. It lies between the one defined by the basis (4.1.26) and the universal differential calculus, which has a free algebra of forms with no relations.

Denote by L_i the Lie derivation with respect to the derivation e_i. Then it is easy to see that

$$L_1\theta^1 = -\theta^2, \qquad L_1\theta^2 = 0,$$
$$L_2\theta^1 = +\theta^1, \qquad L_2\theta^2 = 0,$$
$$L_3\theta^1 = -v^{-1}w\theta^2, \quad L_3\theta^2 = -v^{-1}u\theta^2 - \theta^1.$$

We find then that none of the derivations e_i is a Killing derivation as defined in Equation (3.4.19). The first two form a solvable Lie algebra and g is not defined in terms of the Killing metric. Compare this with the result of Example 3.22.

It is interesting to study the structure of the extended h-deformed quantum plane in the commutative limit. In terms of the commutative limits $\tilde{u}$, $\tilde{v}$ of the generators u, v of the algebra $\mathcal{A}$ and the corresponding commutative limit $\tilde{\theta}^a$ of the frame, the metric is given by the line element

$$ds^2 = (\tilde{\theta}^1)^2 + (\tilde{\theta}^2)^2 = \tilde{v}^{-2}(d\tilde{u}^2 + d\tilde{v}^2).$$

This is indeed the metric of the Poincaré half-plane. The extended h-deformed plane can be considered as a noncommutative version of the Poincaré half-plane.

The derivations e_i define, in the commutative limit, 3 vector fields

$$X_i = \lim_{h \to 0} e_i.$$

If we define $\tilde{w}$ to be the commutative limit of w then

$$X_1 = \tilde{v}\partial_{\tilde{u}}, \qquad X_2 = -\tilde{v}\partial_{\tilde{v}}, \qquad X_3 = -\tilde{w}\partial_{\tilde{u}} + \tilde{u}\partial_{\tilde{v}}.$$

By construction these vector fields form a Lie algebra with the same commutation relations as the e_i. We have seen that the X_i cannot be Killing vector fields; there is in fact no reason for them to be so.

The Poincaré half-plane does have however three Killing vector fields X'_i, given by

$$X'_1 = \partial_{\tilde{u}}, \qquad X'_2 = \tilde{u}\partial_{\tilde{u}} + \tilde{v}\partial_{\tilde{v}}, \qquad X'_3 = \frac{1}{2}(\tilde{v}^2 - \tilde{u}^2)\partial_{\tilde{u}} - \tilde{u}\tilde{v}\partial_{\tilde{v}}.$$

Define a map ϕ of the Poincaré half-plane into itself by

$$\phi(\tilde{u}) = \tilde{u}' = \tilde{u}\tilde{v}^{-1}, \qquad \phi(\tilde{v}) = \tilde{v}' = \tilde{v}^{-1}.$$

This is a regular diffeomorphism. Indeed

$$\phi^2 = \phi \circ \phi = 1.$$

In the spirit of noncommutative geometry we consider $\tilde{u}$ and $\tilde{v}$ as generators of the algebra of functions on the Poincaré half-plane. In ordinary differential geometry a map ϕ of the manifold induces a map ϕ^* of the algebra of differential forms and a map Let ϕ_* of the vector fields. Since we shall not have occasion to refer to the manifold as such we use the notation ϕ to designate the restriction of ϕ^* to the algebra of functions. Since we have

$$\phi_*\partial_{\tilde{u}} = \tilde{v}\partial_{\tilde{u}}, \qquad \phi_*\partial_{\tilde{v}} = -\tilde{u}\tilde{v}\partial_{\tilde{u}} - \tilde{v}^2\partial_{\tilde{v}}$$

it follows that

$$\phi_* X'_i = X_i.$$

The commutative limit of the derivations which defined the differential calculus are related to the Killing vector fields then in a simple way. We have not succeeded in constructing derivations of the algebra whose commutative limits are the Killing vector fields X'_i. The limit $h \to 0$ is a rather singular limit and it need not be true that an arbitrary vector field on the Poincaré half-plane is the limit of a derivation. The action of ϕ^* on the frame is given by

$$\phi^*\tilde{\theta}^1 = \tilde{v}'^{-1}d\tilde{u}' = \tilde{v}\tilde{\theta}^1 + \tilde{u}\tilde{\theta}^2, \qquad \phi^*\tilde{\theta}^2 = -\tilde{v}'^{-1}d\tilde{v}' = -\tilde{\theta}^2.$$

The vector fields X_i are Killing with respect to the metric defined by the line element

$$ds^2 = (\phi^*\tilde{\theta}^1)^2 + (\phi^*\tilde{\theta}^2)^2.$$

We have not really understood the role of the map ϕ nor why it appears but it is certainly connected with the fact that the derivations which define the frame form a solvable Lie algebra and g is not defined in terms of the Killing metric.

The commutation relations (4.1.25) define on the Poincaré half-plane a Poisson structure

$$\{\tilde{u}, \tilde{v}\} = -2\tilde{v}.$$

Since the map ϕ is not a symplectomorphism it cannot be 'lifted' to a morphism of the algebra $\mathcal{A}$. There should be a relation between the Poisson structure and the Riemann curvature. It is not evident from the present example however what this relation is. The geometry of the Poincaré half-plane can be completely globally defined by the action of the $SL(2, \mathbb{R})$ group whose Lie algebra is given by the Killing vectors. From this point of view a complete classification of all Poisson structures on the Poincaré half-plane as well as their possible 'quantum' deformations has been given in detail.   □

The *q-euclidean spaces* or *quantum euclidean spaces* $\mathbb{C}_q^n$ and $\mathbb{R}_q^n$ are formal associative algebras which are covariant under the coaction of the quantum groups $SO_q(n)$ which will be introduced below in Example 4.66. They are defined in terms of generators x^i and quadratic relations of the form

$$P_a{}^{ij}{}_{kl} x^k x^l = 0 \tag{4.1.27}$$

for all i, j. We shall show later in Example 4.66 how $P_a{}^{ij}{}_{kl}$ is defined in terms of the braid matrix $\hat{R}^{ij}{}_{kl}$ which determines the structure of $SO_q(n)$. One obtains the real q-euclidean space by choosing $q \in \mathbb{R}^+$ and by giving the algebra an involution defined by

$$(x^i)^* = x^j g_{ji} \tag{4.1.28}$$

where g_{ij} is the q-deformed euclidean metric. We can use the summation convention if we consider the involution to lower or raise an index. This condition is an $SO_q(n, \mathbb{R})$-covariant condition and n linearly independent, real coordinates can be obtained as combinations of the x^i. The 'length' squared

$$r^2 = g_{ij} x^i x^j = (x^i)^* x^i \tag{4.1.29}$$

is $SO_q(n, \mathbb{R})$-invariant, real and generates the center $\mathcal{Z}(\mathbb{R}_q^n)$ of $\mathbb{R}_q^n$. We can extend $\mathbb{R}_q^n$ by adding to it the square root r of r^2 and the inverse r^{-1}. For various reasons, some mentioned in Example 4.66, we add also an extra generator Λ called the *dilatator* and its inverse Λ^{-1} chosen such that

$$x^i \Lambda = q \Lambda x^i. \tag{4.1.30}$$

We shall choose Λ to be unitary. Since r and Λ do not commute the center of the new extension is trivial. By differentiating Equation (4.1.30) we obtain the condition

$$\xi^i \Lambda + x^i d\Lambda = q d\Lambda x^i + q \Lambda \xi^i$$

on the differential $d\Lambda$. A possible solution is given by the two conditions

$$x^i d\Lambda = q d\Lambda x^i, \qquad \xi^i \Lambda = q \Lambda \xi^i. \tag{4.1.31}$$

In particular one can consistently set $d\Lambda = 0$. We shall do this in Example 4.18 and Example 4.21 although it means that the condition $df = 0$ does not imply that

$f \in \mathbb{C}$. This is not entirely satisfactory since one would like the only 'functions' with vanishing exterior derivative to be the 'constant functions'. It could be remedied by considering a more general solution to Equation (4.1.31). This would however complicate our calculations since it would increase the number of independent forms by one. From (3.1.35) we see that a necessary condition for $d\Lambda = 0$ is that

$$e_a \Lambda = [\lambda_a, \Lambda] = 0.$$

The λ_a must be therefore homogeneous functions of degree zero of the x^i. We are forced then to extend again the algebra to include some inverse elements. We shall see this in the examples.

There are two $SO_q(n)$-covariant differential calculi $\Omega^*(\mathbb{R}_q^n)$ and $\bar{\Omega}^*(\mathbb{R}_q^n)$ over the algebra $\mathbb{R}_q^n$. Let d and $\bar{d}$ be the respective differentials and set $\xi^i = dx^i$ and $\bar{\xi}^i = \bar{d}x^i$. The calculi are determined respectively by the commutation relations

$$x^i \xi^j = q\hat{R}^{ij}{}_{kl} \xi^k x^l \tag{4.1.32}$$

for $\Omega^1(\mathbb{R}_q^n)$ and

$$x^i \bar{\xi}^j = q^{-1}(\hat{R}^{-1})^{ij}{}_{kl} \bar{\xi}^k x^l \tag{4.1.33}$$

for $\bar{\Omega}^1(\mathbb{R}_q^n)$. For neither calculus is it possible to extend the involution (4.1.28). The $SO_q(n)$-invariance is incompatible with the involution. To construct a real calculus we follow the prescription of Example 3.16. The involution can be extended to the direct sum $\Omega^1(\mathcal{A}) \oplus \bar{\Omega}^1(\mathcal{A})$ by setting

$$(\xi^i)^* = \bar{\xi}^j g_{ji}, \tag{4.1.34}$$

since this exchanges relations (4.1.32) and (4.1.33). In the limit $q \to 1$ we recover the standard involution and the differential is real.

From the equations $[r, \theta^a] = 0$ we find that we must choose an Ansatz for the frame of the form

$$\theta^a = \Lambda^{-1} \theta_j^a \xi^j \tag{4.1.35}$$

where the θ_j^a are elements of the extended algebra $\mathbb{R}_q^n$. The conditions $[x^i, \theta^a] = 0$ become

$$x^i \theta_j^a = \hat{R}^{-1ki}{}_{lj} \theta_k^a x^l \tag{4.1.36}$$

in terms of the braid matrix.

Consider the $SO_q(n)$-invariant 1-form

$$\eta = g_{ij} x^i \xi^j = q^{-1} g_{ij} \xi^j x^i.$$

Using the projector decomposition Equation (4.4.19) of the braid matrix as well as the relations (4.4.21) between the braid matrix and its inverse one can easily verify that

$$[\eta, x^i] = -q^{-2}(q-1)r^2 \xi^i.$$

Hence we conclude that

$$\theta = (q-1)^{-1}q^2 r^{-2}\eta \tag{4.1.37}$$

is the 1-form of Equation (3.1.38) which defines the differential. It satisfies the conditions

$$d\theta = 0, \quad \theta^2 = 0$$

and so the coefficients K_{ab} of Equation (3.1.45) vanish. In fact θ is an exact form:

$$d(\log r) = \frac{1}{2}(1 - q^{-2})\theta. \tag{4.1.38}$$

These q-deformations are of especial interest in the cases $n = 1$ and $n = 3$.

Example 4.18 The algebra $\mathbb{R}_q^1$ has only two generators x and Λ which satisfy the commutation relation $x\Lambda = q\Lambda x$. We shall choose x hermitian and $q \in (1, \infty)$. This is a modified version of the Weyl algebra with q real instead of with unit modulus. We write $x = q^y$ as an equality within $\mathbb{R}_q^1$. Then the commutation relations between Λ and y can be written as

$$\Lambda^{-1}y\Lambda = y + 1. \tag{4.1.39}$$

We shall have occasion to renormalize y. We introduce a renormalization parameter z as

$$z = q^{-1}(q - 1) > 0.$$

The renormalization is then given by the substitution

$$zy \mapsto y. \tag{4.1.40}$$

The differential calculus $\Omega^*(\mathbb{R}_q^1)$ is based on the relations

$$xdx = qdxx, \quad dx\Lambda = q\Lambda dx \tag{4.1.41}$$

for the 1-forms. If we choose

$$\lambda_1 = -z^{-1}\Lambda$$

then

$$e_1 x = q\Lambda x, \quad e_1\Lambda = 0$$

and the calculus (4.1.41) is defined by the condition $df(e_1) = e_1 f$ for arbitrary $f \in \mathbb{R}_q^1$. By setting

$$\lambda_2 = z^{-1}x$$

and introducing a second derivation

$$e_2\Lambda = q\Lambda x, \quad e_2 x = 0$$

one could extend the calculus (4.1.41) by the condition $df(e_2) = e_2 f$ for arbitrary $f \in \mathbb{R}_q^1$. One would find $xd\Lambda = qd\Lambda x$. We shall not do so for the reasons given above.

Also it will be seen that Λ is in a sense an element of the phase space associated to x and we are interested in position-space geometry.

The adjoint derivation $e_1^\dagger$ of e_1 is defined by (2.1.33):

$$e_1^\dagger f = (e_1 f^*)^*.$$

The $e_1^\dagger$ on the left-hand side is not an adjoint of an operator e_1. It is defined uniquely in terms of the involution of $\mathbb{R}_q^1$ whereas e_1 acts on this algebra as a vector space.

Since Λ is unitary we have $(\lambda_1)^* \neq -\lambda_1$ and e_1 is not a real derivation. We use therefore the second differential calculus $\bar{\Omega}^*(\mathbb{R}_q^1)$ defined by the relations

$$x\bar{d}x = q^{-1}\bar{d}xx, \qquad \bar{d}x\Lambda = q\Lambda\bar{d}x \qquad (4.1.42)$$

and based on the derivation $\bar{e}_1$ formed from $\bar{\lambda}_1 = -\lambda_1^*$. This calculus is defined by the condition $\bar{d}f(\bar{e}_1) = \bar{e}_1 f$ for arbitrary $f \in \mathbb{R}_q^1$. The derivation $\bar{e}_1$ is also not real. It is easy to see however that

$$e_1^\dagger = \bar{e}_1. \qquad (4.1.43)$$

The frame elements θ^1 and $\bar{\theta}^1$ dual to the derivations e_1 and $\bar{e}_1$ are given by

$$
\begin{aligned}
\theta^1 &= \theta_1^1 dx, & \theta_1^1 &= \Lambda^{-1}x^{-1}, \\
\bar{\theta}^1 &= \bar{\theta}_1^1 \bar{d}x, & \bar{\theta}_1^1 &= q^{-1}\Lambda x^{-1}.
\end{aligned}
\qquad (4.1.44)
$$

To construct a real calculus we shall follow the prescription of Example 3.16 and introduce the algebra $\mathbb{R}_{qr}^1$. Consider the element

$$\lambda_{r1} = (\lambda_1, \bar{\lambda}_1) = z^{-1}(-\Lambda, \Lambda^{-1}) \qquad (4.1.45)$$

of $\mathbb{R}_q^1 \times \mathbb{R}_q^1$. By the definition of the involution (3.2.20) we have $\lambda_{r1}^* = -\lambda_{r1}$ and the associated derivation $e_{r1} = \operatorname{ad}\lambda_{r1}$ is real. We define d_r as in (3.2.22). We have

$$e_{r1}x = (q\Lambda, \Lambda^{-1})x \qquad (4.1.46)$$

from which we conclude that

$$xd_r x = (q, q^{-1})d_r xx, \qquad d_r x\Lambda = q\Lambda d_r x. \qquad (4.1.47)$$

Within $\Omega_r^*(\mathbb{R}_{qr}^1)$ these are the equivalent of the relations (4.1.41) and (4.1.42).

If d_r is to be a differential then the extension to higher-order forms much be such that $d_r^2 = 0$. We have then

$$(d_r x)^2 = 0. \qquad (4.1.48)$$

The algebraic structure of

$$\Omega_r^*(\mathbb{R}_{qr}^1) \subset \Omega^*(\mathbb{R}_q^1) \times \bar{\Omega}^*(\mathbb{R}_q^1) \qquad (4.1.49)$$

is given by the relations (4.1.47) and the condition (4.1.48). It follows that

$$d_r \theta_r^1 = 0, \qquad (\theta_r^1)^2 = 0. \tag{4.1.50}$$

The forms θ^1, $\bar{\theta}^1$ and θ_r^1 are closed. They are also exact. In fact if we define $K \in \mathbb{R}_q^1 \times \mathbb{R}_q^1$ by

$$K = z(\Lambda^{-1}, \Lambda) \tag{4.1.51}$$

then we find that

$$\theta^1 = d(z\Lambda^{-1}y), \qquad \bar{\theta}^1 = \bar{d}(z\Lambda y), \qquad \theta_r^1 = d_r(Ky).$$

We can consider the derivations e_1 and $\bar{e}_1$ also as elements of $\mathbb{R}_q^1$. As such they satisfy the commutation relations

$$e_1 x = q\Lambda x + x e_1, \qquad e_1 \Lambda = \Lambda e_1,$$
$$\bar{e}_1 x = \Lambda^{-1} x + x\bar{e}_1, \qquad \bar{e}_1 \Lambda = \Lambda \bar{e}_1. \tag{4.1.52}$$

If one takes the adjoint of the first two equations one finds the relation

$$e_1^* + \bar{e}_1 = c \tag{4.1.53}$$

for some constant c. This is to be compared with (4.1.43). The equation $e_1 f = [\lambda_1, f]$ relates the derivation e_1 to the operator λ_1. There is an ambiguity

$$\lambda_1 \mapsto \lambda_1 + z^{-1}\alpha$$

in this identification which depends on a complex parameter α. A similar ambiguity $\bar{\alpha}$ exists for $\bar{\lambda}_1$. We shall give c as a function of α and $\bar{\alpha}$ below in Example 4.47. We shall examine some physical problems on this 'space' in Section 7.3.

One can think of the algebra $\mathbb{R}_q^1$ as describing a set of 'lines' x embedded in a 'plane' (x, Λ) and defined by the condition '$\Lambda = $ constant'. To within a normalization the unique local metric is given by

$$g(\theta_r^1 \otimes \theta_r^1) = 1. \tag{4.1.54}$$

Using it we introduce the element

$$g'^{11} = g(d_r x \otimes d_r x) = (e_{r1}x)^2 g(\theta_r^1 \otimes \theta_r^1) = (e_{r1}x)^2 \tag{4.1.55}$$

of the algebra. Then

$$\sqrt{g'^{11}} = e_{r1}x, \qquad \left(\sqrt{g'^{11}}\right)^* = \sqrt{g'^{11}}. \tag{4.1.56}$$

To find the associated linear connection we set as usual

$$D_1\theta^1 = -\omega^1{}_{11}\theta^1 \otimes \theta^1, \qquad \bar{D}_1\bar{\theta}^1 = -\bar{\omega}^1{}_{11}\bar{\theta}^1 \otimes \bar{\theta}^1 \tag{4.1.57}$$

as Ansatz. From the general theory of Section 3.6 these must satisfy a left and right
Leibniz rule

$$D_1(f\theta^1) = df \otimes \theta^1 - f\omega^1{}_{11}\theta^1 \otimes \theta^1, \quad D_1(\theta^1 f) = \sigma(\theta^1 \otimes df) - \omega^1{}_{11}f\theta^1 \otimes \theta^1,$$
$$\bar{D}_1(f\bar{\theta}^1) = \bar{d}f \otimes \bar{\theta}^1 - f\bar{\omega}^1{}_{11}\bar{\theta}^1 \otimes \bar{\theta}^1, \quad \bar{D}_1(\bar{\theta}^1 f) = \bar{\sigma}(\bar{\theta}^1 \otimes \bar{d}f) - \bar{\omega}^1{}_{11}f\bar{\theta}^1 \otimes \bar{\theta}^1,$$

where $f \in \mathbb{R}^1_q$ and the generalized flips σ and $\bar{\sigma}$ can be written as

$$\sigma(\theta^1 \otimes \theta^1) = S\theta^1 \otimes \theta^1, \qquad \bar{\sigma}(\bar{\theta}^1 \otimes \bar{\theta}^1) = \bar{S}\bar{\theta}^1 \otimes \bar{\theta}^1.$$

From the compatibility conditions

$$D_1(\Lambda\theta^1) = D_1(\theta^1\Lambda), \quad D_1(x\theta^1) = D_1(\theta^1 x),$$
$$\bar{D}_1(\Lambda\bar{\theta}^1) = \bar{D}_1(\bar{\theta}^1\Lambda), \quad \bar{D}_1(x\bar{\theta}^1) = \bar{D}_1(\bar{\theta}^1 x)$$

it is easy to see that, to within a multiplicative constant, there are only two solutions
to Equation 4.1.57. The connection compatible with the metric (4.1.54) is given by

$$\omega^1{}_{11} = 0, \quad S = 1, \quad \bar{\omega}^1{}_{11} = 0, \quad \bar{S} = 1.$$

The real, torsion-free covariant derivative compatible with the unique local metric is
given by

$$D_r\theta^1_r = 0. \tag{4.1.58}$$

This can also be written in the form

$$D_r(d_r x) = (q^2\Lambda^2, \Lambda^{-2})x\theta^1_r \otimes \theta^1_r.$$

The generalized flip (3.4.3) is given by $\sigma_r = 1$. This yields by (3.6.8) the involution

$$(\theta^1_r \otimes \theta^1_r)^* = \theta^1_r \otimes \theta^1_r$$

on the tensor product if the covariant derivative is to be real:

$$D_r\xi^* = (D_r\xi)^*.$$

The geometry is 'flat' in the sense that the curvature tensor defined by D_r vanishes.
The interpretation is somewhat unsatisfactory however here because of the existence
of elements in the algebra which do not lie in the center but which have nevertheless
vanishing exterior derivative. These elements play a less important role in the
geometry of the algebras $\mathbb{R}^n_q$ for larger values of n.   □

Example 4.19 The second solution to Equation 4.1.57 is given by

$$\omega^1{}_{11} = \Lambda, \quad S = q^{-1}, \quad \bar{\omega}^1{}_{11} = q\Lambda^{-1}, \quad \bar{S} = q.$$

We set

$$g(\theta^1 \otimes \theta^1) = g^{11}, \qquad g(\bar{\theta}^1 \otimes \bar{\theta}^1) = \bar{g}^{11}.$$

The metric-compatibility condition (3.6.7) can be written

$$dg^{11} = -(1+S)\omega^1{}_{11}g^{11}\theta^1, \qquad d\bar{g}^{11} = -(1+\bar{S})\bar{\omega}^1{}_{11}\bar{g}^{11}\bar{\theta}^1.$$

There is a solution, which necessarily defines a non-local metric, given by

$$g_{11} = (q\Lambda x)^2, \qquad \bar{g}_{11} = (\Lambda^{-1}x)^2.$$

This can also be written in the form

$$g(dx \otimes dx) = 1, \qquad g(\bar{dx} \otimes \bar{dx}) = 1 \qquad (4.1.59)$$

and the corresponding covariant derivative can be written also as

$$D_1 dx = 0, \qquad \bar{D}_1 \bar{dx} = 0.$$

The 'space' now is a discrete subset of the positive real axis with an accumulation point at the origin.

The 'non-locality' means that if f is a 'function' and α a form then the norm of $f\alpha$ cannot be equal to f times the norm of α. To see this we multiply (4.1.59) from the right by x. If we supposed that the metric were left- and right-linear then we would find

$$x = xg(dx \otimes dx) = g(xdx \otimes dx) = q^2 g(dx \otimes dxx) = q^2 g(dx \otimes dx)x = q^2 x.$$

The first and fifth equalities are mathematical trivialities; the third follows directly from (4.1.41) . Therefore either the second or the fourth, or both, must be false. There are no 2-forms and so the curvature and torsion of the non-local metric vanish.

The Ansatz for a covariant derivative on the real calculus is

$$D_r\theta_r^1 = -\omega^1{}_{r11}\theta_r^1 \otimes \theta_r^1. \qquad (4.1.60)$$

The generalized flip is given by $\sigma_r = (q^{-1}, q)$ and the appropriate involution (3.6.8) on the tensor product is given by

$$(\theta_r^1 \otimes \theta_r^1)^* = (q^{-1}, q)(\theta_r^1 \otimes \theta_r^1).$$

The solution to (4.1.60) is given by

$$\omega^1{}_{r11} = (\Lambda, q\Lambda^{-1}), \qquad (\omega^1{}_{r11})^* = (q, q^{-1})\omega^1{}_{r11}.$$

The connection coefficient is not hermitian but the covariant derivative is real. $\quad\square$

Example 4.20 According to the general theory of Section 3.5 one defines a Yang-Mills covariant derivative of an element ψ of a $\mathbb{R}^1_q$-module $\mathcal{H}$ as a map

$$\mathcal{H} \xrightarrow{D_r} \Omega^1_r(\mathbb{R}^1_{qr}) \otimes \mathcal{H}$$

which satisfies the left Leibniz rule

$$D_r(f\psi) = d_r f \psi + f D_r \psi.$$

We have here dropped the tensor product symbol. As in (3.5.13) we define

$$D_r \psi = \theta_r \psi - \psi \phi_r, \qquad \phi_r = A_r - \theta_r$$

and choose the gauge group to act from the right: $\psi \mapsto \psi g$. When the gauge potential vanishes one has then

$$D_r \psi = \theta^1_r e_{r1} \psi, \qquad e_{r1} \psi = ([\lambda_1, \psi], [\bar\lambda_1, \psi]). \tag{4.1.61}$$

With this action e_{r1} is independent of the terms c_r defined below in (4.3.36). We shall return to this covariant derivative in Examples 7.31 and 7.32. □

Example 4.21 Another interesting case is $n = 3$. In this case it is convenient to set $x^a = (x^-, y, x^+)$. For simplicity of notation we introduce also

$$h = \sqrt{q} - 1/\sqrt{q}.$$

We shall suppose that $q > 1$, which implies that $h > 0$. The Equations (4.1.27) can be written in the form

$$
\begin{aligned}
x^- y &= q \, yx^-, \\
x^+ y &= q^{-1} yx^+, \\
[x^+, x^-] &= hy^2.
\end{aligned}
\tag{4.1.62}
$$

The first two equations define two copies of the quantum plane with q and q^{-1} as deformation parameter and a common generator y.

The metric matrix is given by $g_{ij} = g^{ij}$ with

$$g_{ij} = \begin{pmatrix} 0 & 0 & 1/\sqrt{q} \\ 0 & 1 & 0 \\ \sqrt{q} & 0 & 0 \end{pmatrix}. \tag{4.1.63}$$

The condition (4.1.28) can be written in the form

$$(x^-)^* = \sqrt{q}x^+, \qquad y^* = y, \qquad (x^+)^* = \frac{1}{\sqrt{q}}x^-.$$

Three linearly independent, hermitian 'coordinates' can be obtained as combinations of the x^i. We define

$$x^r = \Lambda_i^r x^i, \qquad \Lambda_i^r := \frac{1}{\sqrt{2}} \begin{pmatrix} 1 & 0 & \sqrt{q} \\ 0 & \sqrt{2} & 0 \\ i & 0 & -i\sqrt{q} \end{pmatrix}. \qquad (4.1.64)$$

With respect to the new 'coordinates' the metric is given by

$$g^{rs} = g^{ij}\Lambda_i^r\Lambda_j^s = \frac{1}{2} \begin{pmatrix} q+1 & 0 & i(q-1) \\ 0 & 2 & 0 \\ -i(q-1) & 0 & q+1 \end{pmatrix}.$$

The metric is hermitian but no longer real. In the limit $q \to 1$ one sees that $g^{rs} \to \delta^{rs}$. It is more convenient to remain with the original 'coordinates' and a real metric.

The 'length' squared (4.1.29) can be written also in the forms

$$r^2 = (x^i)^* x^i = (\sqrt{q} + \frac{1}{\sqrt{q}})x^- x^+ + qy^2 = (\sqrt{q} + \frac{1}{\sqrt{q}})x^+ x^- + q^{-1}y^2.$$

The commutation relation between x^+ and x^- can also be written in the form

$$qx^+ x^- - q^{-1}x^- x^+ = hr^2.$$

We shall extend $\mathbb{R}_q^3$ by adding the inverse y^{-1} of y as a new generator with the obvious extra relations.

Written out explicitly the Equations (4.1.32) become

$$x^- \xi^- = q^2 \xi^- x^-,$$
$$y\xi^- = q\,\xi^- y,$$
$$x^+ \xi^- = \xi^- x^+,$$

$$x^- \xi^0 = q\,\xi^0 x^- + (q^2 - 1)\xi^- y,$$
$$y\xi^0 = q\,\xi^0 y - h(q+1)\xi^- x^+,$$
$$x^+ \xi^0 = q\,\xi^0 x^+,$$

$$x^- \xi^+ = \xi^+ x^- - h(q+1)\xi^0 y + h^2(q+1)\xi^- x^+,$$
$$y\xi^+ = q\,\xi^+ y + (q^2 - 1)\xi^0 x^+,$$
$$x^+ \xi^+ = q^2 \xi^+ x^+.$$

By taking the exterior derivative of these equations one finds the relations

$$(\xi^\pm)^2 = 0,$$
$$(\xi^0)^2 = h\,\xi^-\xi^+,$$
$$\xi^-\xi^0 = -q^{-1}\xi^0\xi^-,$$
$$\xi^0\xi^+ = -q^{-1}\xi^+\xi^0,$$
$$\xi^-\xi^+ = -\xi^+\xi^-.$$

The Equations (4.1.36) yield equations for θ_-^a,

$$x^-\theta_-^a = q^{-1}\theta_-^a x^- - q^{-1}(q^2 - 1)\theta_0^a y + h^2(1 + q)\theta_+^a x^+,$$
$$y\theta_-^a = \theta_-^a y + h(1 + q)\theta_0^a x^+,$$
$$x^+\theta_-^a = q\theta_-^a x^+,$$

equations for θ_0^a,

$$x^-\theta_0^a = \theta_0^a x^- + h(1 + q)\theta_+^a y,$$
$$y\theta_0^a = \theta_0^a y - q^{-1}(q^2 - 1)\theta_+^a x^+,$$
$$x^+\theta_0^a = \theta_0^a x^+$$

and equations for θ_+^a,

$$x^-\theta_+^a = q\theta_+^a x^-,$$
$$y\theta_+^a = \theta_+^a y,$$
$$x^+\theta_+^a = q^{-1}\theta_+^a x^+.$$

These equations admit to within a linear transformation a unique solution and from Equation (4.1.35) we find that

$$\theta^- = \Lambda^{-1} y^{-1} \xi^-,$$
$$\theta^0 = \Lambda^{-1} r^{-1} (\sqrt{q}(q + 1)y^{-1}x^+\xi^- + \xi^0),$$
$$\theta^+ = -\Lambda^{-1} r^{-2} (\sqrt{q}q(q + 1)y^{-1}(x^+)^2\xi^- + (q + 1)x^+\xi^0 - y\xi^+).$$

An analogous expression can be found for the frame $\bar\theta^a$ of the differential calculus $\bar\Omega^*(\mathbb{R}_q^3)$. From the relations

$$[(\theta^a)^*, f^*] = -[\theta^a, f]^* = 0, \qquad f \in \mathbb{R}_q^3,$$

it follows that $(\theta^a)^*$ can be written in terms of $\bar\theta^b$. We choose the second frame so that the relation

$$(\theta^a)^* = \bar\theta^b g_{ba} \tag{4.1.65}$$

is satisfied. This is to be compared with Equation (4.1.34).

By direct calculation one finds that

$$P_t{}^{ab}{}_{cd}\theta^c\theta^d = 0, \qquad P_s{}^{ab}{}_{cd}\theta^c\theta^d = 0.$$

Therefore the $P^{ab}{}_{cd}$ of Equation (3.1.43) is given by

$$P^{ab}{}_{cd} = P_a{}^{ab}{}_{cd}.$$

(The two indices a are of different origin and no summation is intended.)   On the right-hand side is the q-deformed antisymmetric projector in the decomposition (4.4.19).

Consider the elements $\lambda_a \in \mathbb{R}_q^3$ with

$$\lambda_- = +h^{-1}q\Lambda y^{-1}x^+,$$
$$\lambda_0 = -h^{-1}\sqrt{q}\Lambda y^{-1}r, \qquad (4.1.66)$$
$$\lambda_+ = -h^{-1}\Lambda y^{-1}x^-.$$

By direct calculation one verifies that they define the derivations dual to the frame and that $\theta = -\lambda_a\theta^a$ is given by the Equation (4.1.37). Since Λ is unitary the hermitian adjoints λ_a^* are given by

$$\lambda_\pm^* = -\Lambda^{-2}g^{\pm b}\lambda_b, \qquad \lambda_0^* = \Lambda^{-2}g^{0b}\lambda_b.$$

The fact that the λ_a are not anti-hermitian is related to the fact that the differential d is not real. We have chosen this rather odd normalization so that the λ_a satisfy the commutation relations

$$\lambda_-\lambda_0 = q\lambda_0\lambda_-,$$
$$\lambda_+\lambda_0 = q^{-1}\lambda_0\lambda_+, \qquad (4.1.67)$$
$$[\lambda_+, \lambda_-] = h(\lambda_0)^2.$$

These equations can be rewritten more compactly in the form

$$P^{ab}{}_{cd}\lambda_a\lambda_b = 0.$$

This is Equation (3.1.50) with

$$C^a{}_{bc} = 0, \qquad F_{ab} = 0.$$

It is easy to check that

$$g^{ab}\lambda_a\lambda_b = qh^{-2}\Lambda^2.$$

Equations (4.1.67) are the same commutation relations as those (4.1.62) satisfied by the x^i. This is a remarkable fact and it underlines how weak a constraint the commutation relations are on the algebra. The λ_a are related in fact to the x^i by a rather complicated nonlinear relation (4.1.66). They differ however from the x^i in that they commute with Λ.

If one introduces the corresponding elements $\bar{\lambda}_a$ which yield the derivations dual to the frame $\bar{\theta}^a$ one finds that the involution on the λ_a can be written

$$\lambda_a^* = -g^{ab}\bar{\lambda}_b.$$

This is to be compared with (4.1.28). It follows from (4.1.65) that $(df)^* = \bar{d}f^*$.

A straightforward but long calculation yields the relations

$$g_{ab}\theta^b_j\theta^a_i = r^{-2}g_{ij}, \qquad g^{ij}\theta^b_j\theta^a_i = r^{-2}g^{ab}. \qquad (4.1.68)$$

The metric then is given by the map

$$g^{ab} = g(\theta^a \otimes \theta^b) = g(\theta^a_i\xi^i \otimes \theta^b_j\xi^j).$$

In the commutative limit we can therefore conclude that

$$g(\xi^i \otimes \xi^j) = r^2 g^{ij}.$$

The frame is necessarily related to the coordinate frame by r^{-1} times a rotation. In terms of the coordinates (4.1.64) we find the line element

$$ds^2 = r^{-2}\delta_{st}x^s x^t. \qquad (4.1.69)$$

The metric is conformally flat with conformal factor r^{-2}. If one uses spherical polar coordinates then one sees immediately that the space is $S^2 \times \mathbb{R}$ with $\log r$ as the preferred coordinate along the line. The radius of the sphere is equal to one.

To write the commutative limit of the frame it is most convenient to use the coordinates $x^r = (x, y, z)$ defined in Equation (4.1.64) and use the components

$$\theta^1 = \frac{1}{\sqrt{2}}(\theta^- + \theta^+), \qquad \theta^2 = \theta^0, \qquad \theta^3 = \frac{i}{\sqrt{2}}(\theta^- - \theta^+)$$

of the frame. A short calculation yields

$$\theta^1 = y^{-1}dx - y^{-1}r^{-1}(x - iz)dr,$$
$$\theta^2 = y^{-1}dr - iy^{-1}r^{-1}(xdz - zdx),$$
$$\theta^3 = y^{-1}dz - iy^{-1}r^{-1}(x - iz)dr$$

and by direct calculation one can verify that in fact the line element is indeed as given by Equation (4.1.69):

$$ds^2 = (\theta^1)^2 + (\theta^2)^2 + (\theta^3)^2 = r^{-2}(dx^2 + dy^2 + dz^2).$$

It is an unsatisfactory aspect of the construction of Example 3.16 that although the differential is real the frame in the commutative limit is not.

The closest one can come to a linear connection compatible with the metric is by choosing $S^{ij}{}_{kl} = q\hat{R}^{ij}{}_{kl}$ or $S^{ij}{}_{kl} = q^{-1}(\hat{R}^{-1})^{ij}{}_{kl}$. The generalized flip is defined by the braid matrix and the metric (4.1.63) is symmetric in the sense of (3.4.2). One finds in these two cases respectively

$$S^{im}{}_{ln}g^{np}S^{jk}{}_{mp} = q^2 g^{ij}\delta^k_l, \qquad S^{im}{}_{ln}g^{np}S^{jk}{}_{mp} = q^{-2}g^{ij}\delta^k_l.$$

instead of (3.6.27). The curvature of the corresponding linear connections (3.6.22) vanishes. We have connections of vanishing curvature but which are not compatible with the metric. This can be understood from the commutative limit. Consider the identity

$$D_{(0)}(r\theta^a) = -\frac{1}{2}(1 + q^{-2})\theta \otimes \theta^a r + r\sigma(\theta^a \otimes \theta)$$

which follows from Equation (4.1.38). The conformal factor r has the effect of introducing a function of q in the expression for the connection. This might be used to arrange the compatibility condition but it is not possible to choose $r\theta^a$ as a frame since it does not commute with Λ. This element, which does not lie in the center but nevertheless has vanishing exterior derivative, is the origin of the problem.
□

The purely algebraic constructions of this section can be made more concrete by adding topology to the algebra. This will be done in Section 4.3 after a short diversion into the commutative limit. We shall comment before Example 4.56 on how one might best extend the topology to the differential calculus.

4.2 Poisson structures

One of the more interesting problems of noncommutative geometry in physics is the study of the commutative, 'classical' limit. Let $\mathcal{A} = \mathcal{C}(V)$ be the algebra of smooth, complex-valued functions on a real space-time V, which for simplicity we shall suppose diffeomorphic to $\mathbb{R}^4$. We shall argue in Chapter 7 that space-time should be more properly described by a noncommutative $*$-algebra $\mathcal{A}_{\hbar}$ over the complex numbers of the type described in the Example 4.7. Let $\mathcal{A}_0$ be the commutative limit of $\mathcal{A}_{\hbar}$. We shall suppose that the algebra $\mathcal{A}_{\hbar}$ is a *deformation* of $\mathcal{A}_0$ in the sense of Gerstenhaber: as vector spaces the two are isomorphic; only the product is deformed. The relation between $\mathcal{A}_{\hbar}$ and the classical space-time V is given by the inclusion relation $\mathcal{A} \subset \mathcal{A}_0$. We shall assume that we can identify $\mathcal{A}_0 = \mathcal{C}(V_0)$ as the algebra of smooth, complex-valued functions on a real parallelizable manifold V_0 and that there is a projection of V_0 onto V. The difference in dimension between V_0 and V is one of the measures of the extent to which the verb 'to quantize' as applied to the coordinates of V is a misnomer; one could *in extremis* 'quantize' the coordinates of V_0. The observables will be some subset of the hermitian elements of $\mathcal{A}_{\hbar}$. We shall not discuss this problem here; we shall implicitly suppose that all hermitian elements of $\mathcal{A}_{\hbar}$ are observables, including the 'coordinates'. We shall not however have occasion to use explicitly this fact. If there is a gravitational field then there must be some source, of characteristic mass μ. If $\mu^2 \hbar$ tends to zero with $\hbar$ then V_0 will be without curvature. We are interested here in the case in which $\mu^2 \hbar$ tends to some finite non-vanishing value as $\hbar \to 0$.

The simplest example of an algebra is the tensor algebra $T(\mathcal{H})$ over the vector space $\mathcal{H}$ spanned by the 4 generators x^μ. This algebra has a natural filtration T^p in powers of $\hbar$. One defines T^0 as the symmetric algebra over the generators. This can be identified with the algebra of polynomials on the classical space-time V and, by an appropriate completion, with the algebra of smooth functions. One defines T^1 as the kernel of the projection of $T(\mathcal{H})$ onto T^0. It is an ideal of $T(\mathcal{H})$ each of whose elements contains at least one commutator. One defines T^p as the ideal of those elements which contain at least p commutators. The most general algebra $\mathcal{A}_\hbar$ is a quotient of $T(\mathcal{H})$ by some ideal $\mathcal{I}$ and the filtration of $T(\mathcal{H})$ defines a corresponding filtration $\{T_\hbar^p\}$ of $\mathcal{A}_\hbar$. There is an intimate connection between the dimension of V_0 and the 'size' of $\mathcal{I}$. We set

$$\tilde{J}^{\mu_1\cdots\mu_p} = \lim_{\hbar\to 0} J^{\mu_1\cdots\mu_p}.$$

For a discussion of the possibility that the limit is singular we refer to Section 7.1.

A set of independent elements of the complete set of the $\tilde{J}^{\mu_1\cdots\mu_p}$ are local coordinates of V_0. If V is Minkowski space-time then the condition of Lorentz invariance in the commutative limit forces $\tilde{x}^\lambda$ and at least 4 of the 6 coordinates $\tilde{J}^{\mu\nu}$ to be independent. In general the $\tilde{J}^{\mu_1\cdots\mu_p}$ for $p \geq 3$ can at least in part be functions of $\tilde{x}^\lambda$ and $\tilde{x}^{\mu\nu}$. This will depend on the structure of the ideal $\mathcal{I}$. If $\mathcal{I} = 0$ then all the $\tilde{J}^{\mu_1\cdots\mu_p}$ for $p \geq 3$ are independent coordinates and $\dim V_0 = \infty$. If on the other hand $\tilde{J}^{\mu_1\cdots\mu_p} = \tilde{J}^{\mu_1\cdots\mu_p}(\tilde{x}^\lambda)$ for all $p \geq 2$ then $V_0 = V$.

Quite generally the commutator of $\mathcal{A}_\hbar$ defines a *Poisson structure* on V_0. Let f and g be elements of $\mathcal{A}_\hbar$ and let f, g designate the corresponding limit functions on V_0. The Poisson bracket is given by

$$\{f, g\} = \lim_{\hbar\to 0} \frac{1}{i\hbar}[f, g]. \tag{4.2.1}$$

The restriction of the bracket to $\mathcal{A}$ is given by

$$\{f, g\} = J^{\mu\nu}\partial_\mu f \partial_\nu g.$$

Let $\Omega^*(\mathcal{A}_\hbar)$ be a differential calculus over $\mathcal{A}_\hbar$. We shall assume that $\Omega^*(\mathcal{A}_\hbar)$ is non-degenerate in the sense that if $df = 0$ then $f \in \mathcal{Z}_\hbar$. This is a natural assumption. It is easy to see that $\Omega^1(\mathcal{A}_\hbar)$ is generated by the dx^λ as a bimodule. We shall suppose that $\Omega^1(\mathcal{A}_\hbar)$ is a free left (and right) $\mathcal{A}_\hbar$-module of rank n with a frame θ^a, $0 \leq a \leq n - 1$. In view of the fact that we suppose that our algebras are 'quantizations' of parallelizable manifolds this is also a natural assumption. It is easy to see that n must be bounded below by the dimension of V_0:

$$n \geq \dim V_0.$$

An interesting question is the relation between the limit

$$\Omega_0^* = \lim_{\hbar\to 0} \Omega^*(\mathcal{A}_\hbar)$$

of a differential calculus over $\mathcal{A}_k$ and the de Rham differential calculus $\Omega^*(V_0)$ over V_0. The universal calculus $\Omega_u^*(\mathcal{A}_k)$ obviously does not have $\Omega^*(V_0)$ as a limit. It would seem that the existence of a frame is a necessary condition for this to be true if V_0 is parallelizable. It is not obvious that the dependence on a set of derivations is a sufficient condition. Suppose that $\Omega_0^* = \Omega^*(V_0)$. The commutator on $\Omega^*(\mathcal{A}_k)$ induces, in the commutative limit, a Poisson bracket on $\Omega^*(V_0)$. Let ξ and η be two elements of $\Omega^*(\mathcal{A}_k)$ which to be specific we shall suppose are 1-forms. We use the same symbol to designate the corresponding de Rham 1-forms. This notation can be misleading since it can happen that as an element of $\Omega^1(\mathcal{A}_k)$ ξ is exact but that as element of $\Omega^1(V_0)$ it is only closed. An example of this is provided by the symplectic form ω of Example 3.2, as is shown in Section 7.2.

Example 4.22 The Poisson bracket of ξ and η is given by the same Formula (4.2.1) as for the functions. Obviously the Jacobi identity remains satisfied but the bracket of two 1-forms need not be an element of $\Omega^*(V_0)$. To see this we write $\xi = \xi_a \theta^a$ and $\eta = \eta_a \theta^a$. Then

$$[\xi, \eta] = [\xi_a, \eta_b]\theta^a\theta^b + \eta_b\xi_a[\theta^a, \theta^b].$$

The first term on the right-hand side tends obviously to the element $\{\xi_a, \eta_b\}\theta^a\theta^b$ of $\Omega^2(V_0)$. The limit of the second term however will be in general an element of $\Omega^1(V_0) \otimes \Omega^1(V_0)$. There is an interesting case in which a 'twisted' bracket $\{\xi, \eta\}_C$ can be defined which takes its values in $\Omega^*(V_0)$. We suppose that the θ^a satisfy relations of the form (3.1.43) which we rewrite using (3.1.51) as

$$\theta^a\theta^b + C^{ab}{}_{cd}\theta^c\theta^d = 0$$

where the $C^{ab}{}_{cd}$ are elements of the center of the algebra such that

$$\lim_{k\to 0} C^{ab}{}_{cd} = \delta^b_c\delta^a_d.$$

Then we can define

$$\{\xi, \eta\}_C = \lim_{k\to 0} \frac{1}{ik}(\xi_a\eta_b - C^{cd}{}_{ab}\xi_c\eta_d)\theta^a\theta^b.$$

$\square$

Let $\mathcal{A} = C(V)$ be the algebra of smooth functions on a manifold V. A *deformation* of $\mathcal{A}$ is a family of algebras $\mathcal{A}_k$ each of which is identical to $\mathcal{A}$ as a vector space but which has a different product. It is traditional to designate the deformed product as a *-product* and to write

$$f * g = fg + k\phi_2(f, g) + o(k^2)$$

where $\phi_2(f, g)$ is a functional of the two elements f and g. One is primarily interested in the product for small values of the expansion parameter k. One requires each of the deformed algebras to be associative, a condition which places strong restrictions on ϕ_2. In fact, from the associativity rule $(f * g) * h = f * (g * h)$ one finds, using again the notation of Example 3.36, the cocycle condition

$$\delta\phi_2 = 0.$$

Every deformation defines a Poisson structure on the limit manifold by the formula

$$\{f, g\} = \lim_{k \to 0} \frac{1}{k}(f * g - g * f).$$

It is given in terms of $\phi_2(f, g)$ obviously by

$$\{f, g\} = \phi_2(f, g) - \phi_2(g, f).$$

Consider a new deformation defined by $\phi_2' = \phi_2 + \delta\phi_1$ for some ϕ_1. It is easy to see that $\{f, g\}$ remains unchanged. There exists ϕ_1 such that $\phi_2'(f, g) = -\phi_2'(g, f)$.

Suppose that $\mathcal{A}$ is a formal algebra with n generators x^i which satisfy commutation relations of the form

$$[x^j, x^k] = i k J^{jk}, \qquad J^{jk} \in \mathcal{A}, \qquad (J^{jk})^* = J^{jk}. \qquad (4.2.2)$$

The case $n = 4$ was mentioned already in Example 4.7. Suppose that J^{ij} is non-degenerate. The center then of $\mathcal{A}$ is trivial and the inverse J_{ij}^{-1} of J^{ij} exists in the sense that

$$J_{ij}^{-1} J^{jk} = \delta_i^k, \qquad J_{ij}^{-1} \in \mathcal{A}.$$

Suppose also that the algebra has as well n generators λ_a which satisfy the quadratic relations (3.1.50). Consider first the case with J^{ij} central elements of the algebra. Consider also the smooth manifold $V = \mathbb{R}^n$ and the algebra $\tilde{\mathcal{A}}$ generated by the coordinates $\tilde{x}^i$ and the conjugate momenta p_j. We shall use the convention of distinguishing between the operator p_j and the result $i\tilde{\partial}_j f$ of the action of p_j on f. There is a simple representation of $\mathcal{A}$ as a subalgebra of the algebra of (pseudo-)differential operators $\tilde{\mathcal{A}}$, given by the identification

$$x^i = \tilde{x}^i + \frac{1}{2} k J^{ij} p_j. \qquad (4.2.3)$$

From this it follows immediately that

$$f(x^i) = f(\tilde{x}^i) + \frac{1}{2} k J^{jk} p_k \partial_j f + o(k^2) = f(\tilde{x}^i) + \frac{1}{2} k J^{jk} \partial_j f p_k + o(k^2)$$

and from this 'Taylor' expansion in phase space we can deduce the commutation relations

$$[f, x^j] = i k J^{ij} \partial_i f + o(k^2)$$

and hence
$$[f,g] = i \hbar J^{ij} \partial_i f \partial_j g + o(\hbar^2).$$

If we have two such *-products with commutators defined by J^{ij} and $\hat{J}^{ij}$ and if we introduce the derivations e_a associated to the λ_a then we must replace the Equation (4.2.3) above by

$$\hat{x}^i = x^i + \frac{1}{2}\hbar\theta^{ia}p_a \qquad (4.2.4)$$

where p_a is the operator which yields $i e_a f$ when acting on f. We shall assume that the θ^{ab} are central and we introduce the quantities θ^{ai} and θ^{ij} defined by the equalities

$$\theta^{ij} = \theta^{ib}e_bx^j = \theta^{ab}e_ax^ie_bx^j.$$

Furthermore we shall assume that

$$\theta^{(ij)} = o(\hbar).$$

These quantities are not necessarily central. It follows that

$$[\hat{x}^i \,; \hat{x}^j] = [x^i, x^j] + i\hbar\theta^{ij} + o(\hbar^2). \qquad (4.2.5)$$

That is we have
$$\hat{J}^{ij} = J^{ij} + \theta^{ij}. \qquad (4.2.6)$$

We deduce the 'Taylor' expansion

$$f(\hat{x}^i) = f(x^i) + \frac{1}{2}\hbar\theta^{ab}e_afp_b + o(\hbar^2)$$

from which follows the difference

$$[f\,;g] - [f,g] = i\hbar\theta^{ab}e_af * e_bg + o(\hbar^2) \qquad (4.2.7)$$

between the commutation relations.

Example 4.23 Using an arbitrary Poisson structure as given by an antisymmetric matrix θ^{ij} one can associate a new product $f*g$ to a commutative algebra of functions, defined by the formula

$$f * g(x) = \lim_{x' \to x} e^{\frac{1}{2}i\hbar\theta^{ij}\partial_i\partial'_j} f(x)g(x') = fg(x) + \frac{1}{2}i\hbar\theta^{ij}\partial_i f\partial_j g + o(\hbar^2). \qquad (4.2.8)$$

This product is known as the *(Moyal) star product*. It corresponds to a definite ordering prescription of the phase-space variables. If the product is associative to first order it is associative to all orders. Below in Example 4.61 we shall show how it can be obtained using a twist. □

Example 4.24 We have seen that a commutation relation gives rise to a Poisson structure in the commutative limit. We have argued also that on a given algebra with a given differential calculus there is essentially a unique metric. If space-time then is an approximation to some noncommutative geometry then it must have a Poisson structure and this structure should be intimately connected with the metric. As a general mathematical problem it is of interest to study on a given manifold what *a priori* relations there could be between a Riemann structure on the one hand and a Poisson structure on the other. □

Example 4.25 Reconsider the algebra homomorphism ρ of Example 3.33 and assume that it can be formally defined by the action

$$x^i = \rho(\hat{x}^i) = \Lambda^i(\hat{x}^j)$$

on the generators. By assumption then

$$\rho(\hat{x}^i \,\hat{*}\, \hat{x}^j) = x^i * x^j = \rho(\hat{x}^i) * \rho(\hat{x}^j).$$

The kernel of ρ is a 2-sided ideal so $\hat{\mathcal{A}}$ cannot in any sense of the word be 'simple'. If $\hat{\mathcal{A}}$ is commutative then so obviously is $\mathcal{A}$; if on the other hand $\mathcal{A}$ is commutative then the kernel of ρ contains necessarily the ideal generated by the commutators. If $\hat{J}$ is non-degenerate then this can again by identified with $\mathcal{A}$ and so $\rho = 0$. In the special case with J and $\hat{J}$ constant non-degenerate matrices we can choose $F^i(\hat{x}^j) = F^i_j \hat{x}^j$ a linear transformation. We have then

$$x^i * x^i = F^i_k F^j_l \hat{x}^k \,\hat{*}\, \hat{x}^l, \qquad J^{ij} = F^i_k F^j_l \hat{J}^{kl}.$$

In general the relation between the generators is much more complicated. If we can write for example $F^i(\hat{x}^j) = \hat{x}^i - \xi^i(\hat{x}^j)$ as a linear perturbation then

$$i k J^{ij} = [x^i, x^j] = [\rho(\hat{x}^j), \rho(\hat{x}^j)] = i k \hat{J}^{ij} - k[x^{[i}, \xi^{j]}],$$

which we write in the form (4.2.6). If we suppose that J^{ij} is constant then writing ξ^i as $\xi^i = i k J^{ia} a_a$ we find that

$$\theta^{ij} = k J^{ia} J^{jb} e_{[a} a_{b]}.$$

In this case the perturbation of the commutation relations is related to the exact form

$$f = da, \qquad a = a_b \theta^b, \qquad f = \frac{1}{2} e_{[a} a_{b]} \theta^a \theta^b.$$

One might be tempted to consider the $F^i(\hat{x}^j)$ as a 'change of coordinates'. But the change is in the 'phase space' of which $\tilde{\mathcal{A}}$ is the structure algebra and so when one looks for a similar transformation in ordinary geometry one must imagine not only a change of coordinates but also a shift in the position because of the term in the definition of the generators which depends on the momentum. What can more properly be considered as a change of coordinates is an automorphism of the algebra, for example the inner automorphism

$$\hat{x}^i = \Lambda^{-1} x^i \Lambda.$$

In this case the product is conserved. It is perhaps preferable to consider ρ as a change of product on one fixed vector space. We drop then the hat on the generators and distinguish the two products by a hat on one of them.

In general one can consider the set S_0 of all products on the vector space $\mathcal{A}$. There is a subset $S_1 \subset S_0$ in which the product is associative; this is the set which interests us here. Let π be a given product and consider the orbit $S_2 \subset S_1$ of π under the group of all possible maps ρ. This group has a subgroup of automorphisms of $\mathcal{A}$, which leave the product invariant. In a formal sense S_2 can be identified with the quotient of the two groups. In general S_1 will be a union of orbits of different products of non-isomorphic algebras. If we assume that there are no relations other than the commutation relations (4.2.2) then the set S_2 will be parameterized by the J^{ij}. To pass from stratum of S_1 to another would require a singular variation in J. A familiar example from the theory of Lie algebras is furnished by the embedding $SU_n \hookrightarrow SO_{n^2-1}$. If $\{\lambda_i\}$ is a set of generators of the Lie algebra of SU_n then so is the set $\{\hat{\lambda}_i\}$ with $\hat{\lambda}_i = g^{-1}\lambda_i g$ for $g \in SU_n$. One can write then $\hat{\lambda}_i = \Lambda_i^j \lambda_j$ where the transformation coefficients are complex numbers. It is the analog of those transformations of SO_n which do not respect the Lie algebra structure which interests us here. As a limiting case with singular ρ one can consider an algebra $\hat{\mathcal{A}}$ with a non-degenerate $\hat{J}$ and an algebra $\mathcal{A}$ with $J = 0$ so that one can identify x^i with $\tilde{x}^i$, the 'space' coordinates of $\tilde{\mathcal{A}}$. The 'lift' by the inverse of ρ is a quantization procedure, a way of associating an operator to a function. One such method is the Weyl-Moyal quantization procedure which furnishes a 'natural' right inverse for ρ which lifts an element $f \in \mathcal{A}$ to an element $\hat{f} \in \hat{\mathcal{A}}$, defined in Example 4.23. This is a map between two different strata of S_1.   □

4.3   Topological algebras

> *Nel mezzo del cammin di nostra vita*
> *mi ritrovai per una selva oscura* *

We mentioned in the Introduction that quantized phase-space is the prototype structure algebra of a noncommutative geometry. One way in which one can acquire intuition about noncommutative geometry in general is to illustrate as far as possible

---

*Dante Alighieri before meeting Virgil

the formalism with examples from quantum mechanics and quantum field theory. The *observables* of both of these theories are the self-adjoint elements of an algebra with involution (the *algebra of observables*) and the result of an observation is an eigenvalue of the observable. The eigenvalue can form part of a continuous spectrum, for example the position or momentum of a particle, or it can be an element of a discrete spectrum, for example the value of a component of spin along a given axis. For technical reasons it is usually supposed that the algebra is a *Banach algebra*, a Banach space such that the norm $\|ab\|$ of the product ab of two elements a and b satisfies the inequality

$$\|ab\| \le \|a\|\|b\|. \tag{4.3.1}$$

The norm defines the *norm topology* on $\mathcal{A}$: a sequence of elements a_i tends to zero in norm if and only if $\|a_i\| \to 0$. Since the lack of commutativity is due to quantum effects we have $\|ab - ba\| = o(\hbar)$. In the classical limit the algebra of observables is commutative. Let $a \mapsto a^*$ be the involution. Then one supposes also that

$$\|aa^*\| = \|a\|^2.$$

A Banach algebra which satisfies this condition is called a C^*-*algebra*. As was mentioned already in the Introduction, an interesting example is the algebra $C^0(V)$ of continuous functions on a compact space V. A norm $\|f\|$ of an element $f \in C^0(V)$ can be defined by the formula

$$\|f\| = \sup_{x \in V} |f(x)|. \tag{4.3.2}$$

As an algebra $C^0(V)$ can be identified with a commutative subalgebra of the algebra of all the bounded operators on itself considered as a vector space. To each $f \in C^0(V)$ we associate the operator $\hat{f}$ defined by $\hat{f} : g \mapsto fg$ for arbitrary $g \in C^0(V)$. It follows that we can write (4.3.2) also as

$$\|\hat{f}\| = \sup_{g}\{\|fg\| : \|g\| \le 1\}.$$

The norm (4.3.2) is a convenient norm which does not use a measure on V but it is not the one which is used on the algebra of observables of classical physics.

The algebra $\mathcal{B}(\mathcal{H})$ of all bounded operators on a Hilbert space $\mathcal{H}$ is a noncommutative C^*-algebra and conversely every C^*-algebra can be realized as a closed subalgebra of $\mathcal{B}(\mathcal{H})$ for some $\mathcal{H}$. The norm $\|a\|$ of an element $a \in \mathcal{B}(\mathcal{H})$ is given in terms of the norm $\|\psi\|$ of an element $\psi \in \mathcal{H}$ by the formula

$$\|a\| = \sup_{\psi}\{\|a\psi\| : \|\psi\| \le 1\}.$$

The inequality (4.3.1) follows directly. The algebra $\mathcal{B}(\mathcal{H})$ is therefore a Banach algebra. In particular it follows that $\|aa^*\| \le \|a\|^2$ since $\|a^*\| = \|a\|$. To prove the reverse inequality we must use the fact that $\mathcal{H}$ is a Hilbert space with an inner product (ψ, ψ') which satisfies the Schwarz inequality $|(\psi, \psi')| \leq \|\psi\| \|\psi'\|$. We have then

$$\|a\|^2 = \sup_{\psi}\{(a^*\psi, a^*\psi) : \|\psi\| \leq 1\} \leq \sup_{\psi}\{\|aa^*\psi\| : \|\psi\| \leq 1\} = \|aa^*\|. \qquad (4.3.3)$$

It follows that $\mathcal{B}(\mathcal{H})$ is a C^*-algebra. The simplest examples of noncommutative C^*-algebras are the algebras M_n of $n \times n$ complex matrices.

The dual $\mathcal{A}^*$ of a C^*-algebra $\mathcal{A}$ is the space of continuous linear functionals on $\mathcal{A}$. The norm $\|\omega\|$ of an arbitrary element ω in $\mathcal{A}^*$ is defined to be

$$\|\omega\| = \sup_{a \in \mathcal{A}}\{|\omega(a)| : \|a\| \leq 1\}.$$

A linear functional is positive if $\omega(aa^*) \geq 0$ for all $a \in \mathcal{A}$. A *state* is a positive linear functional with unit norm. The set of all states is a convex set. In fact if ω and ω' are states then it is easy to see that $\alpha\omega + (1 - \alpha)\omega'$ is also a state if α lies between 0 and 1. The extremal elements of a convex set are those elements which cannot be decomposed as a sum of two other elements. For example the extremal elements of the unit disc are its boundary and the extremal elements of a triangle are its 3 vertices. If ω is an extremal element of the set of states then it is called a *pure state*. Otherwise the state is a *mixed state*.

If $\mathcal{A} = C^0(V)$ then a state is equivalent to a probability measure μ on V and we can write

$$\omega_\mu(f) = \int f d\mu.$$

The state is pure if and only if the measure is a Dirac measure concentrated on some point $x \in V$, in which case we can write

$$\omega_x(f) = f(x).$$

The pure states can be identified then with the points of V and we can say that $\mathcal{A}$ is an algebra of functions on the space of its pure states. The space can be endowed with a topology with respect to which the functions are continuous. Each measure on V makes $C^0(V)$ into a Hilbert space $L^2(V, \mu)$ with inner product

$$(g, h) = \int g^* h d\mu. \qquad (4.3.4)$$

We have then

$$(g, fh) = \int g^* f h d\mu$$

and $C^0(V)$ as an algebra can be identified with a commutative algebra $L^\infty(V, \mu)$ of bounded operators on itself considered as a Hilbert space. Each $g \in C^0(V)$ defines a state by the formula

$$\omega_g(f) = \int |g|^2 f d\mu.$$

The original measure defines the state ω_1. The norms which are used on the algebras of observables of classical physics are derived from inner products of the form (4.3.4).

If $\mathcal{A}$ is a general C^*-algebra of operators on a Hilbert space $\mathcal{H}$, then $\mathcal{H}$ is *semisimple* as an $\mathcal{A}$-module: every submodule $\mathcal{H}_1$ has a complement, a submodule $\mathcal{H}_2$ such that $\mathcal{H} = \mathcal{H}_1 \oplus \mathcal{H}_2$. Since $\mathcal{A}$ is an algebra with involution, $\mathcal{H}_2$ will in fact be the orthogonal complement of $\mathcal{H}_1$. We can decompose then $\mathcal{H}$ as a direct sum of simple modules, modules which have no submodules. For example, consider $\mathcal{A} = M_r \times M_s$ as a subalgebra of M_{r+s} with the natural identifications. Then $\mathbb{R}^{r+s}$ is an $\mathcal{A}$-module which splits as the direct sum $\mathbb{R}^{r+s} = \mathbb{R}^r \oplus \mathbb{R}^s$ of two $\mathcal{A}$-modules each of which is simple. Considered as an M_{r+s}-module however, $\mathbb{R}^{r+s}$ is simple.

Each unit vector $\psi \in \mathcal{H}$ defines a state ω_ψ by the formula

$$\omega_\psi(a) = (\psi, a\psi). \tag{4.3.5}$$

The vector ψ is the *state vector*. Suppose that $\mathcal{H}$ can be written as the direct sum of two $\mathcal{A}$-modules and that the state vector can be written $\psi = (\psi_1 + \psi_2)/\sqrt{2}$ as the sum of two unit vectors $\psi_1 \in \mathcal{H}_1$ and $\psi_2 \in \mathcal{H}_2$. Then ω_ψ can be written $\omega_\psi = (\omega_1 + \omega_2)/\sqrt{2}$ as the sum of two states and it is not a pure state. Conversely, if we accept the fact that every state on $\mathcal{A}$ is of the form (4.3.5) for some ψ, then it follows that ω is a pure state if the corresponding state vector ψ lies within a simple component of $\mathcal{H}$.

If $\mathcal{A}$ is the algebra $\mathcal{B}(\mathcal{H})$ of all bounded operators on $\mathcal{H}$ then P_ψ, the projector onto the vector ψ in $\mathcal{H}$, is in $\mathcal{A}$ and we can write (4.3.5) as

$$\omega_\psi(a) = \mathrm{Tr}(P_\psi a).$$

A more general state is one which can be written in the form

$$\omega_\rho(a) = \mathrm{Tr}(\rho a) \tag{4.3.6}$$

where the statistical *density matrix* ρ is a positive bounded operator on $\mathcal{H}$ of unit trace. The state ω_ρ is a pure state if and only if ρ is a projector of rank 1.

Example 4.26 The notion of a pure state comes from statistical mechanics. To illustrate it we can consider the example of an elementary spin system. Let n be an ensemble of particles of spin 1/2 to each of which we associate the three Pauli matrices σ_i^a, $1 \leq i \leq n$. The Hilbert space of states $\mathcal{H}_i$ associated to particle i is equal to $\mathbb{C}^2$. The total Hilbert space $\mathcal{H}$ of the system is therefore given by

$$\mathcal{H} = (\mathbb{C}^2)^{\otimes n}.$$

The algebra of observables is the tensor product

$$\mathcal{A} = \bigotimes_{i=1}^{n} \mathcal{A}_i$$

where $\mathcal{A}_i \subseteq M_2$. Let $a_i \in \mathcal{A}_i$. The action of the element $a = a_1 \otimes \cdots \otimes a_n \in \mathcal{A}$ on the element $\psi = \psi_1 \otimes \cdots \otimes \psi_n \in \mathcal{H}$ is given by $a\psi = a_1\psi_1 \otimes \cdots \otimes a_n\psi_n$.

Consider first of all the classical case. Each factor $\mathcal{A}_i$ is commutative and can be chosen to be equal to the algebra of diagonal matrices, generated by σ_3. The algebra $\mathcal{A}$ is commutative. The space V of pure states consists of the 2^n possible spin configurations and $\mathcal{A}$ can be identified with the space of functions on V. The only simple submodules of $\mathcal{H}$ are of dimension 1 and each gives rise to a single pure state. In the quantum case each algebra $\mathcal{A}_i$ is equal to the complete algebra M_2 and $\mathcal{A}$ is no longer commutative. The space $\mathcal{H}$ is simple as an $\mathcal{A}$-module. Every unit vector ψ gives rise to a pure state by (4.3.5).   □

The algebra $\mathcal{B}(\mathcal{H})$ of bounded operators on a Hilbert space $\mathcal{H}$ is a C^*-algebra with a state given by (4.3.5). The *GNS theorem* (Gelfand-Naimark-Segal) states that conversely any C^*-algebra $\mathcal{A}$ with a state ω can be considered as a closed subalgebra of the algebra $\mathcal{B}(\mathcal{H})$ for some $\mathcal{H}$. The proof involves an explicit construction of $\mathcal{H}$ from $\mathcal{A}$ using ω. Consider in fact $\mathcal{A}$ as a vector space and define an inner product (a, b) on it by

$$(a, b) = \omega(a^* b). \tag{4.3.7}$$

Let $\mathcal{H}_0$ be the set of vectors of norm zero:

$$\mathcal{H}_0 = \{b : (b, b) = 0\}. \tag{4.3.8}$$

Then the quotient $\mathcal{H} = \mathcal{A}/\mathcal{H}_0$ is a Hilbert space. To each $a \in \mathcal{A}$ we associate the operator $\hat{a}$ defined by

$$\hat{a} : b \mapsto ab \tag{4.3.9}$$

for arbitrary $b \in \mathcal{H}$. It can be shown that for any two elements $a, b \in \mathcal{A}$, $\omega(b^* a^* a b) \leq \|a\|^2 \omega(b^* b)$. Therefore if $b \in \mathcal{H}_0$ then so is ab and (4.3.9) is well defined. It is in fact an isomorphism of $\mathcal{A}$ into a closed subalgebra of $\mathcal{B}(\mathcal{H})$. Let ψ be the equivalence class in $\mathcal{H}$ of the identity in $\mathcal{A}$. Then $\omega(a) = \omega_\psi(a) = (\psi, \hat{a}\psi)$. In general the Hilbert space $\mathcal{H}$ is not separable.

We have shown that a commutative C^*-algebra can be considered as an algebra of functions on its pure states. Let V be the space of pure states of a noncommutative C^*-algebra $\mathcal{A}$. Then of course $\mathcal{A}$ and $C^0(V)$ cannot be isomorphic as algebras, but we can define an embedding of $\mathcal{A}$ as a vector space in $C^0(V)$ by the map $a \mapsto f_a$ where f_a is defined by $f_a(\omega) = \omega(a)$. A pure state is a natural noncommutative generalization of a point.

Example 4.27 The matrix algebra M_n is a C^*-algebra which acts on the Hilbert space $\mathbb{C}^n$. The latter is a simple M_n-module. Therefore any unit vector in $\mathbb{C}^n$ gives rise to a pure state. One easily sees in fact that the space of pure states can be identified with the projective space $\mathbb{P}^{n-1}(\mathbb{C})$. In particular the space of pure states of M_2 is the 2-sphere S^2. In Example 2.1 we gave a multipole expansion of the functions on S^2. In Section 7.2 we shall see that M_2 can be identified with the monopole and dipole terms of the expansion and therefore it can be embedded as a vector space in $\mathcal{C}^0(S^2)$.   □

Example 4.28 A single spinning particle in ordinary euclidean space can be described by three hermitian operators $\hbar J_a$, $1 \leq a \leq 3$. If there is to be no preferred direction the description must be invariant under space rotations and the J_a must form an irreducible representation of the Lie algebra of the rotation group. The commutation relations mean that one cannot measure all three components of the spin simultaneously. The representation is determined by its dimension n and the value of the Casimir operator is given by $J_a J^a = (n^2 - 1)/4$. The operators J_a are then elements of M_n. The state of the system is described by a vector in $\mathbb{P}^{n-1}(\mathbb{C})$ and if, for example, J_3 is measured it can take half-integer values between $-(n-1)/2$ and $(n-1)/2$. The classical limit is given by $\hbar \to 0$ and $n \to \infty$ in such a way that $\hbar n$ remains finite. We shall use this limit in Section 7.2 to give a finite noncommutative version of the geometry of the 2-sphere.   □

Example 4.29 Let $\tilde{x}$ and $\tilde{p}$ be the position and momentum of a particle moving on a line. Under quantization they become elements x, p of a noncommutative algebra which satisfy the commutation relation $[x, p] = i$. Set

$$a = \frac{1}{\sqrt{2}}(x + ip).$$

Then the commutation relation becomes $[a, a^*] = 1$, considered in Example 4.1. Introduce the elements

$$u = e^{ix}, \qquad v = e^{ip}.$$

Then u and v satisfy the commutation relation

$$uv = qvu \tag{4.3.10}$$

with $q = e^{-i}$. The *Weyl algebra*, defined in Example 4.2 as a formal algebra, is the C^*-algebra $\mathcal{A}$ generated by u and v. Let $s, t \in \mathbb{R}$. If we replace ix by isx and ip by itp in the definitions of u and v then by a theorem of von Neumann there is a unique irreducible representation of $\mathcal{A}$ as the algebra $\mathcal{B}(\mathcal{H})$ of bounded operators on a Hilbert space $\mathcal{H}$, with the images of u and v unitary operators strongly continuous in s and t. We can choose the representation

$$x = \tilde{x}, \qquad p = -i\frac{d}{d\tilde{x}}$$

given in Example 4.10. The pure states of $\mathcal{A}$ are given by the Schrödinger wave functions ψ of unit norm in $\mathcal{H} = L^2(\mathbb{R})$. Each ψ gives rise to a state by

Formula (4.3.5) and since $\mathcal{H}$ is a simple $\mathcal{A}$-module the state is pure. The fact that every pure state is obtained in this manner follows from the GNS theorem. The hypothesis of strong continuity was essential to obtain all of the operators on $\mathcal{H}$. If it is dropped the C^* algebra $\mathcal{A}$ is a proper subalgebra of $\mathcal{B}(\mathcal{H})$, which, for example, need not contain the square roots of u and v. We shall see below that the strongly continuous representation of the Weyl algebra is the unique representation as a type-I_∞ factor. We shall refer to this also as the *Heisenberg algebra*. It was shown by von Neumann that the Weyl algebra has an infinite number of inequivalent representations as a von Neumann algebra.

Consider the derivations e_1 and e_2 of $\mathcal{A}$ defined by

$$e_1 = -\frac{1}{i}\mathrm{ad}\,p, \qquad e_2 = \frac{1}{i}\mathrm{ad}\,x,$$

Then we have exactly as in Example 2.2

$$e_1 u = iu, \quad e_1 v = 0, \qquad e_2 u = 0, \quad e_2 v = iv,$$

and $[e_1, e_2] = 0$. Therefore if $a \in \mathcal{A}$ is such that $e_i a = 0$ for $i = 1, 2$ then a is proportional to the identity. The Weyl algebra has sufficient derivations. As in the matrix case one can formally define then dx and similarly dp by the condition that $dx(X) = Xx$ where X is a derivation. In the classical limit dx and dp would tend to the ordinary 1-forms $d\tilde{x}$ and $d\tilde{p}$ on classical phase space.

If the particle is an harmonic oscillator it has as hamiltonian $\tilde{H} = (\tilde{p}^2 + \tilde{q}^2)/2$. Under quantization this becomes

$$H = \frac{1}{2}(p^2 + x^2) = \frac{1}{2}(aa^* + a^*a)$$

and we see that

$$[H, a] = -a, \qquad [H, a^*] = a^*.$$

If then ψ is a state vector of energy E, $a^*\psi$ ($a\psi$) describes a state of energy $E + 1$ ($E - 1$). One says that a^* (a) creates (annihilates) a quantum mode. The *ground state* or *vacuum state* is defined by the vector $|0\rangle$ which minimizes H. Necessarily then $a|0\rangle = 0$. Using the commutation relations we can rewrite the hamiltonian as

$$H = (a^*a + \frac{1}{2})\hbar\nu.$$

The constant term on the right-hand side is called the *vacuum energy*. To have the correct dimensions we have explicitly included the frequency ν of the harmonic oscillator, set equal to one by a choice of units. If we subtract the vacuum energy from the hamiltonian we obtain an operator for which the vacuum is an eigenvector with eigenvalue zero. The process of using the commutation relations to subtract the vacuum energy is called *normal ordering*. The spectrum of H is given by

$$E_j = (j + \frac{1}{2})\hbar\nu, \qquad j = 0, 1, 2, \cdots,$$

with each level of unit multiplicity.

The a describes a particle of one degree of freedom. A quantum system described by n degrees of freedom would be given by a set of a_j and their adjoints a_j^* which satisfy the *canonical commutation relations* (CCR's)

$$[a_j, a_k^*] = \delta_{jk}, \qquad [a_j, a_k] = 0, \tag{4.3.11}$$

for $1 \leq j, k \leq n$. We have here again chosen units with Planck's constant equal to one. □

Example 4.30 Let $\mathcal{T}$ be the C^*-algebra freely generated by n elements a_i and their adjoints a_i^*. A *quasi-free state* ω is determined by its values on the 2-element words, the *2-point functions*, and by the normalization condition $\omega(1) = 1$. We define ω to be zero on words of odd length and on words of even length to be given by the product of its values on the successive 2-word factors. For example

$$\omega(a_{i_1} \ldots a_{i_k} a_{i_{k+1}}^* \ldots a_{i_{2k}}^*) = \omega(a_{i_k} a_{i_{k+1}}^*)\omega(a_{i_{k-1}} a_{i_{k+2}}^*) \cdots \omega(a_{i_1} a_{i_{2k}}^*), \tag{4.3.12}$$

for $1 \leq k \leq n$. Let $\mathcal{A}$ be the quotient algebra defined by the *canonical anticommutation relations* (CAR's)

$$a_i a_j^* + a_j^* a_i = \delta_{ij}, \qquad a_i a_j + a_j a_i = 0,$$

for $1 \leq i, j \leq n$. The algebra $\mathcal{A}$ is of finite dimension and can be identified with the Clifford algebra in dimension n; the Dirac matrices are given by $\gamma_i = a_i + a_i^*$. Equation (4.3.12) cannot be used to define a state on $\mathcal{A}$ since it is not consistent with the quotient structure. But consider the state ω whose non-vanishing values are of the form

$$\omega(a_1 \ldots a_k a_k^* \ldots a_1^*) = \det \omega(a_i a_j^*)$$

where here $\omega(a_i a_j^*)$ is considered as a $k \times k$ matrix. Then ω is a *quasi-free state* on $\mathcal{A}$. If the 2-point functions are of the form (4.3.5) then ω is a pure state. It is then called a *free state*. □

The description of a quantum system in terms of creation and annihilation operators is convenient only when the phase space of the system can in some approximation at least be considered as a euclidean space. This does not apply to the spinning particle of Example 4.28, which is a system with a compact phase space. We recall that *elementary particles* are assumed to be described by vectors in an irreducible representation of the Lie algebra of the Poincaré group. One concludes that they are either *bosons* or *fermions*. The former must have integer spin and be quantized using CCR's; the latter must have half-integer spin and be quantized using CAR's. The C^*-algebra $\mathcal{A}$ generated by the creation and annihilation operators is called the *algebra of observables* and the Hilbert space $\mathcal{F}$ whereon it operates is called *Fock space*. Any arbitrary vector in $\mathcal{F}$ can be obtained by applying an element of $\mathcal{A}$ to the vacuum state $|0\rangle$; $\mathcal{F}$ is a simple $\mathcal{A}$-module. One can consider $\mathcal{F}$ as the space given by the GNS construction from the algebra of observables and the vacuum state.

Example 4.31 In Example 2.14 we defined a real classical scalar field on a manifold V. If V has a metric with Minkowski signature and can be decomposed into a time direction and a space V_S which is compact and time-independent then the quantized scalar field can be expressed in terms of a countable number of creation and annihilation operators. The Laplace operator Δ_S on V_S has a discrete unbounded spectrum p_j^2. Let $\{\phi_{Sj}\}$ be the corresponding eigenvectors normalized by the condition

$$\frac{1}{\mathrm{Vol}(V_S)} \int \phi_{Sj}^* \phi_{Sk} dx = \delta_{jk}.$$

Here dx is the invariant volume element of V_S with respect to the metric induced from V. Set $E_j^2 = p_j^2 + \mu^2$. Then we can expand any solution ϕ to the field equation

$$(\Delta + \mu^2)\phi = (\frac{\partial^2}{\partial t^2} + \Delta_S + \mu^2)\phi = 0$$

in the form

$$\phi = \sum_j (\tilde{a}_j \phi_j^+ + \tilde{a}_j^* \phi_j^-) \tag{4.3.13}$$

where

$$\phi_j^+ = e^{-iE_j t} \phi_{Sj}, \qquad \phi_j^- = \phi_j^{+*}$$

and the $\tilde{a}_j$ are complex numbers. The first terms on the right-hand side are called the positive modes since they correspond to the positive roots of the equation $E_j^2 = p_j^2 + \mu^2$; the second terms are correspondingly called negative modes. The latter were reinterpreted by Dirac as anti-particles of positive energy. Only positive modes with $p_j^2 \ll \mu^2$ are excited in the non-relativistic limit and therefore $E_j \simeq \mu + p_j^2/2\mu$. As applied to these modes

$$(\Delta_S + \mu^2)^{1/2} \simeq \mu + \frac{1}{2\mu}\Delta_S.$$

So we see that the positive-energy states are eigenvectors of the operator $H \simeq \mu + H_S$ where

$$H_S = \frac{1}{2\mu}\Delta_S$$

is the hamiltonian of a non-relativistic particle of mass μ. In the non-relativistic limit the field equation reduces to Schrödinger's equation

$$i\frac{\partial \phi}{\partial t} = H_S \phi.$$

The map

$$\tilde{a}_j \mapsto a_j$$

which consists in replacing the $\tilde{a}_j$ by operators a_j on a Fock space $\mathcal{F}$ which satisfy the CCR's (4.3.11) is called *second quantization*. Each a_j^* (a_j) creates (annihilates) a particle mode. The index j plays the role of the momentum coordinates. In the simplest case with V_S a torus, j would designate a Fourier mode. If the points of the torus are restricted to a lattice then the sum is finite. The vacuum state $|0\rangle$ is given by $a_j|0\rangle = 0$. The number operators N_j are defined as $N_j = a_j^* a_j$. They count the number of modes in a given vector in $\mathcal{F}$. Using them one can write the hamiltonian H as

$$H = \sum_j E_j N_j.$$

This expression has been normal-ordered, written with the creation operators to the left. From the definitions it follows that the 1-particle states $a_j^*|0\rangle$ are eigenstates of H with eigenvalue E_j.   □

Example 4.32 Let $\mathcal{H}$ be the Hilbert space of square-integrable Dirac spinors on the space V_S of the preceding example and expand a solution to the Dirac equation

$$\psi = \sum_j (\tilde{b}_j \psi_j^+ + \tilde{d}_j^* \psi_j^-)$$

in terms of an orthonormal basis $\psi_j^\pm$ of the Weyl spinors associated to the Dirac operator (2.2.18) which are eigenfunctions of the hamiltonian. Associated to this basis one has a set of creation operators (b_j^*, d_j^*) and annihilation operators (b_j, d_j) of respectively positive- and negative-energy modes and therefore a fermionic Fock space $\mathcal{F}$, equipped with a vacuum state vector $|0\rangle$. The (b_j, d_j) are related to the Dirac spinors in the same way as the a_j is related to the scalar field in the preceding example but they satisfy CAR's. There are twice as many because the Dirac field is complex-valued. We have suppressed the helicity index on the b and d. The map

$$(\tilde{b}_j, \tilde{d}_j) \mapsto (b_j, d_j)$$

is called *second quantization*. The vacuum $|0\rangle$ is defined to be the vector in $\mathcal{F}$ with

$$b_j|0\rangle = 0, \qquad d_j^*|0\rangle = 0. \tag{4.3.14}$$

The second equality was interpreted by Dirac as meaning that the *Dirac sea* of negative-energy states is full. To each $a \in \mathcal{B}(\mathcal{H})$ we associate formally a set of 4 operators on Fock space which we write as a matrix

$$Q(a) = \begin{pmatrix} Q_{++}(a) & Q_{+-}(a) \\ Q_{-+}(a) & Q_{--}(a) \end{pmatrix}$$

with

$$Q_{++}(a) = \sum_{jk} (\psi_j^+, a\psi_k^+) b_j^* b_k, \quad Q_{+-}(a) = \sum_{jk} (\psi_j^+, a\psi_k^-) b_j^* d_k,$$
$$Q_{-+}(a) = \sum_{jk} (\psi_j^-, a\psi_k^+) d_j^* b_k, \quad Q_{--}(a) = \sum_{jk} (\psi_j^-, a\psi_k^-) d_j^* d_k.$$

Using the canonical anticommutation relations and the rules of matrix multiplication one finds that

$$Q([a, b]) = [Q(a), Q(b)].$$

We introduce also the normal-ordered form $q(a)$ of $Q(a)$ obtained by subtracting the vacuum energy. Let P_- be the projection onto the negative-energy states of $\mathcal{H}$. Then from the definition of normal ordering and the definition (4.3.14) of the fermionic vacuum one finds that

$$q(a) = Q(a) - \text{Tr}(P_- a P_-).$$

The normal ordering does not in general respect the Lie-algebra structure defined by the commutator. The difference

$$s(a, b) = [q(a), q(b)] - q([a, b]) \tag{4.3.15}$$

is known as a *Schwinger term*. We have encountered it already in Example 2.24. A straightforward calculation yields

$$s(a, b) = \text{Tr}(P_- [a, b] P_-). \tag{4.3.16}$$

Introduce the projector P_+ complementary to P_-: $P_- + P_+ = 1$. Then one can also write $s(a, b)$ in the form

$$s(a, b) = \text{Tr}(P_- a P_+ b P_- - P_- b P_+ a P_-). \tag{4.3.17}$$

The above is quite formal and $s(a, b)$ is not properly defined unless restrictions are placed on the operators a and b. We shall mention these below in Example 6.13.

Let $\psi_j^{\pm\prime}$ be a second orthonormal basis of $\mathcal{H}$ obtained using a second Dirac operator coupled, for example, to a different external potential A or different metric on V. To it also are associated a Fock space $\mathcal{F}'$ and a vacuum state vector $|0\rangle'$. The change of basis

$$\psi_j^\pm \mapsto \psi_j^{\pm\prime}$$

defines an element $U \in \mathcal{B}(\mathcal{H})$ called a *Bogoliubov transformation*. The thereto associated map

$$|0\rangle \mapsto |0\rangle', \qquad b_j \mapsto b_j', \qquad d_j \mapsto d_j'$$

of $\mathcal{F}$ to $\mathcal{F}'$ we shall denote by $Q(U)$. If

$$P_+ U P_- = 0, \qquad P_- U P_+ = 0$$

then $\mathcal{F} = \mathcal{F}'$ and $|0\rangle' = |0\rangle$.

As a special case consider a hamiltonian of the form $H = H_0 + H_I$ with H_0 the free Dirac hamiltonian on V_S and H_I an interaction term. Choose the basis $\psi_j^\pm$ to consist of the eigenfunctions of H_0 and the basis $\psi_j^{\pm\prime}$ to consist of the eigenfunctions of H. In this case $Q(U)$ is called the *dressing operator*. One can show that it can only exist in a formal sense. A second case arises when the field take its values in the representation space of some group and the U is a gauge transformation. A third case arises when $\mathcal{F}' = \mathcal{F}$ is the Fock space associated to a free hamiltonian corresponding to incoming and outgoing particle states and the Bogoliubov transformation is the result of a change of basis from the incoming modes to the outgoing modes with the evolution determined by a hamiltonian describing an interaction with an external field.   □

Example 4.33 One would usually like to suppose that a change of basis leave the Fock space invariant and that the only consequence be a change of modes. In this case one can consider each choice of basis as a point in an abstract space and each change of basis as an arrow between two points. Similarly one can consider each choice of set of modes as a point in a different space and the Bogoliubov transformations as the corresponding arrows. The Q is a map then from the first set into the second which respects the arrows. 'First quantization is a mystery; second quantization is a functor.'   □

Example 4.34 Consider an ensemble of non-relativistic electrons constrained to move on a plane in the presence of a constant perpendicular magnetic field B. If one neglects the interaction between the electrons the single-particle hamiltonian H which governs their behaviour is given by

$$H = -\frac{1}{2\mu} D_\lambda D^\lambda.$$

It depends on a covariant derivative D which contains the electromagnetic potential A. The corresponding momentum operators $p_\lambda = -iD_\lambda$ (written here with $\hbar = 1$) do not commute and we have $[p_1, p_2] = ieB$. Because of the rotational symmetry the system is analogous to that described in Example 4.29 and the energy spectrum is similar. It is given by

$$E_j = (j + \frac{1}{2})eB\mu^{-1}, \qquad j = 0, 1, 2, \cdots,$$

where μ is the electron mass. Each value of E_j is known as a *Landau level*. The fact that the system is 2-dimensional manifests itself in a degeneracy of the levels.

Using the potential one can define parallel transport along a displacement a^λ in each coordinate x^λ but since H does not commute with p_λ, parallel transport does not leave it invariant. The magnetic field is nonetheless constant and it is possible to introduce modified momentum operators t_λ which commute with H. They can in fact be chosen so that

$$[p_\lambda, t_\mu] = 0, \qquad [t_1, t_2] = ieB.$$

The 'magnetic translations'

$$u = e^{-ia^1 t_1}, \qquad v = e^{-ia^2 t_2}$$

leave then the hamiltonian invariant. They satisfy the commutation relations (4.3.10) but with

$$q = e^{-2\pi i\alpha}, \qquad \alpha = \frac{1}{2\pi} eBa^1 a^2.$$

Suppose that the plane in which the electrons move is taken as a square of length L with periodic boundary conditions. Then magnetic translations along the entire edge can have no effect and the corresponding value of α must be equal to an integer. That is,

$$\frac{1}{2\pi} eBL^2 = l \in \mathbb{Z}. \tag{4.3.18}$$

One can show that the degeneracy of each Landau level is equal to this integer.

We add now a uniform electric field E parallel to one of the edges of the square, that is, perpendicular to B. This induces a collective motion of the electrons and from Maxwell's equations one sees that in a stationary situation the velocity is perpendicular to E and to B and has a magnitude equal to E/B. The corresponding current J is given by

$$J = e\frac{N}{L^2}\frac{E}{B} = \sigma_H E, \qquad \sigma_H = \frac{eN}{BL^2}$$

where N is the total number of electrons. The parameter σ_H is known as the Hall conductivity.

In experiments the plane is a thin strip of metal and the electron number N can be varied. At low temperatures as electrons are forced into the sheet they fill up progressively the states of lowest energy with exactly one electron per state because of the Pauli exclusion principle. The energy of the highest occupied state is the *Fermi level* μ_F. Variation of the number of electrons is equivalent to the variation of μ_F. Suppose that μ_F is just at the j-th Landau level. There are then $N = jl$ electrons in the system and because of the quantization condition on B one finds that

$$\sigma_H = \frac{el}{BL^2} j = \alpha_0 j. \tag{4.3.19}$$

The Hall conductivity is quantized in units of $\alpha_0 = e^2/2\pi$. This fact is known as the integer *quantum Hall effect*.

In real samples all of this is modified. There are no periodic boundary conditions and the electrons are not free to move only under the influence of the electromagnetic field. Due to the interactions with the lattice of atoms which constitutes the metallic strip the degeneracy of the Landau levels is lifted. Each level becomes a band whose width depends on the strength of the binding of the electrons to the lattice and which can overlap to form a continuum of levels. Moreover due to a phenomenon known as 'Anderson localization' the electron states whose energies are on the edges of each band are not mobile and they do not contribute to the current. They can be thought of as trapped by the impurities of the lattice. There is a gap, the 'mobility gap', in the spectrum of the conduction electrons between each Landau level. If the Fermi level is varied in the region of a gap there is no change in the conductivity and a plateau appears in the Hall conductivity as a function of the electron number. The extraordinary fact is that in spite of the complication of the system and to within considerable accuracy the value of the Hall conductivity on the plateaux remains quantized. The charge-carrying elements need not be electrons. It was observed originally by Hall that the direction of the current depended on the metal which was used. The charge can be carried by holes as well as electrons. In more refined measurements plateaux appear at rational numbers which are not integers. This is thought to be due to conductivity by collective modes with fractional charge.   □

Example 4.35 The operators x and p of the phase space of a single particle do not commute and usually one diagonalizes either one or the other. This situation is not satisfactory in noncommutative geometry where one would like all 'coordinates' to play a symmetric role. There is a *coherent-state* formalism however which allows more symmetry between x and p. It gives rise to a complete but over-determined set of state vectors which are in a sense closer to the classical wave functions; to each point in the classical phase space corresponds a vector in the Hilbert space of states. There are then obviously too many states.

Let $|0\rangle$ be a state concentrated at the origin of phase space. To be precise we choose the state which minimizes the operator $x^2 + p^2$ and we define $|0\rangle$ as the corresponding eigenvector. The '0' refers to the origin; it does not refer to the fact that the solution minimizes $x^2 + p^2$. From the solution to the harmonic-oscillator problem we know that the eigenvalue is equal to 1 and that, for example, in the representation of Example 4.29

$$|0\rangle = \frac{1}{\pi^{1/4}} e^{-\tilde{x}^2/2}.$$

We wish to associate a similar state to each point in phase space.

Using the notation of Example 4.29 and in analogy with the definition of a given there we define

$$z = \frac{1}{\sqrt{2}}(\tilde{x} + i\tilde{p}).$$

This is here just a convenient representation of a point in phase space by one complex number. Define the set of operators

$$T(z) = e^{za^* - \bar{z}a} = e^{i(\hat{p}x - \tilde{x}p)}.$$

These are the translation operators from the origin to the point z. It is easy to see that the adjoint T^* of T is given by $T^*(z) = T(-z)$. The normal-ordered form is

$$T(z) = e^{-|z|^2/2} e^{za^*} e^{-\bar{z}a}.$$

The $T(z)$ satisfy the composition law

$$T(z_1)T(z_2) = e^{i\omega_{12}} T(z_1 + z_2)$$

and so, apart from the here irrelevant phase factor given by $2i\omega_{12} = z_1 \bar{z}_2 - z_2 \bar{z}_1$, the $T(z)$ form a representation of the translation group in the phase-space plane.

We define the state $|z\rangle$ to be the image of $|0\rangle$ under $T(z)$:

$$|z\rangle = T(z)|0\rangle.$$

The states $|z\rangle$ are eigenvectors of a,

$$a|z\rangle = z|z\rangle,$$

and they can be expressed in terms of the eigenvectors $|n\rangle$ of the number operator N:

$$|z\rangle = e^{-|z|^2/2} \sum_{n=0}^{\infty} \frac{z^n}{n!} |n\rangle.$$

If we designate by $\langle z|$ the conjugate state then

$$|\langle z_1 | z_2 \rangle|^2 = e^{-|z_1 - z_2|^2}.$$

So the set $|z\rangle$ is not orthonormal. It is, however, complete. The unit operator can be written in the form

$$1 = \frac{1}{2\pi} \int d\tilde{x} d\tilde{p} |z\rangle\langle z|.$$

As far as we are concerned the most important properties of the set of states $|z\rangle$ lie in the equations

$$\langle z|x|z\rangle = \tilde{x}, \qquad \langle z|p|z\rangle = \tilde{p}. \tag{4.3.20}$$

They furnish a map between the generators (x, p) of the quantum algebra and the points $(\tilde{x}, \tilde{p})$ of classical phase space. There is a one-to-one correspondence between the set of states and the set of points. $\qquad \square$

In certain contexts it is of interest to consider an unbounded operator a on a Hilbert space $\mathcal{H}$. We have done so in Example 4.29 and we shall do so below in Examples 4.45 and 4.46. In such cases the domain $\mathcal{D}(a)$ of definition of a is a proper subspace of $\mathcal{H}$, which we shall suppose to be dense in $\mathcal{H}$. It is possible for an unbounded hermitian operator to be not exactly identifiable with its adjoint. By definition $\mathcal{D}(a^*) \supset \mathcal{D}(a)$ but the converse need not be true. If $\mathcal{D}(a^*) = \mathcal{D}(a)$ then a is said to be *self-adjoint*. If a can be extended to an operator $\tilde{a}$ defined on a larger domain $\mathcal{D}(\tilde{a})$ such that $\mathcal{D}(\tilde{a}^*) \simeq \mathcal{D}(\tilde{a})$ then a is said to be essentially self-adjoint. Introduce the vector spaces $\mathcal{D}_\pm$ by

$$\mathcal{D}_\pm = \{\psi \in \mathcal{H} : a^*\psi = \pm i\psi\}.$$

Since an hermitian operator can have only real eigenvalues it is obviously necessary that $\mathcal{D}_\pm = 0$ for a to be self-adjoint. This is also a sufficient condition. In fact one can show that

$$\mathcal{D}(a^*) = \mathcal{D}(a) \oplus \mathcal{D}_+ \oplus \mathcal{D}_-.$$

The *deficiency indices* are defined to be the integers

$$n_\pm = \dim \mathcal{D}_\pm.$$

If $n_+ = n_-$ then we can extend a to $\tilde{a}$ defined on $\mathcal{D}(\tilde{a}) = \mathcal{D}(a) \oplus \mathcal{D}_+$ by setting $\tilde{a}\psi = -i\psi$ for all $\psi \in \mathcal{D}_+$. The adjoint $\tilde{a}^*$ can be defined on $\mathcal{D}(\tilde{a}^*) = \mathcal{D}(a) \oplus \mathcal{D}_-$ by $\tilde{a}^*\psi = -i\psi$ for all $\psi \in \mathcal{D}_-$. Since now $\mathcal{D}(\tilde{a}^*) \simeq \mathcal{D}(\tilde{a})$ the extension is self-adjoint and the original operator was essentially self-adjoint. If $n_+ \neq n_-$ then there can be no such extension.

Example 4.36 Typically the problem of an unbounded hermitian operator which is not self-adjoint is due to a boundary condition. As an example consider the operator a on the Hilbert space $L^2([0,1])$ defined to be $a = i\partial_x$ on the dense subspace of smooth functions $f(x)$ which satisfy the boundary conditions $f(1) = f(0) = 0$. The Hilbert-space norm is given by

$$(g, f) = \int_0^1 g^* f \, dx.$$

Partial integration yields the formula

$$(i\partial_x g, f) - (g, i\partial_x f) = i(g^*(1)f(1) - g^*(0)f(0)).$$

The right-hand side must vanish since we have supposed that $i\partial_x$ is hermitian. Because of the conditions on f this will be the case whatever the values of $g(0)$ and $g(1)$. The domain of a^* contains then the two functions $e^{\pm x}$ which belong to $\mathcal{D}_\mp$: $\dim \mathcal{D}_+ = \dim \mathcal{D}_- = 1$. It is strictly greater than the domain of a and a

is not self-adjoint. Consider now an extension $\tilde{a}$ of the same operator with $\mathcal{D}(\tilde{a})$ defined by the condition $f(1) = e^{-1}f(0)$. Then it follows that $g(1) = e^{+1}g(0)$. Therefore $\mathcal{D}(\tilde{a}^*) \simeq \mathcal{D}(\tilde{a})$ and the extension is self-adjoint. The original operator a was essentially self-adjoint. The extension is an extension of it to $\mathcal{D}_+$. The same operator on the space $L^2([0, \infty))$ would be not essentially self-adjoint. This is because now e^{-x} belongs to the space but e^x does not. $\qquad\qquad\qquad\square$

A *representation* of a C^*-algebra $\mathcal{A}$ is a morphism π of $\mathcal{A}$ into $\mathcal{B}(\mathcal{H})$ which respects the involution. It is *faithful* if the kernel of π is equal to zero. Two representations π and π' are said to be *equivalent* if there exists a unitary operator U such that $\pi'(f) = U^{-1}\pi(f)U$ for every f in $\mathcal{A}$. The isomorphism between $\pi(\mathcal{A})$ and $\pi'(\mathcal{A})$ is called a *spatial isomorphism* or *unitary equivalence*. It is much stronger than an *algebraic isomorphism* between $\mathcal{A}$ and some other algebra. As in the commutative case each state gives rise in general to an inequivalent representation of the abstract algebra as an algebra of bounded operators on a Hilbert space. This fact is important to have in mind when one studies the noncommutative geometry of formal algebras. From a formal point of view, two isomorphic algebras are to be identified. But from a physical point of view two algebras should perhaps be identified only if they are unitarily equivalent as algebras of operators on some Hilbert space and perhaps two operators should be considered as physically identical if they coincide on every finite-dimensional subspace of the Hilbert space. For this reason it is of interest to consider the *weak topology* on $\mathcal{B}(\mathcal{H})$: a sequence of elements f_i tends to zero weakly if and only if $(\phi, f_i\psi) \to 0$ for arbitrary $\phi, \psi \in \mathcal{H}$. The weak closure is then a subalgebra of $\mathcal{B}(\mathcal{H})$ which is in general strictly larger than the original algebra and contains some information about the representation. Essentially different states give rise to different weak closures of $\mathcal{A}$. A weakly closed subalgebra of $\mathcal{B}(\mathcal{H})$ is a *von Neumann algebra*.

Example 4.37 On a separable Hilbert space with basis $|k\rangle$ we introduce the sequence of operators f_j by $f_j|k\rangle = \delta_{jk}|k\rangle$. As a matrix f_j has a 1 in the j-th diagonal position and zero elsewhere. Then $\|f_j\| = 1$ but $f_j \to 0$ weakly. $\qquad\square$

Example 4.38 Besides the weak and the norm topology there is the *strong topology* on $\mathcal{B}(\mathcal{H})$: a sequence of elements f_j tends to zero strongly if and only if $\|f_j\psi\| \to 0$ for arbitrary $\psi \in \mathcal{H}$. On a separable Hilbert space with basis $|k\rangle$ consider the shift operator Λ defined by $\Lambda|k\rangle = |k+1\rangle$. (We use here the notation of Example 4.47.) Then the sequence $\Lambda^n \to 0$ weakly but $\|\Lambda^n|k\rangle\|$ is independent of n. In the strong topology $\|\Lambda\| = 1$. $\qquad\qquad\qquad\square$

Let $\mathcal{M}$ be a $*$-subalgebra of $\mathcal{B}(\mathcal{H})$ for some separable Hilbert space $\mathcal{H}$ and designate by $\mathcal{M}'$ the algebra of all elements of $\mathcal{B}(\mathcal{H})$ which commute with $\mathcal{M}$. The algebra $\mathcal{M}'$ is said to be the *commutant* of $\mathcal{M}$. It is a weakly closed subalgebra of $\mathcal{B}(\mathcal{H})$ and therefore a C^*-algebra. It is obvious that $\mathcal{M} \subset \mathcal{M}''$ and one can show that the equality holds if and only if $\mathcal{M}$ is weakly closed. From the GNS theorem mentioned above it follows that under certain restrictions one can consider a von Neumann algebra as being equivalent to a C^*-algebra plus a state. We saw above that in the commutative case a state was the same thing as a measure. One can consider then von Neumann algebras as the noncommutative versions of measure spaces just as one can consider C^*-algebras as the noncommutative versions of topological spaces.

If $\mathcal{H} = \mathcal{H}_1 \oplus \mathcal{H}_2$ and $\mathcal{M} = \mathcal{M}_1 \oplus \mathcal{M}_2$ with $\mathcal{M}_1 \subset \mathcal{B}(\mathcal{H}_1)$ and $\mathcal{M}_2 \subset \mathcal{B}(\mathcal{H}_2)$ both von Neumann algebras then the center $\mathcal{Z}(\mathcal{M}) = \mathcal{M} \cap \mathcal{M}'$ is non-trivial; it contains the identity of each of the components. Conversely, if the center is non-trivial then $\mathcal{M}$ can be written as a direct sum, a process which can be continued until one arrives at a direct sum (or direct integral) of algebras each with trivial center. Such a von Neumann algebra is called a *factor*. Consider the Hilbert space $\mathcal{H} \otimes \mathbb{C}^2$ and let I_2 be the unit element of $\mathcal{M}_2$. Then if $\mathcal{M} \subset \mathcal{B}(\mathcal{H})$ is a factor so is $\mathcal{M} \otimes I_2$. Since the latter is reducible as representation of $\mathcal{M}$ one sees that a factor is slightly more general than an irreducible representation. Factors were classified by Murray and von Neumann into types I, II and III. The classification is based on the properties of the set of elements of $\mathcal{M}$ which satisfy

$$e^* = e, \quad e^2 = e. \tag{4.3.21}$$

Such elements e are called *idempotents* or *projections* or *projectors*. They are considered in another context in the Section 5.1. The range of an idempotent is the noncommutative analogue of a measurable set.

Let $\mathcal{M}$ be a factor and let e and e' be idempotents. An element $u \in \mathcal{M}$ such that

$$e = u^*u, \qquad e' = uu^* \tag{4.3.22}$$

is called a *partial isometry*. It is a unitary operator from the range of e onto the range of e'. If e and e' satisfy (4.3.22) they are said to be *equivalent*. A similar notion is defined in (5.1.6). We denote by $[e]$ the equivalence class of e.

Let e_1 and e_2 be idempotents of $\mathcal{M}$ and e an idempotent such that the range of e is a subspace of the range of e_2. If $[e_1] = [e]$ one writes $[e_1] \leq [e_2]$; if $[e_1] \neq [e_2]$ one writes $[e_1] < [e_2]$. Suppose the range of e_1 is a proper subspace of the range of e_2. If $[e_1] = [e_2]$ then e_1 is said to be *infinite*; otherwise it is said to be *finite*. Since $\mathcal{H}$ is separable the only possible infinite equivalence class is $[1]$. Either $[1]$ is in fact infinite, in which case $\mathcal{M}$ is said to be *infinite*, or $[1]$ is finite, in which case $\mathcal{M}$ is said to be *finite*. The equivalence classes are totally ordered and for each $e \neq 0$ we have

$$[0] < [e] \leq [1].$$

If e_2, e_2 are idempotents with $e_1 e_2 = 0$ then $e_1 + e_2$ is an idempotent and one can define

$$[e_1] + [e_2] = [e_1 + e_2].$$

If further $[e_1] = [e_2]$ then one can write $[e_1 + e_2] = 2[e_1]$. As we shall see in the examples, in general $m[e]$ does not exist for arbitrary $[e]$ and arbitrary integer m. However, if $[e_2]$ is finite and $[e_1] < [e_2]$ then it is always possible to write $[e_2]$ in the form

$$[e_2] = m[e_1] + [r], \qquad [r] < [e_1],$$

for some integer m.

There is a trace function $\tau[e]$, also known as the *relative dimension*, with values in the real numbers, which satisfies the properties

$$\tau[0] = 0,$$
$$\tau[e] > 0 \qquad \text{if } [e] \neq 0,$$
$$\tau[e] = \infty \qquad \text{if } [e] \text{ is infinite},$$
$$\tau([e_1] + [e_2]) = \tau[e_2] + \tau[e_2].$$

If $\mathcal{M}$ is finite one can fix the normalization by setting

$$\tau[1] = 1.$$

As an algebra a factor $\mathcal{M}$ is generated by its idempotents. If $\mathcal{M}$ is finite then τ has a unique extension to all elements $f \in \mathcal{M}$ which defines a *trace state* τ on $\mathcal{M}$, a state which satisfies

$$\tau(ab) = \tau(ba).$$

If $\mathcal{M}$ is of type II_∞ then there exist finite idempotents and the relative dimension τ can be extended to a semifinite trace on $\mathcal{M}$, but which is not a state.

If there exists a minimal $[e_0] \neq [0]$, with $[e] \geq [e_0]$ for all $[e]$, then $\mathcal{M}$ is said to be of type I. If there exists no minimal $[e_0]$ but there exists some finite $[e]$ then $\mathcal{M}$ is said to be of type II. If there are no finite $[e]$, which means that $[1]$ is infinite and $[0]$ and $[1]$ are the only classes, then $\mathcal{M}$ is said to be of type III. If $\mathcal{M}$ is of type I then any finite $[e]$ can be written $[e] = m[e_0]$ for some integer m. If $\mathcal{M}$ is finite then $m \leq n$ for some integer n and $\mathcal{M}$ is of type I_n. From the normalization condition one finds that

$$\tau[e_0] = \frac{1}{n}.$$

Note that e_0 could very well have an infinite trace as an element of $\mathcal{B}(\mathcal{H})$. We shall see this in Example 4.41 below. If $m[e_0]$ exists for all integer m then $\mathcal{M}$ is of type I_∞. The complete algebra $\mathcal{B}(\mathcal{H})$ is of type I and the minimal $[e_0]$ is the equivalence class of all 1-dimensional subspaces. The classification is complete in the type-I case; any $\mathcal{M}$ of type I is isomorphic to $\mathcal{B}(\mathcal{H})$ for some $\mathcal{H}$.

If $\mathcal{M}$ is of type II and finite then the relative dimension can take all values in the interval $[0, 1]$; the algebra is of type II_1. If $[1]$ is infinite then the algebra is of type II_∞. A factor $\mathcal{M}$ is of type II_∞ if and only if $\mathcal{M} = \mathcal{N} \otimes \mathcal{B}(\mathcal{H})$ where $\mathcal{N}$ is of type II_1 and $\dim \mathcal{H} = \infty$.

Example 4.39 The matrix algebra M_{mn} is a factor of type I_{mn} which can be factorized $M_{mn} = M_m \otimes M_n$ as the tensor product of a factor M_m of type I_m and a factor M_n of type I_n. The factors M_m and M_n are subfactors of M_{mn} and it is straightforward to see that

$$M'_m = M_n, \qquad M'_n = M_m.$$

There is a sort of *duality* here. If one varies m so that M_m becomes larger within M_{mn} then M_n becomes correspondingly small. The relative dimension of M_m within M_{nm} is the integer n^2. Factors of type II can have subfactors of the same type with non-integer relative dimension.   □

Example 4.40 Consider the C^*-algebra $\mathcal{A} = M_n$ acting on $\mathcal{H} = \mathbb{C}^n$ and the state $\tau(a) = n^{-1}\text{Tr}(a)$. According to the GNS theorem one can consider $\mathcal{A} \simeq \mathcal{H} \otimes \mathcal{H}$ as a Hilbert space with inner product given by (4.3.7). In this case $\mathcal{H}_0 = 0$ and the map

$$a \mapsto \hat{a} = a \otimes 1$$

yields a faithful representation of $\mathcal{A}$ as a subalgebra $\mathcal{M}$ of $\mathcal{B}(\mathcal{H} \otimes \mathcal{H})$. Choose an orthonormal basis $|j\rangle$ of $\mathcal{H}$ and let $|j, k\rangle = |j\rangle \otimes |k\rangle$ be a basis for $\mathcal{H} \otimes \mathcal{H}$. Then

$$a|i\rangle = \sum_{j=1}^{n} a_{ij}|j\rangle, \qquad \hat{a}|j, k\rangle = a \otimes 1|j, k\rangle = \sum_{i=1}^{n} a_{ji}|i, k\rangle.$$

Introduce the vector

$$|0\rangle = \frac{1}{\sqrt{n}} \sum_{j=1}^{n} |j, j\rangle$$

in $\mathcal{H} \otimes \mathcal{H}$. The identification of $\mathcal{A}$ with $\mathcal{H} \otimes \mathcal{H}$ is given by

$$a \mapsto \frac{1}{\sqrt{n}} \sum_{j,k=1}^{n} a_{jk}|j, k\rangle.$$

In particular, $|0\rangle$ corresponds to the unit element of $\mathcal{A}$. From the identities

$$\tau(a) = n^{-1}\text{Tr}(a) = \langle 0|\hat{a}|0\rangle = \langle 0|a \otimes 1|0\rangle$$

one sees that $|0\rangle$ is the state vector corresponding to the state τ. It is obvious that the commutant $\mathcal{M}'$ of $\mathcal{M}$ is the set of elements of $\mathcal{B}(\mathcal{H} \otimes \mathcal{H})$ of the form $1 \otimes a$. The algebra $\mathcal{A}$ is a factor of type I_n with $\mathcal{A}' = \mathbb{C}$. The algebra $\mathcal{M}$ is also of type I_n but $\mathcal{M}' \simeq \mathcal{M}$. As algebras $\mathcal{A}$ and $\mathcal{M}$ are isomorphic but they have different commutants since they are realized on different Hilbert spaces. The representation $\mathcal{M}$ of $\mathcal{A}$ is a reducible representation.   □

Let $\mathcal{A}$ be a C^*-algebra of operators on a Hilbert space $\mathcal{H}$ and consider $\mathcal{A}$ as a subspace of $\mathcal{H} \otimes \mathcal{H}$ with inner product (4.3.7) defined by a state of the form

$$\omega(a) = \mathrm{Tr}(\rho a) \tag{4.3.23}$$

where ρ is a density matrix. If we assume that ω is a *faithful state* then ρ is invertible and $\mathcal{H}_0 = 0$. Choose an orthonormal basis $|j\rangle$ of $\mathcal{H}$ of eigenvectors of ρ. Then

$$\rho|j\rangle = \rho_j|j\rangle, \qquad \sum_{j=1}^{\infty} \rho_j = 1.$$

Introduce the vector

$$|0\rangle = \sum_{j=1}^{\infty} \sqrt{\rho_j}|j,j\rangle$$

in $\mathcal{H} \otimes \mathcal{H}$. Then

$$\hat{a}|0\rangle = \sum_{j,k=1}^{\infty} \sqrt{\rho_j}a_{jk}|k,j\rangle. \tag{4.3.24}$$

From the identities

$$\omega(a) = \mathrm{Tr}(\rho a) = \langle 0|\hat{a}|0\rangle = \langle 0|a \otimes 1|0\rangle$$

one sees that $|0\rangle$ is the state vector corresponding to the state ω. If $\mathcal{A}$ is a factor then so is $\mathcal{M}$ and $\mathcal{M}$ is called the *standard form* of $\mathcal{A}$. The commutant $\mathcal{M}'$ of $\mathcal{M}$ is the set of elements of $\mathcal{B}(\mathcal{H} \otimes \mathcal{H})$ of the form $1 \otimes a$. Since $\mathcal{M}|0\rangle$ is dense in $\mathcal{M}$ the vector $|0\rangle$ is said to be a *cyclic vector*; since $a \otimes 1|0\rangle = 0$ only if $a = 0$ the vector $|0\rangle$ is said to be a *separating vector*. The vector $|0\rangle$ is cyclic and separating also for the commutant $\mathcal{M}'$. The notation $|0\rangle$ for the cyclic and separating vector is taken from physics; the vacuum state vector is generically supposed to be cyclic and separating. It is certainly not separating in the situation discussed in Example 4.29 since it is annihilated by the operator a; compare however Example 4.59.

A factor is said to be *hyperfinite* if it can be realized as the limit of a sequence of matrix algebras. The factor of type I_∞ is hyperfinite. There is also a unique hyperfinite factor of type II_1 which can be constructed as a limit of a tensor product of factors of type I_2.

Example 4.41 Let the Hilbert space $\mathcal{H}$ be written

$$\mathcal{H} = \bigoplus_{i=1}^{n} \mathcal{H}_i$$

as a direct sum of n (isomorphic) Hilbert spaces $\mathcal{H}_i$ of infinite dimension. Then one can consider the algebra $\mathcal{M} \subset \mathcal{B}(\mathcal{H})$, of type I_n, consisting of those elements of $\mathcal{B}(\mathcal{H})$

which reduce to the multiple of the identity on each subspace $\mathcal{H}_i$. Let $[e_0]$ be the equivalence class of the identities. Then $[e_0] < [1]$ is finite in $\mathcal{M}$ although $[e_0] = [1]$ is infinite in $\mathcal{B}(\mathcal{H})$. In particular choose $n = 2$. The resulting algebra $\mathcal{M}_1$ is of type I_2. Decompose once more

$$\mathcal{H}_1 = \mathcal{H}_{11} \oplus \mathcal{H}_{12}, \qquad \mathcal{H}_2 = \mathcal{H}_{21} \oplus \mathcal{H}_{22}.$$

The resulting algebra $\mathcal{M}_2$ is of type I_4. Continuing the decomposition one obtains a sequence $\{\mathcal{M}_n\}$ of factors with $\mathcal{M}_n \subset \mathcal{M}_{n+1}$ and $\mathcal{M}_n$ of type I_{2^n}. We define $\mathcal{M}$ to be the weak closure of the union of the $\mathcal{M}_n$ in $\mathcal{B}(\mathcal{H})$. By construction $\mathcal{M} \neq \mathcal{B}(\mathcal{H})$. Define $\mathcal{N}_1 = \mathcal{M}_1$ and $\mathcal{N}_j = \mathcal{M}_j \cap \mathcal{M}'_{j-1}$ for $n \geq 2$. Then each $\mathcal{N}_j$ is of type I_2 and

$$\mathcal{M}_n = \bigotimes_{j=1}^{n} \mathcal{N}_j.$$

Each $\mathcal{N}_j$ has one non-trivial idempotent e. The 2^n idempotents of $\mathcal{M}_n$ are obtained by using all the different ways one can replace 1 by e in the n-fold tensor product $1 \otimes \cdots \otimes 1$. They are all the finite idempotents of $\mathcal{M}$. From this construction it is clear that $\mathcal{M}$ has no minimal idempotent. We define a trace on $\mathcal{M}_n$ by the condition that

$$\tau[e] = \frac{1}{2}$$

for the non-trivial idempotent e of each $\mathcal{N}_j$. Since the trace is multiplicative with respect to the tensor product of matrices we have

$$\tau[e_0] = \frac{1}{2^n}, \qquad \tau[1] = 1$$

where $e_0 = e \otimes \cdots \otimes e$ is the minimal element of $\mathcal{M}_n$. It follows that $\mathcal{M}$ is of type II_1. The hyperfinite factor is a subalgebra of every type-II_1 factor.   □

Example 4.42 A generalization of the above construction can be given which yields in the limit hyperfinite factors of each possible type. We discussed in Example 4.26 the statistical mechanics of an n-particle quantum spin system. Let

$$\mathcal{A} = M_2^{\otimes n}$$

be the algebra of observables. An element of $\mathcal{A}$ is a sum of elements of the form

$$a = a_1 \otimes \cdots \otimes a_n, \qquad a_i \in M_2. \tag{4.3.25}$$

The algebra is therefore a C^*-algebra with involution and norm defined by those of M_2; the norm of a is the product of the norm of each matrix factor. In the infinite-volume or *thermodynamic limit* when $n \to \infty$ the limit algebra is well defined as a C^*-algebra. Consider now states on $\mathcal{A}$ defined for each finite n and for each state consider the representation of $\mathcal{A}$ as an algebra of operators on the (finite-dimensional) Hilbert space given by the GNS construction. We denote the weak closure again by $\mathcal{A}$. It will depend on the choice of state.

As examples consider product states, defined on the product elements (4.3.25) by

$$\omega(a) = \prod_{i=1}^{n} \text{Tr}(\rho a_i)$$

where ρ is a density matrix on M_2. If ρ is of rank 1 then $\mathcal{A}$ is a factor of type I_∞; the minimal idempotent e_0 is as in the preceding example with e the projection onto the image of ρ but now $\text{Tr}[e_0] = \omega(e_0) = 1$. If $\rho = 1/2$ then $\mathcal{A}$ is of type II_1; $\text{Tr}[1] = \omega(1) = 1$. If ρ is of rank 2 but not proportional to the identity then $\mathcal{A}$ is of type III; a finer classification in this case can be obtained by considering the ratio of the eigenvalues of ρ. In the first case the vector space $\mathcal{H}_0$ defined by Equation (4.3.8) does not vanish and $\mathcal{H} = \mathbb{C}^n$. In the second and third cases the state is faithful and $\mathcal{H}_0 = 0$. Suppose for each n that the state ω is a *thermal state* defined by the density matrix

$$\rho = Z^{-1} e^{-\beta H}, \qquad Z = \text{Tr}(e^{-\beta H}).$$

We have here replaced as usual the product kT of Boltzmann's constant and the temperature by β^{-1}. In general the limiting hamiltonian will have a continuous spectrum and the corresponding state cannot be of the form (4.3.23); the von Neumann algebra is necessarily of type III. In the limiting case with infinite temperature, with $\beta = 0$, then the state is a trace state with $\omega(1) = 1$. This will remain so in the thermodynamic limit and the algebra will be of type II_1. In the opposite limit with $\beta = \infty$ the state is a pure state, the projection onto the ground state and the limiting algebra will be of type I_∞. $\square$

We recall that the group algebra of a finite group is semi-simple. This result has a natural extension to groups with a countable number of elements. Let G be a countable discrete group and let $f \in L^2(G)$. Then the operators L_g and R_g defined by

$$(L_g f)(h) = f(g^{-1}h), \qquad (R_g f)(h) = f(hg)$$

are unitary on $L^2(G)$. We have

$$(L_{g_1}(L_{g_2} f))(h) = (L_{g_1} f)(g_2^{-1}h) = f(g_2^{-1}g_1^{-1}h) = (L_{g_1 g_2} f)(h).$$

This representation of G on $L^2(G)$ is called the left-regular representation. Similarly R_g defines the right-regular representation. By choosing f a polynomial one sees that the algebra generated by the set of L_g or R_g can be identified with the group algebra. Let $\mathcal{M}_L$ ($\mathcal{M}_R$) be the weak closure of the algebra generated by the L_g (R_g)

for all $g \in G$. Then $\mathcal{M}_L$ and $\mathcal{M}_R$ are both type-II$_1$ von Neumann algebras and $\mathcal{M}_L = \mathcal{M}'_R$. If G has the property that the equivalence classes

$$C_g = \{h^{-1}gh : h \in G\}$$

are infinite for all $g \neq 1$ then they are both factors. An example of such a group is the free group on two generators. If G is the union of an increasing family of finite subgroups then $\mathcal{M}_L$ and $\mathcal{M}_R$ are hyperfinite.

The vector $\delta \in L^2(G)$ defined by $\delta(1) = 1$ and $\delta(g) = 0$ for $g \neq 1$ is a unit vector; it is cyclic and separating. Therefore the factors $\mathcal{M}_L$ and $\mathcal{M}_R$ are in standard form. The relative dimension in $\mathcal{M}_L$ is defined by the state $\tau = \omega_\delta$. If e is an idempotent

$$\tau[e] = \omega_\delta(e) = (e\delta, \delta).$$

Let $\mathcal{M}$ be a von Neumann algebra acting on a Hilbert space $\mathcal{H}$ and let G be a topological group of automorphisms of $\mathcal{M}$. We shall suppose that G is a countable discrete group. Let $L^2(G)$ be the Hilbert space of square-integrable functions on G with respect to the Haar measure. For each $g \in G$ define $v_g \in \mathcal{B}(L^2(G))$ by $v_g = L_g$ and as above let $\mathcal{M}_L$ be the von Neuman algebra generated by the set of v_g. We let $\alpha_g(u)$ denote the action of $g \in G$ on $u \in \mathcal{M}$. Consider the Hilbert space $\mathcal{H} \otimes L^2(G)$ and let $\psi \in \mathcal{H}$ and $f \in L^2(G)$. One can define a new von Neumann algebra $\mathcal{M} \times_\alpha G \subset \mathcal{B}(\mathcal{H} \otimes L^2(G))$ generated by the actions u and v_g on the elements of the form $\psi \otimes f$ given by

$$u \circ (\psi \otimes f)(h) = \alpha_h(u)\psi f(h),$$

$$v_g \circ (\psi \otimes f)(h) = (\psi \otimes f)(g^{-1}h) = \psi f(g^{-1}h).$$

This is the *crossed-product* construction. From the definitions it follows that

$$v_{g^{-1}}uv_g \circ (\psi \otimes f)(h) = \alpha_{g^{-1}h}(u)\psi f(h) = \alpha_{g^{-1}}(u) \circ (\psi \otimes f)(h) \qquad (4.3.26)$$

and so

$$uv_g = v_g\alpha_{g^{-1}}(u). \qquad (4.3.27)$$

If the action of G is trivial then $\mathcal{M} \times_\alpha G = \mathcal{M} \otimes \mathcal{M}_L$. The commutation relations (4.3.27) generalize those which define the Weyl algebra.

Example 4.43 A simple example of the crossed-product construction is obtained by choosing $\mathcal{M} = M_n$ and $G = \mathbb{Z}_n$. The Hilbert space $L^2(G)$ is equal to $\mathbb{C}^n$ and the algebra $\mathcal{M}_L$ is equal to $\mathbb{C} \times \cdots \times \mathbb{C}$, with n factors. Let t be the generator of $\mathbb{Z}_n$ and associate to it a unitary matrix $U(t)$ obtained by a cyclic permutation of the rows (or columns) of the unit element of M_n. The action α_t is defined by $\alpha_t(u) = U(t)^{-1}uU(t)$. The new von Neumann algebra $M_n \times_\alpha \mathbb{Z}_n$ is block-diagonal with n copies of M_n along the diagonal; it is a subalgebra of M_{n^2} of dimension n^3 which is not a factor. $\square$

Example 4.44 Let $\mathcal{H} = L^2(S^1)$ where S^1 is identified as the set of elements $z \in \mathbb{C}$ of unit modulus and let $q \in \mathbb{C}$ be fixed of unit modulus. Introduce the operators u and v by their action on $\psi(z) \in \mathcal{H}$:

$$(u\psi)(z) = \psi(qz), \qquad (v\psi)(z) = z\psi(z).$$

The functions

$$|j\rangle = \frac{1}{\sqrt{2\pi}} z^j$$

form an orthonormal basis for $\mathcal{H}$ and the action of u and v can be given by

$$u|j\rangle = q^j|j\rangle, \qquad v|j\rangle = |j+1\rangle.$$

These are infinite-dimensional versions of the operators introduced in (3.1.24) to study the Weyl algebra when q is a root of unity. It is easy to see that the operators u and v are unitary. The von Neumann algebra $\mathcal{M}$ they generate is a representation of the Weyl algebra. If q is not a root of unity then the weak closure is a factor of type II_1.

One can construct another representation of the Weyl algebra using the crossed-product construction. Let $\mathcal{M} \subset \mathcal{B}(L^2(S^1))$ be the commutative von Neumann algebra generated by the element u. Let $\psi \otimes f(z, k)$ be an element of the Hilbert space $L^2(S^1) \otimes L^2(\mathbb{Z})$. We introduce an action α_k of $\mathbb{Z}$ on $\mathcal{M}$ by

$$(\alpha_k(u)\psi)(z) = \psi(q^k z)$$

and we introduce the elements $u, v_l \in \mathcal{B}(L^2(S^1) \otimes L^2(\mathbb{Z}))$ by the actions

$$u \circ (\psi \otimes f)(z, k) = \psi(q^k z) f(k),$$
$$v_l \circ (\psi \otimes f)(z, k) = \psi(z) f(k + l).$$

If we use the basis $|j\rangle$ and introduce a similar basis $|k\rangle(i) = \delta_{ik}$ for $L^2(\mathbb{Z})$ then we can define $|j, k\rangle = |j\rangle \otimes |k\rangle$ as basis for the crossed-product representation. The action of the generators becomes then

$$u|j, k\rangle = q^{j+k}|j, k\rangle, \qquad v_l|j, k\rangle = |j, k + l\rangle.$$

One verifies that $uv_1 = qv_1u$. The commutant of $\mathcal{M} \times_\alpha \mathbb{Z}$ in $\mathcal{B}(L^2(S^1) \otimes L^2(\mathbb{Z}))$ is not trivial. In fact it is easy to see that the operators u' and v' defined by

$$u'|j, k\rangle = q^j|j, k\rangle, \qquad v'|j, k\rangle = |j + 1, k - 1\rangle$$

belong to $(\mathcal{M} \times_\alpha \mathbb{Z})'$. It is also easy to see that they generate a second representation of the Weyl algebra.

When $q^n = 1$ for some integer n then the vector

$$\delta = \frac{1}{\sqrt{n}} \sum_{j=1}^{n} |j, 0\rangle$$

is the state vector and the relative dimension of an idempotent e is given by

$$\tau[e] = \omega_\delta(e) = (\delta, e\delta).$$

It is easy to see that

$$\omega_\delta(1) = 1, \qquad \omega_\delta(u^j v^k) = 0, \quad 0 \le j, k \le n.$$

When q is not a root of unity one defines a state by

$$\omega_\delta(1) = 1, \qquad \omega_\delta(u^j v^k) = 0, \quad j, k \ge 0.$$

The characteristic function of a measurable set of the circle defines an idempotent whose relative dimension is given by the measure of the set. The relative dimension of this set of idempotents obviously takes all values in the interval $[0, 1]$. The relative dimension of the other idempotents is impossible to calculate explicitly but from the general theory we know that each is equivalent to some characteristic function of a measurable set of the circle. $\square$

Example 4.45 We now consider a variation of the above construction which will be of interest in Example 4.47. Fix $q > 1$. Let S be the discrete subset of the real line given by

$$S = \{q^j : j \in \mathbb{Z}\}.$$

Let $\mathcal{H} = L^2(S)$ be the Hilbert space of square-summable functions on S and choose $\mathcal{M} = L^\infty(S)$. Let $G = \mathbb{Z}$ with the action

$$(\alpha_k(u)\psi)(q^j) = \psi(q^{j+k})$$

on $u \in \mathcal{M}$. We have then

$$u \circ (\psi \otimes f)(q^j, k) = \psi(q^{j+k})f(k),$$
$$v_l \circ (\psi \otimes f)(q^j, k) = \psi(q^j)f(k + l).$$

In particular if $\{|k\rangle\}$ is the basis of $L^2(S)$ given by $|k\rangle(i)^\bullet = \delta_{ik}$ we have

$$u \circ (\psi \otimes |k\rangle)(q^j) = \psi(q^{j+k})|k\rangle,$$
$$v_l \circ (\psi \otimes |k\rangle)(q^j) = \psi(q^j)|k + l\rangle.$$

Consider the identity function as the singular limit $\psi(q^j) \to q^j$. Then we have

$$u|k\rangle = q^k|k\rangle, \qquad v_l|k\rangle = |k + l\rangle. \tag{4.3.28}$$

For reasons to become apparent in Example 4.47 we introduce $x = u$ and $\Lambda = v_1$. The commutation relations which follow from (4.3.28) can be written then $x\Lambda = q\Lambda x$.

For each integer j introduce the idempotent e_j in $\mathcal{M} \times_\alpha \mathbb{Z}$ which projects onto $|j\rangle$. These idempotents are all equivalent and the equivalence class $[e_0]$ is minimum. Within the commutative algebra generated by the element u this is obvious. It follows that $\mathcal{M} \times_\alpha \mathbb{Z}$ is of type I_∞. Since the action of $\mathbb{Z}$ is effective and with no nontrivial fixed points one can show that the algebra is a factor.   □

Example 4.46 Let now $S = \mathbb{Z}$. We have then

$$u \circ (\psi \otimes f)(j, k) = \psi(j + k)f(k),$$
$$v_l \circ (\psi \otimes f)(j, k) = \psi(j)f(k + l).$$

In particular if $\{|k\rangle\}$ is the basis of $L^2(S)$ given by $|k\rangle(i) = \delta_{ik}$ we have

$$u \circ (\psi \otimes |k\rangle)(j) = \psi(j + k)|k\rangle,$$
$$v_l \circ (\psi \otimes |k\rangle)(j) = \psi(j)|k + l\rangle.$$

Consider the identity function as the singular limit $\psi(j) \to j$. Then we can write as above

$$u|k\rangle = k|k\rangle, \qquad v_1|k\rangle = |k + 1\rangle. \qquad (4.3.29)$$

For reasons to become apparent in Example 4.47 we introduce $y = u$ and $\Lambda = v_1$. The commutation relations which follow from (4.3.29) can be written then $\Lambda^{-1}y\Lambda = y+1$. The operator y is related to the x of the preceding example by $x = q^y$. The algebra $\mathcal{M} \times_\alpha \mathbb{Z}$ is also a factor of type I_∞.   □

Example 4.47 By the previous two examples we can represent the algebra $\mathbb{R}_q^1$ of Example 4.18 on a Hilbert space $\mathcal{R}_q$ with basis $|k\rangle$ by

$$x|k\rangle = q^k|k\rangle, \qquad \Lambda|k\rangle = |k + 1\rangle. \qquad (4.3.30)$$

This explains the origin of the expression 'dilatator'. The spectrum of Λ is continuous. The element y has the representation

$$y|k\rangle = k|k\rangle \qquad (4.3.31)$$

on the basis elements. With the renormalized value (4.1.40) of y the spacing between the spectral lines vanishes with z. We shall refer to the old units as Planck units and the new ones as laboratory units.

As operators on $\mathcal{R}_q$ one finds the representations

$$e_1|k\rangle = -z^{-1}|k + 1\rangle + z^{-1}\beta|k\rangle, \qquad \bar{e}_1|k\rangle = z^{-1}|k - 1\rangle + z^{-1}\bar{\beta}|k\rangle$$

for the derivations with two arbitrary complex parameters β and $\bar{\beta}$. Recall the parameters α and $\bar{\alpha}$ introduced in Example 4.18. If we choose

$$\beta = \alpha, \qquad \bar{\beta} = \bar{\alpha}$$

we can write

$$e_1 = z^{-1}\alpha + \lambda_1, \qquad \bar{e}_1 = z^{-1}\bar{\alpha} + \bar{\lambda}_1. \qquad (4.3.32)$$

In terms of α and $\bar{\alpha}$ we find the expression

$$c = z^{-1}(\alpha^* + \bar{\alpha}) \qquad (4.3.33)$$

for the constant c of Equation (4.1.53). From (4.1.46) we are prompted to introduce the anti-hermitian element e_{r1} of $\mathbb{R}^1_{qr}$ with the commutation relations

$$[e_{r1}, x] = (q\Lambda, \Lambda^{-1})x, \qquad [e_{r1}, \Lambda] = 0. \qquad (4.3.34)$$

From the definition of e_{r1} as derivation one sees that the solution is given by

$$e_{r1} = \lambda_{r1} + c_r. \qquad (4.3.35)$$

If $\alpha = \pm 1$, $\bar{\alpha} = \mp 1$ then

$$c_r = \pm z^{-1}(1, -1). \qquad (4.3.36)$$

The limit $q \to 1$ is rather difficult to control. From the relations of the algebra and the two differential calculi one might expect $\Lambda \to 1$. This is consistent with the limiting relations $e_1 x = \bar{e}_1 x = x$ and the intuitive idea that x is an exponential function on the line. However the representation (4.3.30) of the algebra becomes quite singular; one has rather $x \to 1$. This would imply that the parameters α and $\bar{\alpha}$ must tend to zero as $q \to 1$. If one renormalizes according to (4.1.40) then one finds that the relation (4.1.39) is consistent with the limit $\Lambda \to 1$ as $q \to 1$. We shall assume this to be the case. We have then

$$\lim_{q \to 1} \mathcal{A}_r = \mathcal{A}$$

and the real differential calculus coincides with the diagonal elements of the product in (3.2.19). $\qquad \Box$

Example 4.48 The algebra $\mathbb{R}^1_q$ is a subalgebra of the graded algebra of forms $\Omega^*(\mathbb{R}^1_q)$ and the representation (4.3.30) can be extended to a representation of the latter. In fact since $\Omega^1(\mathbb{R}^1_q)$ and $\bar{\Omega}^1(\mathbb{R}^1_q)$ are free $\mathbb{R}^1_q$-modules of rank one with respectively the special basis θ^1 and $\bar{\theta}^1$ we can identify

$$\Omega^*(\mathbb{R}^1_q) = \mathbb{R}^1_q \otimes \bigwedge{}^*, \qquad \bar{\Omega}^*(\mathbb{R}^1_q) = \mathbb{R}^1_q \otimes \bigwedge{}^*$$

where $\bigwedge^*$ is the exterior algebra over $\mathbb{C}^1$ and the extension is trivial; we can represent $\Omega^*(\mathbb{R}_q^1)$ and $\bar{\Omega}^*(\mathbb{R}_q^1)$ on $\mathcal{R}_q$. For the two elements dx and $\bar{d}x$ we have respectively

$$dx|k\rangle = \gamma q^{k+1}|k+1\rangle, \qquad \bar{d}x|k\rangle = \bar{\gamma}q^k|k-1\rangle, \qquad (4.3.37)$$

with two arbitrary real parameters γ and $\bar{\gamma}$. One sees that $(dx)^* = \bar{d}x$ if and only if $\gamma^* = \bar{\gamma}$. It is easy to see that the commutation relations (4.1.41) and (4.1.42) are satisfied. The above representations are certainly not unique.

Considered as derivations e_1 and $\bar{e}_1$ cannot be implemented on $\mathcal{R}_q$. However e_1 can be considered as an 'annihilation' operator which maps $\Omega^1(\mathbb{R}_q^1)$ into $\Omega^0(\mathbb{R}_q^1) \equiv \mathbb{R}_q^1$. Similarly θ^1 has an interpretation as a 'creation' operator which takes $\Omega^1(\mathbb{R}_q^1)$ into $\Omega^2(\mathbb{R}_q^1) \equiv 0$. In this respect $\Omega^*(\mathbb{R}_q^1)$ resembles a fermionic Fock space. The same of course holds for $\bar{e}_1$ and $\bar{\theta}^1$ and $\bar{\Omega}^*(\mathbb{R}_q^1)$. An analogous situation was described in Example 3.11. On $\mathcal{R}_q$ the frame elements θ^1 and $\bar{\theta}^1$ become the operators

$$\theta^1 = \gamma, \qquad \bar{\theta}^1 = \bar{\gamma} \qquad (4.3.38)$$

proportional to the unit element. They were so constructed. The representation of the 1-forms of the product of the two differential calculi can be considered as given on the direct sum $\mathcal{R}_q \oplus \mathcal{R}_q$ of two separate and distinct copies of $\mathcal{R}_q$, one for dx and one for $\bar{d}x$. On $\mathcal{R}_q \oplus \mathcal{R}_q$ the involution of Example 3.16 is given by the map $\gamma \mapsto \bar{\gamma}$. We shall choose

$$\gamma = 1, \qquad \bar{\gamma} = 1 \qquad (4.3.39)$$

so that the map simply exchanges the two terms of $\mathcal{R}_q \oplus \mathcal{R}_q$.

The algebra $\mathbb{R}_q^1$ is also a subalgebra of the graded algebra of forms $\Omega_r^*(\mathbb{R}_{qr}^1)$ of (4.1.49) and the representation (4.3.30) can be extended to a representation of it as well. Again since $\Omega_r^1(\mathbb{R}_{qr}^1)$ is a free $\mathbb{R}_q^1$-module of rank one with the special basis θ_r^1 we can identify

$$\Omega_r^*(\mathbb{R}_{qr}^1) = \mathbb{R}_{qr}^1 \otimes \bigwedge{}^*$$

where $\bigwedge^*$ is the exterior algebra over $\mathbb{C}^1$ and therefore the extension is again trivial. From (4.3.37) one sees that $d_r x$ can be represented by the operator

$$d_r x|k\rangle = q^k(q\gamma|k+1\rangle + \bar{\gamma}\overline{|k-1\rangle}).$$

We have placed a bar over the second term to underline the fact that it belongs to the second copy of $\mathcal{R}_q$. On $\mathcal{R}_q \oplus \mathcal{R}_q$ we have the representation

$$\theta_r^1 = 1. \qquad (4.3.40)$$

The θ_r^1 here is to be interpreted as an element on the $\bigwedge^*$ and the equality gives its representation as the unit in $\mathbb{R}_q^1$. The second of Equations (4.1.50) is to be interpreted then as the equation $1 \wedge 1 = 0$ in the exterior algebra. $\square$

Example 4.49 In the Example 4.18 it is possible to give an intuitive interpretation of the metric (3.4.1) in terms of observables since we have a representation of x and $d_r x$ on the Hilbert space $\mathcal{R}_q$. In this representation the distance s along the 'line' x is given by the expression

$$ds(k) = \| \sqrt{g_{11}'} d_r x (|k\rangle + \overline{|k\rangle}) \| = \| \theta_r^1 (|k\rangle + \overline{|k\rangle}) \|. \qquad (4.3.41)$$

This comes directly from the original definition of dx as an 'infinitesimal displacement'. Using (4.3.40) we find that

$$ds(k) = \| \, |k\rangle + \overline{|k\rangle} \, \| = 1.$$

The 'space' is discrete and the spacing between 'points' is uniform. The distance operator s can be identified with the element y represented in (4.3.31). This means that if we measure y using laboratory units, introduced in Equation (4.1.40), then we shall do the same with s. In these units then the distance between neighbouring 'points' is given by

$$ds(k) = z.$$

It is of interest to stress that the structure of this 'space' endowed with the two different metrics we have considered is the same. With both of them we saw in Examples 4.45 and 4.46 that the weak completion of the algebra $\mathbb{R}_q^1$ was a type-I_∞ factor. The choice of metric does not influence the algebraic structure of the space. However the 'space' with the non-local metric is incomplete; the origin and all points to the right are missing. In this case it would be natural either to add the origin to obtain a 'space' with boundary or to add the origin and another copy of the 'space' to obtain again the entire real line. In both of these cases the algebra is no longer a factor. □

Example 4.50 The algebra of Example (4.17) has a representation given on the Hilbert space $L^2(\mathbb{R})$ by

$$(u\psi)(\alpha) = -2h\alpha\partial_\alpha\psi(\alpha) - h\psi(\alpha), \qquad (v\psi)(\alpha) = \alpha\psi(\alpha)$$

on smooth functions. If we set $\Lambda = e^{iu}$, $x = v$ and $q = e^{2ih}$ then we find that $x\Lambda = q\Lambda x$, the relation which defines $\mathbb{R}_q^1$. □

Example 4.51 The algebra $\mathbb{R}_q^3$ of Example 4.21 has a representation given on the Hilbert space $L^2(\mathbb{R}^3)$ by

$$(x^-\psi)(\alpha, \beta, \gamma) = \frac{1}{\sqrt{1+q}}(\alpha - i\gamma)\psi(\alpha, q^{-1}\beta, \sqrt{(1-q^2)(q^{-1}\beta)^2 + \gamma^2}),$$

$$(y\psi)(\alpha, \beta, \gamma) = \sqrt{q}\beta\psi(\alpha, \beta, \gamma),$$

$$(x^+\psi)(\alpha,\beta,\gamma) = \frac{1}{\sqrt{1+q^{-1}}}(\alpha + i\sqrt{(q^2-1)\beta^2+\gamma^2})\psi(\alpha, q\beta, \sqrt{(q^2-1)\beta^2+\gamma^2}),$$

$$(r\psi)(\alpha,\beta,\gamma) = \sqrt{\alpha^2 + q^2\beta^2 + \gamma^2}\psi(\alpha,\beta,\gamma),$$

$$(\Lambda\psi)(\alpha,\beta,\gamma) = \psi(q^{-1}\alpha, q^{-1}\beta, q^{-1}\gamma)\, q^{-3/2}.$$

There is here a delicate domain problem which we shall ignore: strictly speaking x^- is defined only on those functions $\psi(\alpha,\beta,\gamma)$ which vanish in the strip $q^2\gamma^2 \le (q^2-1)\beta^2$. In the commutative limit the unitary operator Λ tends to 1 and the other, unbounded, operators tend to their expected values. The representation has been chosen with y and r diagonal. If it is restricted to the Hilbert space of functions of β alone then the subalgebra generated by y and Λ becomes the representation already considered of $\mathbb{R}^1_q$. ☐

Example 4.52 The construction of Example 4.44 can be modified to be valid for real values of q. Choose $q > 1$ and a positive measure μ_0 on the interval $[1, q)$. Extend μ_0 to a measure μ on the positive real axis $\mathbb{R}_+$ so that $\mu(q^k\mathcal{U}) = \mu(\mathcal{U})$ for each measurable subset $\mathcal{U}$ of $[1, q)$ and each integer k. Let $\mathcal{H} = L^2(\mathbb{R}_+, \mu)$. Introduce the operators u and v by their action on $\psi(\alpha) \in \mathcal{H}$:

$$(u\psi)(\alpha) = \alpha\psi(\alpha), \qquad (v\psi)(\alpha) = \psi(q^{-1}\alpha)\, q^{-1/2}.$$

Then the von Neuman algebra $\mathcal{M}$ generated by the unbounded self-adjoint operator $x = u$ and the unitary operator $\Lambda = v$ is a representation of the algebra $\mathbb{R}^1_q$. In the limit $q \to 1$ the operator Λ tends to the identity and if the measure is the Lebesgue measure the operator x describes the positive real axis. All functions periodic with respect to the q-shift, that is all $\psi(\alpha) \in \mathcal{H}$ such that $\psi(q\alpha) = \psi(\alpha)$, obviously belong to the center of $\mathcal{M}$. If the measure μ_0 is not a Dirac measure this center is not trivial and the von Neumann algebra is not a factor. ☐

In Examples 4.45 and 4.46 we remarked that the von Neumann algebra generated by x and Λ or by y and Λ is a factor of type I$_\infty$. We have in fact applied the crossed-product construction to a von Neumann algebra $\mathcal{M}$ of operators on the Hilbert space of functions on the real line which are square integrable with respect to a certain measure. Under conditions which generalize the condition that the action be effective and have no fixed points, that it be *ergodic* and *free*, one can show that the algebra

$\mathcal{M} \times_\alpha G$ is a factor. If the measure is concentrated on a discrete countable set of points, as in the two examples, it is of type I; if it is such that all points have measure zero then it is of type II or III. Two measures which are not equal will in general yield von Neumann algebras which are not unitarily equivalent.

In certain contexts the Weyl algebra is referred to as the *noncommutative torus* or the *rotation algebra*. To motivate this terminology we briefly mention *foliations*.

A manifold V is foliated by a set of manifolds called *leaves* if every point $x \in V$ has a neighbourhood which can be identified as the cartesian product of a neighourbood of a leaf L and some open set $\mathcal{U}$ of $\mathbb{R}^q$. The integer q is the codimension of the foliation. A manifold T (of dimension q) is transverse to the foliation if whenever $x \in T$ the set $\mathcal{U}$ can be considered as a neighbourhood of T. If L cuts T exactly once then one can identify V with $L \times T$. In general this is not the case. In fact the space of leaves of a foliation of a manifold V is in general not a manifold; it is often esoteric even as a topological space and its algebra of continuous or measurable functions furnishes little information about its structure. A more interesting algebra $\mathcal{A}$ can be constructed by using T but instead of associating to each point $x \in T$ the value of a function f on T, considering f as a function on V and taking a sort of mean value over the leaf L through x. The product $f * g(\gamma)$ of two elements of $f, g \in \mathcal{A}$ is a convolution product along a path γ in L through x using to integrate a measure induced from V. If T is compact then the integral can be replaced by a sum over the intersection points of γ with T. This product is in general noncommutative.

Example 4.53 The torus $\mathbb{T}^2$ of Example 2.2 can be foliated by the family of curves

$$\tilde{x} = \tilde{x}_0 + 2\pi n t, \qquad \tilde{y} = \tilde{y}_0 + 2\pi l t$$

where $t \in \mathbb{R}$ and (l, n) are mutually prime with $l < n$. If we choose T parallel to the y-axis, each leaf L cuts T exactly n times and the space of leaves is the quotient of T by the action of $\mathbb{Z}_n$, which is again S^1. The convolution sum of the previous paragraph is a sum from 1 to n and algebra $\mathcal{A} = \mathcal{A}_{l/n}$ is the cross-product of the algebra of continuous or measurable functions on S^1 by the group $\mathbb{Z}_n$. Let $|j\rangle$ be a basis of the square summable functions on $\mathbb{Z}_n$ and identify $|n\rangle$ with $|0\rangle$ as we did in (3.1.24). Then $\mathcal{A}_{l/n}$ is generated by the functions u and v defined by

$$u|j\rangle = q^y|j\rangle, \qquad v|j\rangle = |j + 1\rangle$$

where we have set as usual $q = e^{2\pi i l/n}$. Since $v^n = 1$ we see that $\mathcal{A}_{l/n}$ can be identified with $L^2(S^1) \otimes M_n$. If we take the limit $n \to \infty$ so that $l/n \to \alpha$, an irrational number, then the set of points where L cuts T becomes dense in T. In the limit it can be shown that $\mathcal{A} = \mathcal{A}_\alpha$ is the cross-product of the algebra of continuous or measurable functions on S^1 by the integers; it is the Weyl algebra with $q = e^{2\pi i \alpha}$. $\qquad\square$

Example 4.54 The algebra $\mathcal{A}_\alpha$ of the previous example has two special derivations e_1 and e_2, formally identical to those (2.1.26) of the ordinary torus, and a unique trace τ which is invariant under the action of the derivations and which is normalized so that $\tau(1) = 1$. By the GNS construction it can be identified as an algebra of operators acting on itself by left multiplication and completed to form a factor of type II$_1$. This we saw in Example 4.44. Using the same notation, we embed the algebra of functions on the unit circle S^1 into $\mathcal{A}_\alpha$ by $z \mapsto u$. The derivations defined by Equations (4.1.1) become inner in the enlarged algebra of unbounded operators. If we write $z = e^{2\pi i \alpha t}$ then to be consistent with the notation of Chapter 3 we find that we must set

$$2\pi \alpha t = r^{-1} x, \qquad 2\pi \alpha = r^{-2} k.$$

Were it rational, the inverse of the parameter α would be the volume of the 'torus' in units of the elementary cell volume; the 'coordinate' t is considered to be without dimension. It is convenient to normalize the derivations so that the corresponding λ_a are given by

$$\lambda_1 = \frac{d}{dt}, \quad \lambda_2 = 2\pi i \alpha^{-1} t, \qquad [\lambda_1, \lambda_2] = 2\pi i \alpha^{-1}.$$

According to our general formula (3.1.10) the Dirac operator θ is given then by

$$\theta = -\frac{d}{dt} \cdot \theta^1 - 2\pi i \alpha^{-1} t \cdot \theta^2.$$

If we use the embedding to pull θ back to S^1 then

$$(\theta f)(t) = -\frac{df}{dt} \theta^1 - 2\pi i \alpha^{-1} t f(t) \theta^2.$$

We showed in Section 3.3 that the map $\psi \mapsto -\theta \psi$ defined a covariant derivative on a module associated to the algebra. The corresponding curvature $\Omega = d\theta + \theta^2$ is

$$\Omega = (-e_{[1}\lambda_{2]} + [\lambda_1, \lambda_2]) \theta^1 \theta^2 = -2\pi i \alpha^{-1} \theta^1 \theta^2$$

and the K_{12} introduced in Section 3.1 is given by

$$K_{12} = 2\pi i \alpha^{-1}.$$

For arbitrary smooth $f(t)$ and $g(t)$ we define the element $e \in \mathcal{A}_\alpha$ by

$$e = f(u) + v g(u) + g(u) v^{-1}.$$

One can show that for appropriate explicit choices e is a projector and that

$$\mathrm{Tr}(e) = \alpha.$$

This can be compared with the discussion given below in Section 5.2. Furthermore Equation (5.1.8) holds. The integer is obtained as a product of a curvature times an 'area'. This is also the structure of Equation (3.1.27). □

Example 4.55 Another noncommutative algebra which can be associated to a set of foliations is the *quantum disc*, the C^*-algebra generated by an element z and its adjoint z^* which satisfy the commutation relations

$$z^* z - q z z^* = 1 - q, \qquad q < 1.$$

On the Hilbert space $\mathcal{H} = L^2(S^1)$ this algebra has representations of dimension 1, given by the action

$$(z\psi)(\alpha) = e^{i\theta}\psi(\alpha), \qquad 0 \leq \theta \leq \pi$$

of z on $\psi(\alpha) \in \mathcal{H}$. There is also a representation of infinite dimension given by the action

$$z|k\rangle = \sqrt{1 - q^{k+1}}|k + 1\rangle, \qquad z^*|k\rangle = \sqrt{1 - q^k}|k - 1\rangle$$

on the basis vectors $|k\rangle = e^{ik\alpha}$ of $\mathcal{H}$.   □

In the examples we have given in this section the representation of the algebra has a trivial extension to the differential calculus. In none of them does the existence of the calculus place restrictions on the algebra. This is not an entirely satisfactory result; one would prefer that the existence of the extension place restrictions at least on the representation of the algebra. In ordinary geometry the existence of the de Rham differentials forces the structure algebra to be the algebra of smooth functions. By analogy one is forced to conclude that the algebra $\mathbb{R}_q^1$ is such that every representation of it is 'smooth'. From this point of view the construction given below in Section 5.3 is superior. The algebra even is defined in terms of the differential calculus.

Example 4.56 In Chapter 5 we shall introduce the notion of a p-summable Fredholm module $(\mathcal{H}, F)$. We shall define then the algebra $\mathcal{A}$ to be the subset of the even elements of $\mathcal{B}(\mathcal{H})$ such that the commutator $[f, a]$ lies in a predetermined ideal $\mathcal{L}^p$. One defines a differential by the formula

$$da = i[F, a]$$

where by assumption d maps $\mathcal{A}$ into the odd elements of $\mathcal{L}^p$. This is the restriction to $\mathcal{A}$ of Equation (5.3.4). The above construction is a noncommutative generalization of a possible definition of the de Rham differential on compact manifolds with metrics of euclidean signature, using the Dirac operator. Compare with Example 2.10. It forms the basis of the more mathematical approaches to noncommutative geometry. The set $\{\mathcal{A}, \mathcal{H}, F\}$ is called a K-*cycle* or *spectral triple*. It is discussed briefly in Chapter 5. In Example 3.9 we constructed the spectral triple $\{M_2^+, \mathbb{C}^2, i\eta\}$. In this case, as in every finite-dimensional case, $\mathcal{L}^p = \mathcal{B}(\mathcal{H})$ for all p.   □

We terminate this section with an indication of the applications of the theory of von Neumann algebras to quantum field theory.

Example 4.57 Return to the Example 4.40 and consider the anti-linear map of $\mathcal{M}$ into itself given by complex conjugation. Define the anti-linear *modular conjugation* operator J by

$$J\hat{a}|0\rangle = \hat{a}^*|0\rangle, \qquad J(ca) = \bar{c}Ja, \qquad c \in \mathbb{C}.$$

Obviously $J^2 = 1$. More important, because the state ω is a trace state, we have $\omega(ab) = \omega(ba)$ and therefore

$$\langle 0|, (J\hat{a})^* J\hat{a}|0\rangle = \omega(\hat{a}\hat{a}^*) = \omega(\hat{a}^*\hat{a}) = \langle 0|, \hat{a}^*\hat{a}|0\rangle.$$

Therefore J is anti-unitary:

$$J^*J = 1.$$

Since we defined $\hat{a}$ by left multiplication $J\hat{a}$ will be defined by right multiplication and we find

$$J\hat{a}|0\rangle = J(a \otimes 1)|0\rangle = 1 \otimes a^*|0\rangle.$$

The a^* here is considered to act from the right. One can consider J then to be the anti-linear extension of the flip

$$J|j, k\rangle = |k, j\rangle$$

and from this one sees that

$$J\mathcal{M}J = \mathcal{M}'. \tag{4.3.42}$$

□

Let now $\mathcal{M}$ be a factor in standard form, with a cyclic and separating state vector $|0\rangle$. Define an (anti-linear) operator S by

$$Sa|0\rangle = a^*|0\rangle.$$

In general S is not unitary. However we can define $\Delta = S^*S$ and since S^*S is a positive operator we can take the square root and write $S = J\Delta^{1/2}$ for some anti-unitary operator J. This is the *polar decomposition* of S. If the state is of the form (4.3.23) for some invertible ρ then one sees that

$$\Delta = \rho \otimes \rho^{-1}.$$

In fact from (4.3.24) it follows that

$$S\hat{a}|0\rangle = J\Delta^{1/2}\hat{a}|0\rangle = J\sum_{j,k=1}^{n} \sqrt{\rho_k} a_{jk}|k, j\rangle = \sum_{j,k=1}^{n} \sqrt{\rho_k} \bar{a}_{jk}|j, k\rangle = \hat{a}^*|0\rangle.$$

The *Tomita-Takesaki theorem* states that (4.3.42) remains true and that

$$U(t)^*\mathcal{M}U(t) = \mathcal{M}, \qquad U(t) = \Delta^{-it}, \quad t \in \mathbb{R}. \tag{4.3.43}$$

The operator Δ is called the *modular operator*. We saw in the Example 4.57 that it reduces to the identity if the state vector defines a trace state. In general it is unbounded. The automorphism defined by $U(t)$ plays an important role in the finer classification of von Neumann algebras, especially those of type III, and an important group one can form is the set of values of t for which it is an inner automorphism. We mention only that if this group reduces to the identity then the factor is a special type-III factor called III$_1$. This means in particular that every non-vanishing idempotent is equivalent to the identity, with $\tau[1] = \infty$. There is a unique hyperfinite factor of type III$_1$. In Example 4.42 we outlined the construction of some of the hyperfinite types (but not that of III$_1$). For finite n the modular automorphism is always inner since the algebra is isomorphic to a matrix algebra. It generically does not remain so in the thermodynamic limit.

Example 4.58 Let H be a hamiltonian and $\mathcal{M}$ an algebra of observables. Then the time evolution of an element $a \in \mathcal{M}$ is given by the formula

$$a \mapsto U(t)^* a U(t), \qquad U(t) = e^{-iHt}.$$

If the state ω is a thermal state then, because of the cyclic properties of the trace, for any observables a and b one sees that

$$\omega(U(t)^* a U(t) b) = \omega(b U(t + i\beta)^* a U(t + i\beta)).$$

This is the KMS (Kubo-Martin-Schwinger) condition. It implies that physical observables in a thermal bath are periodic in imaginary time with period β. We mentioned in Example 4.42 the thermodynamic limit of an n-particle spin system, in particular one in a thermal state. The Tomita-Takesaki theorem permits one in principle to generalize the KMS condition to these more general states.   ☐

Example 4.59 One of the basic assumptions of quantum field theory in flat space-time is *locality*: the values of a scalar field at two space-like-separated points are commuting operators. Let $\mathcal{U}$ be an open region of space-time and denote by $\mathcal{U}'$ the set of all points which are space-like separated from all points of $\mathcal{U}$. Consider an arbitrary quantum field ϕ, the free scalar field for example, and set

$$\phi(f) = \int \phi(x) f(x)$$

where f is a smooth function with support in $\mathcal{U}$. Then $\phi(f)$ is an unbounded operator on a Hilbert space $\mathcal{H}$. Let $\mathcal{M}(\mathcal{U})$ be the von Neumann algebra generated by all such field configurations. One assumes that $\mathcal{M}(\mathcal{U})$ is a factor in standard form with the vacuum $|0\rangle$ as cyclic and separating vector: $\mathcal{M}(\mathcal{U})|0\rangle$ is dense in $\mathcal{H}$ and $\mathcal{M}(\mathcal{U})$ contains no element which annihilates $|0\rangle$ for any $\mathcal{U}$ with $\mathcal{U}' \neq \emptyset$. The postulate of *locality* can be formulated as

$$\mathcal{U}_2 \subset \mathcal{U}_1' \implies \mathcal{M}(\mathcal{U}_2) \subset \mathcal{M}(\mathcal{U}_1)'.$$

A stronger condition is the assumption of *Haag duality*: $\mathcal{M}(\mathcal{U}) = \mathcal{M}(\mathcal{U}')'$. If $\mathcal{U}$ is a bounded region enclosed by an advanced and retarded light cone then it can be shown that $\mathcal{M}(\mathcal{U})$ is the unique hyperfinite factor of type III_1. If $\mathcal{U}$ is all of Minkowski space then the corresponding algebra is invariant under the action of the Poincaré group. It is the set of all bounded operators on $\mathcal{H}$, a factor of type I_∞. Because of this universal behaviour the algebra type contains no information about the particle content or the interaction hamiltonian. In fact it does not even depend on the dimension of space-time. This could be considered as a motivation for introducing a differential structure over the algebra of observables quite independently of any consideration of 'noncommutative space-time'. Especially if formulated as in Example 4.56 a differential structure would open the possibility of a finer analysis of the algebra of observables than the 'measure-theoretic' one we have just briefly outlined.   □

4.4   Quantum groups

The examples of differential calculi we gave in Chapter 3 are invariant under the action of a Lie group. We shall see in the last chapter that the invariance gives rise in fact to a multiplet of gauge potentials and Higgs fields. This is perhaps not entirely justified since symmetries in physics have a tendency to be broken. It could be argued that if one is to 'make space noncommutative' one should do the same with the invariance group. This logic leads naturally to the notion of a *quantum group*. Quantum groups are examples of noncommutative geometries and they can be studied as such in their own right. Here our interest in them lies in the fact that they act on particular deformations of ordinary euclidean space called quantum spaces examples of which we have already introduced in Section 4.1.

If a compact manifold is also a Lie group then the algebra of functions defined on it has additional structure. Let e be the identity element. From a function $f(x)$ of one variable there is a natural way of constructing a function $(\Delta f)(x, y) = f(xy)$ of two variables, there is a natural map $\epsilon : f \mapsto f(e)$ of the algebra into the complex numbers and there is a natural map $S(f)(x) = f(x^{-1})$ of the algebra into itself. These operations define respectively the *coproduct*, the *counit* and the *antipode*. The group operation on the manifold yields thus additional operations on the algebra of functions which make it into a *Hopf algebra*. If the group is not commutative then the coproduct is not commutative. The group analogue of a 'noncommutative manifold' then is a general Hopf algebra whose product also is noncommutative. Since the expression 'noncommutative group' had been already used to designate something different, these general Hopf algebras have been referred to as 'quantum groups', but they are not groups and the noncommutative structure is not necessarily related to quantum mechanics.

A Hopf algebra structure can be thought of as the minimal additional structure which is necessary to define a product of two representations of an associative algebra.

Let π_1 and π_2 be two representations of the elements of the algebra $\mathcal{A}$ as linear operators on the vector spaces $\mathcal{H}_1$ and $\mathcal{H}_2$ and suppose that one wishes to construct a product $\pi = \pi_1 \odot \pi_2$ of π_1 and π_2, a representation of $\mathcal{A}$ on $\mathcal{H}_1 \otimes \mathcal{H}_2$. Two possibilities immediately come to mind. One could define π by the equation $\pi(a) = \pi_1(a) \otimes \pi_2(a)$, for $a \in \mathcal{A}$. This definition would not yield a linear representation; if α is a complex number then $\pi(\alpha a) = \alpha^2 \pi(a)$. One could also dxefine π by $\pi(a) = \pi_1(a) \otimes 1 + 1 \otimes \pi_2(a)$. This is not a representation of the associative algebra structure since

$$\pi(a)\pi(b) - \pi(ab) = \pi_1(a) \otimes \pi_2(b) + \pi_1(b) \otimes \pi_2(a)$$

but it would at least yield a representation of the underlying Lie-algebra structure since $[\pi(a), \pi(b)] = \pi([a,b])$. Suppose however that $\mathcal{A}$ is equipped with a coproduct $\Delta : \mathcal{A} \to \mathcal{A} \otimes \mathcal{A}$. One can then define the product $\pi = \pi_1 \odot \pi_2$ by

$$\pi_1 \odot \pi_2 = (\pi_1 \otimes \pi_2) \circ \Delta.$$

The restrictions which we shall place below on the definition of Δ are just what are required to make π into a sensible representation of $\mathcal{A}$. From the point of view of representation theory a Hopf algebra is a natural generalization of the algebra of functions on a group.

As part of the formal definition of a Hopf algebra it is useful to recall that an associative algebra $\mathcal{A}(m, i)$ over the complex numbers is a complex vector space with an associative *product m*

$$\mathcal{A} \otimes \mathcal{A} \xrightarrow{\ m\ } \mathcal{A}$$

and a *unit i*

$$\mathbb{C} \xrightarrow{\ i\ } \mathcal{A}.$$

Only if $\mathcal{A}$ is commutative is m an algebra homomorphism. Let I designate the identity map. Define $m \otimes I$ and $I \otimes m$ by the maps

$$(\mathcal{A} \otimes \mathcal{A}) \otimes \mathcal{A} \xrightarrow{m \otimes I} \mathcal{A} \otimes \mathcal{A}, \qquad \mathcal{A} \otimes (\mathcal{A} \otimes \mathcal{A}) \xrightarrow{I \otimes m} \mathcal{A} \otimes \mathcal{A}.$$

Then the associativity law can be written as

$$m \circ (I \otimes m) = m \circ (m \otimes I).$$

The fact that i is the unit is expressed by the relations

$$m \circ (I \otimes i) = m \circ (i \otimes I) = I.$$

A *coalgebra* $\mathcal{A}(\Delta, \epsilon)$ over the complex numbers is a complex vector space with an associative coproduct Δ

$$\mathcal{A} \xrightarrow{\ \Delta\ } \mathcal{A} \otimes \mathcal{A}$$

and a counit ϵ

$$A \xrightarrow{\epsilon} \mathbb{C}.$$

Both Δ and ϵ are algebra homomorphisms. If we define $\Delta \otimes I$ and $I \otimes \Delta$ as above then the coassociativity law can be written as

$$(I \otimes \Delta) \circ \Delta = (\Delta \otimes I) \circ \Delta.$$

The fact that ϵ is the counit is expressed by the relations

$$(I \otimes \epsilon) \circ \Delta = (\epsilon \otimes I) \circ \Delta = I.$$

Define a map τ of $A \otimes A \otimes A \otimes A$ into itself by

$$\tau(a \otimes b \otimes c \otimes d) = a \otimes c \otimes b \otimes d.$$

A *bialgebra* $A(m, i, \Delta, \epsilon)$ over the complex numbers is a complex vector space which is an algebra and a coalgebra such that the compatibility conditions

$$\begin{aligned} \Delta \circ m = (m \otimes m) \circ \tau \circ (\Delta \otimes \Delta), \quad & \epsilon \circ m = m \circ (\epsilon \otimes \epsilon), \\ \Delta \circ i = i \otimes i, \quad & \epsilon \circ i = I \end{aligned} \qquad (4.4.1)$$

are satisfied. These conditions are easily seen to be satisfied if A is the algebra of smooth functions on a group. For example to understand the presence of the permutation τ consider $f, g \in A$. The tensor product of f and g is the function of 2 variables defined by $f \otimes g(x, y) = f(x)g(y)$. Therefore $m(f \otimes g)(x) = f(x)g(x)$ and

$$\Delta \circ m(f \otimes g)(x, y) = f(xy)g(xy). \qquad (4.4.2)$$

On the other hand $\Delta \otimes \Delta(f \otimes g)(x, y, z, t) = f(xy)g(zt)$ and therefore $\tau \circ \Delta \otimes \Delta(f \otimes g)(x, y, z, t) = f(xz)g(yt)$. Therefore

$$(m \otimes m) \circ \tau \circ (\Delta \otimes \Delta)(f \otimes g)(x, y) = f(xy)g(xy). \qquad (4.4.3)$$

Equations (4.4.2) and (4.4.3) imply the first of Equations (4.4.1).

A Hopf algebra $A(m, i, \Delta, \epsilon, S)$ is a bialgebra with an antipode, a linear map

$$A \xrightarrow{S} A$$

of A into itself which satisfies the compatibility relations

$$m \circ (I \otimes S) \circ \Delta = m \circ (S \otimes I) \circ \Delta = i \circ \epsilon.$$

One readily verifies that the algebra of smooth functions on a Lie group is a Hopf algebra with a commutative product.

Consider the group $G = SL(n, \mathbb{C})$. An element $x \in G$ can be written in the form of a matrix $x = (a^i_j)$ with determinant equal to 1 and the algebra $\mathcal{A}$ generated by the functions $a^i_j(x)$ is dense in the algebra of smooth functions on G. From the general discussion $\mathcal{A}$ is a Hopf algebra with coproduct given by

$$(\Delta a^i_j)(x, y) = a^i_j(xy) = \sum_k a^i_k(x) a^k_j(y) = \sum_k a^i_k \otimes a^k_j(x, y).$$

That is we have

$$\Delta a^i_j = \sum_k a^i_k \otimes a^k_j. \tag{4.4.4}$$

The counit is given by

$$\epsilon(a^i_j) = \delta^i_j \tag{4.4.5}$$

and the inverse matrix defines the antipode:

$$S(a^i_j) = (a^{-1})^i_j. \tag{4.4.6}$$

Suppose now that the a^i_j are the n^2 basis elements of a vector space. The tensor algebra $\mathcal{T}$ generated by the a^i_j can be made into a bialgebra by defining the coproduct and counit by (4.4.4) and (4.4.5). The coproduct of an arbitrary element of $\mathcal{T}$ is obtained by using the fact that Δ is an algebra homomorphism. For example

$$\Delta(a^i_j \otimes a^k_l) = \sum_{p,q} a^i_p \otimes a^p_j \otimes a^k_q \otimes a^q_l.$$

There is however no antipode. Let $\hat{R}^{i_1 \ldots i_r}{}_{j_1 \ldots j_s}$ be a set of complex numbers, where r and s are arbitrary integers greater than or equal to zero. Consider the quotient algebra $\mathcal{A}$ of $\mathcal{T}$ defined by the relations

$$\hat{R}^{i_1 \ldots i_r}{}_{k_1 \ldots k_s} a^{k_1}_{j_1} \otimes \cdots \otimes a^{k_s}_{j_s} = a^{i_1}_{k_1} \otimes \cdots \otimes a^{i_r}_{k_r} \hat{R}^{k_1 \ldots k_r}{}_{j_1 \ldots j_s}. \tag{4.4.7}$$

Then $\mathcal{A}$ is also a bialgebra with the coproduct (4.4.4) and counit (4.4.5). There is still in general no antipode.

Suppose that $r = s = 2$ and define

$$R^{ij}{}_{kl} = \hat{R}^{ji}{}_{kl}.$$

In this case the matrix $\hat{R}$ is known as the *braid matrix*. Consider the case

$$R^{ij}{}_{kl} = \delta^i_k \delta^j_l, \qquad \hat{R}^{ij}{}_{kl} = \delta^j_k \delta^i_l. \tag{4.4.8}$$

Then (4.4.7) is equivalent to the relation

$$a^i_j a^k_l = a^k_l a^i_j$$

and the algebra $\mathcal{A}$ is the polynomial algebra in the a_j^i. The product here is the product in $\mathcal{A}$ induced from the tensor product in $\mathcal{T}$.

Define the determinant $\det(a_j^i)$ by the usual formula. It is an element of $\mathcal{A}$ which is a linear combination of products of n generators. Require that the elements of $\mathcal{A}$ satisfy the additional relation

$$\det(a_j^i) = 1.$$

The inverse of the matrix (a_j^i) can be defined then as an element of $\mathcal{A}$. We can define an antipode by (4.4.6) and $\mathcal{A}$ is a Hopf algebra. It can be identified with the algebra of functions on $SL(n, \mathbb{C})$.

Example 4.60 Let G be a finite group with n elements. The algebra $\mathcal{C}(G)$ of complex functions on G is the product of n copies of $\mathbb{C}$. The group structure of G is entirely encoded in the coalgebra structure of $\mathcal{C}(G)$. In general the algebra structure of $\mathcal{C}(G)$ contains only information concerning G as a topological set or manifold. $\square$

A difficult problem is to find other forms of $R^{ij}{}_{kl}$ which yield Hopf algebras with a noncommutative product. We shall not attempt a general discussion of this problem but we shall exhibit a family of deformations of (4.4.8) which yield Hopf algebras $SL_q(n, \mathbb{C})$ called the quantum deformations of $SL(n, \mathbb{C})$. It is convenient for this to adopt a matrix notation. Consider $R^{ij}{}_{kl}$ as an $n^2 \times n^2$ matrix by grouping the indices i, j and k, l. Define the $n^2 \times n^2$ matrices A_1 and A_2 by

$$A_1{}^{pq}_{rs} = a_r^p \delta_s^q, \qquad A_2{}^{pq}_{rs} = \delta_r^p a_s^q.$$

Then $(A_1 A_2)^{pq}_{rs} = a_r^p \otimes a_s^q$ and, with $r = s = 2$, Equation (4.4.7) can be written as an equation for $n^2 \times n^2$ matrices:

$$RA_1 A_2 = A_2 A_1 R. \tag{4.4.9}$$

In this representation (4.4.8) is the identity matrix. Introduce the $n \times n$ matrices e_{ij} which have 1 in the intersection of the i-th row and j-th column and zeros everywhere else. Then (4.4.8) considered as an $n^2 \times n^2$ matrix equation can be written

$$R = \sum_{i,j} e_{ii} \otimes e_{jj}.$$

Define for each complex number $q \neq 0$ the R-matrix

$$R_q = q \sum_i e_{ii} \otimes e_{ii} + \sum_{i \neq j} e_{ii} \otimes e_{jj} + (q - q^{-1}) \sum_{i>j} e_{ij} \otimes e_{ji}.$$

The determinant

$$\det_q(a_j^i) = \sum_\sigma (-q)^{|\sigma|} a_{\sigma(1)}^1 \cdots a_{\sigma(n)}^n \tag{4.4.10}$$

lies in the center of $\mathcal{A}$ by virtue of the relations (4.4.9). The sum is taken over all permutations σ and $|\sigma|$ is the parity of σ. The quotient algebra of $\mathcal{T}$ defined by the relations (4.4.9) and the relation

$$\det_q(a_j^i) = 1 \qquad (4.4.11)$$

is the Hopf algebra $SL_q(n, \mathbb{C})$.

The R-matrix acts on the 2-fold tensor product of the original vector space. Introduce the matrix R_{12} which acts on the 3-fold tensor product by R_q on the first 2 factors and the identity on the third. Introduce similarly R_{13} and R_{23}. Then R_q satisfies the equation

$$R_{12}R_{13}R_{23} = R_{23}R_{13}R_{12}. \qquad (4.4.12)$$

This is the *Yang-Baxter equation* without its spectral parameter. The Yang-Baxter equation arises as one of the integrability conditions for the solution of inverse scattering problems. Written in terms of the braid matrix $\hat{R}_q$ Equation (4.4.12) becomes the *braid relation*

$$\hat{R}_{12}\hat{R}_{23}\hat{R}_{12} = \hat{R}_{23}\hat{R}_{12}\hat{R}_{23}. \qquad (4.4.13)$$

We have already encountered this relation in Section 3.7 and we shall encounter it again in Section 6.1.

Example 4.61 Let $\mathcal{A}$ be a Hopf algebra and $\mathcal{M}$ a *Hopf module*. Write the action of the coproduct Δ of $\mathcal{A}$ as

$$\Delta(a) = a_{(1)} \otimes a_{(2)}$$

Here we have used what is called the *Sweedler notation*; the right-hand side is actually a short-hand notation for a sum. We have then

$$\Delta(a)(f \otimes g) = a_{(1)} \otimes a_{(2)}(f \otimes g) = a_{(1)}(f) \otimes a_{(2)}(g), \qquad f, g \in \mathcal{M}.$$

If $\mathcal{M}$ is also an algebra then the action of $\mathcal{A}$ will be compatible with the product m if one imposes the condition

$$am = m \circ \Delta(a) \qquad (4.4.14)$$

on the coproduct Δ. Using the Sweedler notation this becomes

$$a(fg) = a_{(1)}(f)a_{(2)}(g).$$

A coproduct Δ' is related to a coproduct Δ by a *twist* if there is an element $\tau \in \mathcal{A} \otimes \mathcal{A}$ such that

$$\Delta'(a) = \tau^{-1} \circ \Delta(a) \circ \tau.$$

Below in Example 4.69 we mention a cohomological description of this condition. If the action of $\mathcal{A}$ is to remain compatible with the product of $\mathcal{M}$ after the twist then the latter must also be deformed to a product m' such that

$$am' = m' \circ \tau^{-1} \circ \Delta(a) \circ \tau,$$

a condition which can be written in the form

$$am' \circ \tau^{-1} = m' \circ \tau^{-1} \circ \Delta(a),$$

which coincides with the previous condition (4.4.14) if one choose

$$m' = m \circ \tau.$$

The Hopf algebra $\mathcal{A}$ generated by the differential operators (∂_x, ∂_y) acts on the algebra $\mathcal{M} = \mathcal{C}(\mathbb{R}^2)$. The latter is a Hopf module. If one twist the coproduct of $\mathcal{A}$ by

$$\tau = e^{\frac{1}{2}ik\theta^{12}\partial_x \otimes \partial_y}$$

then in order to remain a Hopf module, the product in $\mathcal{M}$ must the deformed to the $*$-product given in Example 4.23. □

Example 4.62 Consider two relations of the form (4.4.7) with respectively $r = 2$, $s = 0$ and $r = 0$, $s = 2$. Rewrite them as

$$Q_{ij}a_k^i a_l^j = Q_{kl}, \qquad Q^{ij}a_i^k a_j^l = Q^{kl}.$$

Suppose that $\det Q_{ij} \neq 0$ and that $Q^{ij} = (Q_{ij})^{-1}$. Then the quotient algebra $\mathcal{A}$ of $\mathcal{T}$ with respect to these relations is a Hopf algebra. The antipode is given by

$$S(a_j^i) = Q^{ik}Q_{lj}a_k^l.$$

The matrix $\hat{R}^{ij}_{kl} = Q^{ij}Q_{kl} + a\delta_k^i \delta_l^j$ satisfies the braid relation if and only if

$$a + a^{-1} + Q_{ij}Q^{ij} = 0.$$

□

Example 4.63 Consider $SL_q(2, \mathbb{C})$ and write

$$(a_j^i) = \begin{pmatrix} a & b \\ c & d \end{pmatrix}.$$

If we reorder the indices $(11, 12, 21, 22) = (1, 2, 3, 4)$ then the matrices R_q and $\hat{R}_q$ can be written as 4×4 matrices

$$R_q = \begin{pmatrix} q & 0 & 0 & 0 \\ 0 & 1 & 0 & 0 \\ 0 & q - q^{-1} & 1 & 0 \\ 0 & 0 & 0 & q \end{pmatrix}, \qquad \hat{R}_q = \begin{pmatrix} q & 0 & 0 & 0 \\ 0 & q - q^{-1} & 1 & 0 \\ 0 & 1 & 0 & 0 \\ 0 & 0 & 0 & q \end{pmatrix}.$$

The braid matrix is of the form of the previous example with

$$Q_{ij} = \begin{pmatrix} 0 & -1 \\ q & 0 \end{pmatrix}.$$

If we set $x^i = (x, y)$ then the relation which defines the algebra of Example 4.3 can be written in the form

$$Q_{ij}x^i x^j = 0, \tag{4.4.15}$$

an equation from which one can deduce the fact that the Hopf algebra $SL_q(2, \mathbb{C})$ coacts on the q-deformed plane. In fact if $x^{i'} = a_k^i \otimes x^k$ then

$$Q_{ij}x^{i'}x^{j'} = Q_{ij}a_k^i a_l^j \otimes x^k x^l = Q_{kl}x^k x^l = 0.$$

The matrix $\hat{R}_q$ has eigenvalues q and $-q^{-1}$ and satisfies the *Hecke relation*

$$(\hat{R}_q - q)(\hat{R}_q + q^{-1}) = 0. \tag{4.4.16}$$

From Equation (4.4.9) follow the commutation relations

$$\begin{aligned} ab &= qba, \quad ac = qca, \quad bc = cb, \\ bd &= qdb, \quad cd = qdc, \quad ad - da = (q - q^{-1})bc. \end{aligned} \tag{4.4.17}$$

The product here is the tensor product. From (4.4.10) we find that the determinant is given as

$$\det{}_q(a_j^i) = ad - qbc \ (= 1).$$

With the condition (4.4.11) the inverse of the matrix (a_j^i) is given by

$$(a_j^i)^{-1} = \begin{pmatrix} d & -q^{-1}b \\ -qc & a \end{pmatrix}.$$

$\square$

Example 4.64 Consider $SL_q(2, \mathbb{C})$ and set

$$\bar{a} = d, \qquad \bar{c} = -q^{-1}b.$$

Suppose that the bar extends to an involution of the algebra. Then q must be real. From the condition (4.4.10) on the determinant and the commutation relations (4.4.17) we see that

$$a^{*i}_{\ j}a_k^j = a_j^i a^{*j}_{\ k} = \delta_k^i.$$

This Hopf algebra is known as $SU_q(2)$.

$\square$

Example 4.65 Consider the braid matrix $\hat{R}^{ij}{}_{kl}$ defined by

$$\hat{R}_h = \begin{pmatrix} 1 & -h & h & h^2 \\ 0 & 0 & 1 & h \\ 0 & 1 & 0 & -h \\ 0 & 0 & 0 & 1 \end{pmatrix}. \tag{4.4.18}$$

If we introduce

$$a^i_j = \begin{pmatrix} a & b \\ c & d \end{pmatrix}$$

then they are the generators of the *jordanian deformation* $SL_h(2,\mathbb{C})$ of $SL(2,\mathbb{C})$ if (4.4.7) with $r = s = 2$ is satisfied, or equivalently if the associated R-matrix satisfies Equation (4.4.9). The braid matrix is of the form of Example 4.62 with

$$Q_{ij} = \begin{pmatrix} 0 & -1 \\ 1 & h \end{pmatrix}.$$

If we set $x^i = (x, y)$ then the relation which defines the algebra of Example 4.4 can be written in the form (4.4.15), an equation from which one can now deduce the fact that the Hopf algebra $SL_h(2,\mathbb{C})$ coacts on the h-deformed plane. □

Example 4.66 We refer to the literature for the precise expression of the braid matrix which defines the Hopf algebras $SO_q(n)$ as well as for the definition of the q-deformed euclidean metric. Suffice it to remark here that $\hat{R}$ admits the projector decomposition

$$\hat{R} = qP_s - q^{-1}P_a + q^{1-n}P_t \tag{4.4.19}$$

where the P_s, P_a, P_t are $SO_q(n)$-covariant q-deformations of respectively the symmetric trace-free, antisymmetric and trace projectors. They are mutually orthogonal and their sum is equal to the identity:

$$P_s + P_a + P_t = 1.$$

The trace projector is 1-dimensional and its matrix elements can be written in the form

$$P_t{}^{ij}{}_{kl} = (g^{mn}g_{mn})^{-1}g^{ij}g_{kl} \tag{4.4.20}$$

where g_{ij} is the q-deformed euclidean metric. The braid matrix and the q-deformed metric satisfy the relations

$$g_{il}\,\hat{R}^{\pm 1\,lh}{}_{jk} = \hat{R}^{\mp 1\,hl}{}_{ij}\,g_{lk}, \qquad g^{il}\,\hat{R}^{\pm 1\,jk}{}_{lh} = \hat{R}^{\mp 1\,ij}{}_{hl}\,g^{lk}. \tag{4.4.21}$$

The introduction of a new generator Λ is necessary in the inhomogeneous extension of the homogeneous quantum groups $SL_q(n)$, $SO_q(n)$ and q-Lorentz; more precisely, Λ appears in the coproduct of the translation generators. □

Example 4.67 Several q-deformed versions of the Lorentz and Poincaré groups have been proposed, some based on the braid matrix mentioned in Example 4.66 (with $n = 4$) and some based on a deformation of the de Sitter group. We shall make a few remarks concerning a q-deformed version of Minkowski space later in Example 7.10. □

In physical applications an important role is played by the representations of the Lie algebra of a group. A Lie algebra g can be embedded in an associative algebra $\mathcal{U} = \mathcal{U}(g)$ called its *enveloping algebra* which has the property that any representation of g in another associative algebra $\mathcal{A}$ can be extended to a morphism of $\mathcal{U}$ into $\mathcal{A}$. One can define $\mathcal{U}$ to be the quotient algebra of the tensor algebra $\mathcal{T}(g)$ over g by the ideal generated by the relations

$$R(X, Y) = [X, Y] - X \otimes Y + Y \otimes X, \qquad X, Y \in g.$$

Of special interest here is the fact that $\mathcal{U}$ is a Hopf algebra with coproduct, counit and antipode obtained from (4.4.4), (4.4.5) and (4.4.6) by setting $a^i_j = \delta^i_j + \epsilon X^i_j$ and retaining terms linear in ϵ. The coproduct is commutative. A quantum deformation can be defined also as a deformation $\mathcal{U}_q$ of $\mathcal{U}$. This is referred to as the dual point of view since a left-invariant vector field on a group can be considered as an element of the dual algebra to the algebra of functions on the group. In general the coproduct of $\mathcal{U}_q$ will not be commutative.

Example 4.68 A left-invariant vector field on a Lie group can be considered as an element of the dual algebra to the algebra of functions on the group. Let $\mathcal{A}$ be the algebra of functions on a group. Let L be a linear map of $\mathcal{A}$ into $\mathcal{A}$ with kernel $L(x, y)$ and let ω be an element of the dual $\mathcal{A}^*$ to $\mathcal{A}$ with kernel $\omega(x)$. The map L is left-invariant if $L(x, y) = L(zx, zy)$ for all elements z of the group. We can define then $\omega \in \mathcal{A}^*$ by $\omega(x) = L(e, x)$. Conversely L defined by $L(x, y) = \omega(x^{-1}y)$ is left-invariant. Let dx be the Haar measure on the group. Since

$$Lf(y) = \int L(y, x) f(x) dx = \int \omega(y^{-1}x) f(x) dx$$

$$= \int \omega(x) f(yx) dx = (I \otimes \omega) \circ \Delta f(y)$$

and

$$\omega(f) = \int \omega(x) f(x) dx = \int L(e, x) f(x) dx = \epsilon \circ Lf$$

the relation between ω and L can be written using the Hopf-algebra structure of $\mathcal{A}$ as

$$L = (I \otimes \omega) \circ \Delta, \qquad \omega = \epsilon \circ L.$$

The condition that L be left-invariant can be written

$$\Delta \circ L = (I \otimes L) \circ \Delta.$$

$\square$

Example 4.69 In Example 2.23 we discussed the deformations of a Lie algebra g and we mentioned that semi-simple Lie algebras have no nontrivial deformations. They have vanishing 1- and 2-cohomology. A deformation of g will entail a deformation of $\mathcal{U}$ and since the latter remains an enveloping algebra the deformed coproduct will remain commutative. What is deformed in fact is the product structure of $\mathcal{U}$. The quantum deformation $\mathcal{U}_q$ is the enveloping algebra of a Lie algebra but the coproduct is also deformed. It is no longer commutative. The quantum deformations of a semi-simple algebra remain however also in a sense trivial. Let Δ' be a first-order deformation of a coproduct Δ which we write as $\Delta' = \Delta + \epsilon t$ where t is a map of $\mathcal{U}$ into $\mathcal{U} \otimes \mathcal{U}$. In particular then t is a 1-cochain of g with values in $\mathcal{U} \otimes \mathcal{U}$. If we require that the product remain unchanged then there is a consistency condition

$$\Delta'([X, Y]) = [\Delta'(X), \Delta'(Y)]$$

which yields the condition that t be a cocycle. It defines therefore an element of $H^1(g; \mathcal{U} \otimes \mathcal{U})$ as it is defined in Section 2.3. The action of g on $\mathcal{U} \otimes \mathcal{U}$ is the adjoint action defined using Δ. If t is a coboundary then $t = \delta r$, where r is an element of $\mathcal{U} \otimes \mathcal{U}$ and

$$\Delta' = \Delta - \epsilon[r, \Delta].$$

The modification of the coproduct can be absorbed into the transformation r. $\square$

Representations of $\mathcal{U}_q$ can be used for example to construct solutions to problems in statistical mechanics and in conformal field theory. When $R = 1$ the relations (4.4.9) define Bose statistics. When $R \neq 1$ they define braid statistics, which have been proposed as alternative statistics for particles in 2-dimensional systems.

Example 4.70 Let $(J_\pm, J_3)$ be the standard basis of the Lie algebra $\underline{sl}_2(\mathbb{C})$ considered in Example 3.3. For any operator F define

$$[F]_q = \frac{q^F - q^{-F}}{q - q^{-1}}.$$

Then the corresponding deformation $\underline{sl}_q(2, \mathbb{C})$ of the universal enveloping algebra can be defined by the relations

$$[J_3, J_\pm] = \pm J_\pm, \qquad [J_+, J_-] = [2J_3]_q. \qquad (4.4.22)$$

The coproduct is defined by

$$\Delta(J_3) = 1 \otimes J_3 + J_3 \otimes 1, \qquad \Delta(J_\pm) = q^{-J_3} \otimes J_\pm + J_\pm \otimes q^{J_3}.$$

The counit ϵ and the antipode S are defined by

$$\epsilon(J_3) = 0, \quad \epsilon(J_\pm) = 0, \qquad S(J_3) = -J_3, \quad S(J_\pm) = -q^{\pm 1} J_\pm.$$

The q-deformed Casimir operator is given by

$$C_q(J_a) = J_+ J_- + [J_3]_q [J_3 - 1]_q.$$

There are several other forms of this q-deformation of $\underline{sl}_2(\mathbb{C})$ which can be found in the literature. Some, but perhaps not all, can be implemented by taking a representation of $\underline{sl}_2(\mathbb{C})$ like the one given in Example 4.11 and replacing the ordinary derivative ∂_x with the Jackson derivative defined by

$$[\partial]_q f(x) = \frac{f(qx) - f(q^{-1}x)}{qx - q^{-1}x}.$$

$\square$

Example 4.71 In Section 2.1 we defined a principal bundle. It is a manifold P with an action of a group G such that the orbits of the action form a manifold V. There are therefore a projection of P onto V and a map of $P \times G$ onto P which satisfy the consistency relation $(ug)g' = u(gg')$ for every $u \in P$ and $g \in G$. The noncommutative version is therefore a pair of algebras $V \subset P$ and a Hopf algebra $\mathcal{G}$ with

$$\mathcal{P} \xrightarrow{\Delta'} \mathcal{P} \otimes \mathcal{G}$$

which satisfy the consistency relation

$$(\Delta' \otimes I) \circ \Delta' = (I \otimes \Delta) \circ \Delta'.$$

The fact that the action of G is trivial on V is expressed as the condition that the elements of $\mathcal{V}$ are those elements $u \in \mathcal{P}$ such that $\Delta' u = u \otimes 1$; the fact that the action of G is free on P is expressed as the condition that the map

$$\mathcal{P} \otimes \mathcal{P} \to \mathcal{P} \otimes \mathcal{G},$$

given by $u \otimes v \mapsto \Delta'(u)(v \otimes 1)$, is surjective. $\square$

We briefly mention how the quantum plane of Example 4.13 behaves under the coaction of the associated quantum group. First we discuss the case with n generators. Let x^i be the standard coordinates of complex euclidean space $\mathbb{C}^n$ and let $\mathcal{A}$ be the algebra of polynomials in x^i. Introduce n new symbols ξ^i as the exterior derivatives of the x^i:

$$\xi^i = dx^i.$$

The algebra $\Omega^*(\mathcal{A})$ generated by the x^i and the ξ^i is dense in the algebra of smooth differential forms on $\mathbb{C}^n$. Consider the tensor algebra $\mathcal{T}$ over the $2n$-dimensional vector space spanned by the (x^i, ξ^i). Then we can consider $\Omega^*(\mathcal{A})$ as the quotient algebra of $\mathcal{T}$ defined by the relations

$$x^i x^j = x^j x^i, \qquad x^i \xi^j = \xi^j x^i, \qquad \xi^i \xi^j = -\xi^j \xi^i. \tag{4.4.23}$$

We have here written the relations in the tensor algebra as equalities in the quotient algebra. They are invariant under the action

$$x^i \mapsto a^i_j x^j, \qquad \xi^i \mapsto a^i_j \xi^j$$

of the group $SL(n, \mathbb{C})$. If $\mathcal{A}$ is the Hopf algebra of functions on $SL(n, \mathbb{C})$ then the action on $\mathbb{C}^n$ can be written as a *coaction*

$$\Omega^*(\mathcal{A}) \to \mathcal{A} \otimes \Omega^*(\mathcal{A})$$

on the algebra $\Omega^*(\mathcal{A})$. If we replace (4.4.23) by relations which are invariant under the coaction of a quantum deformation $SL_q(n, \mathbb{C})$ of $\mathcal{A}$ we are led naturally to define a q-deformed algebra of forms as we did, for instance, in Example 4.15. In general a *quantum space*, with the corresponding algebra of forms, is defined as an appropriate quotient algebra of $\mathcal{T}$ which is invariant under the coaction of a quantum group.

Consider again $\hat{R}$ as a $n^2 \times n^2$ matrix and let $f(\hat{R})$, $g(\hat{R})$ and $h(\hat{R})$ be polynomials in $\hat{R}$. From the relations (4.4.7) for a quantum group with $r = s = 2$ it follows immediately that a quantum space can be defined by relations of the form

$$f(\hat{R})^{ij}{}_{kl} x^k x^l = 0, \qquad x^i \xi^j - g(\hat{R})^{ij}{}_{kl} \xi^k x^l = 0, \qquad h(\hat{R})^{ij}{}_{kl} \xi^k \xi^l = 0. \tag{4.4.24}$$

For example, if $x'^i = a^i_j x^j$, then we have

$$f(\hat{R})^{ij}{}_{kl} x'^k x'^l = f(\hat{R})^{ij}{}_{kl} a^k_p a^l_q x^p x^q = a^i_k a^j_l f(\hat{R})^{kl}{}_{pq} x^p x^q = 0.$$

Of course the definition of ξ^i in terms of x^i will impose consistency conditions on the functions f, g and h.

Consider a quotient algebra defined by quadratic relations of the form

$$x^i x^j - B^{ij}{}_{kl} x^k x^l = 0, \tag{4.4.25}$$

$$x^i \xi^j - C^{ij}{}_{kl} \xi^k x^l = 0. \tag{4.4.26}$$

The commutative case (4.4.23) is given by the values

$$B^{ij}{}_{kl} = C^{ij}{}_{kl} = \delta^j_k \delta^i_l.$$

It follows necessarily from the requirement $d^2 = 0$ that the ξ^i satisfy the relations

$$\xi^i \xi^j + C^{ij}{}_{kl} \xi^k \xi^l = 0.$$

If we take the exterior derivative of Equation (4.4.25) and use (4.4.26) then we find the constraint

$$(1 - B)(1 + C) = 0.$$

The quotient algebra is the algebra of forms on the *quantum hyperplane*.

Example 4.72 The quantum plane with $n = 2$ is especially interesting. Let $\hat{R} = \hat{R}_q$ of Example 4.63. Because of the Hecke relation (4.4.16) we can choose

$$B = q^{-1} \hat{R}_q, \qquad C = q \hat{R}_q \tag{4.4.27}$$

and the three functions f, g and h introduced in (4.4.24) are given by

$$f(t) = 1 - q^{-1} t, \qquad g(t) = qt, \qquad h(t) = 1 + qt.$$

Consider Equation (4.4.25) and multiply it on the right by ξ^m. In the matrix notation which was used in Equation (4.4.12) the resulting equation can be written

$$(1 - B)_{12} x_1 x_2 \xi_3 = 0.$$

If we use (4.4.26) twice to place ξ^m to the left we find the equation

$$(1 - B)_{12} C_{23} C_{12} x_2 x_3 = 0$$

after canceling ξ^m. Using Equation (4.4.25) again this can be written in the more symmetric form

$$(C_{23} C_{12} B_{23} - B_{12} C_{23} C_{12}) x_2 x_3 = 0.$$

This equation is identically satisfied by virtue of (4.4.27) and the braid relation (4.4.13).

The corresponding algebra of forms

$$\Omega^*(\mathcal{A}) = \Omega^0(\mathcal{A}) \oplus \Omega^1(\mathcal{A}) \oplus \Omega^2(\mathcal{A})$$

has 4 generators $x^i = (x, y)$ and $\xi^i = dx^i = (\xi, \eta)$ which satisfy the Weyl relation plus the relations (4.1.8) and (4.1.9):

$$xy = qyx,$$
$$x\xi = q^2 \xi x, \qquad\qquad x\eta = q\eta x + (q^2 - 1)\xi y,$$
$$y\xi = q\xi y, \qquad\qquad y\eta = q^2 \eta y,$$
$$\xi^2 = 0, \quad \eta^2 = 0, \quad \eta\xi + q\xi\eta = 0.$$

These relations follow directly from Equations 4.4.25 and (4.4.26) using the expression (4.4.27) for B and C. The differential calculus is identical to that obtained by derivations in Example (4.15). If $x = u$ and $y = v$ are unitary and $|q| = 1$ then we see that the first of the above commutation relations is identical to that defined in Example 4.2.

Equation (4.1.8) and (4.1.9) are the conditions that must be satisfied if the differential calculus of the quantum plane is to be invariant under the coaction of the quantum group $SL_q(2,\mathbb{C})$. Introduce the trivial differential calculus on $SL_q(2,\mathbb{C})$ with the differential d given by

$$da^i_j = 0.$$

The result of the coaction of $SL_q(2,\mathbb{C})$ on x^i and ξ^i is then

$$x^{i\prime} = a^i_j \otimes x^j, \quad \xi^{i\prime} = a^i_j \otimes \xi^j$$

and it follows that $x^{i\prime}$ and $\xi^{i\prime}$ satisfy the same relations as x^i and ξ^i. ☐

Example 4.73 The relations which define the differential calculus of Example 4.16 can be expressed in a form similar to (4.4.24) where now the braid matrix is given by $\hat{R}_h$ of Equation (4.4.18). Introduce the trivial differential calculus on $SL_h(2,\mathbb{C})$ with the differential d given by

$$da^i_j = 0.$$

The result of the coaction of $SL_h(2,\mathbb{C})$ on x^i and ξ^i is then

$$x^{i\prime} = a^i_j \otimes x^j, \quad \xi^{i\prime} = a^i_j \otimes \xi^j$$

and it follows that $x^{i\prime}$ and $\xi^{i\prime}$ satisfy the same relations as x^i and ξ^i. We saw in Example 4.17 that here again the covariant differential calculus can be defined using derivations. ☐

Example 4.74 We noticed in Section 4.3 that on a topological space a large number of measures can be constructed. This was stated in algebraic language as the fact that a von Neumann algebra could be considered as a C^*-algebra plus a state. On a topological group there are special measures known as Haar measures which are invariant under the action of the group on itself. We use in Section 7.3 the Lebesgue measure dx on Minkowski space which is uniquely defined to within a normalization as the measure which is invariant under the translation subgroup of the Poincaré group. It is in fact invariant under the entire Poincaré group. If the group is noncommutative there are to within normalization a unique left-invariant and a unique right-invariant measure. If the group is compact these two measures coincide and the normalization can be chosen so that the volume of the group is equal to one. We have used Haar measures in the crossed-product constructions of Section 4.3 as

well as in Example 4.68. When the group is a Lie group the Haar measure can
be expressed as a differential form of maximal degree and the (left and/or right)
invariance allows one to identify the Lie algebra as the tangent space at the origin
and to show that the group manifold is completely parallelizable. In Section 2.2 we
notice that the information in this form is contained in the Maurer-Cartan form,
which we write as $\theta = h^{-1}dh$. The left invariance of the 1-form dh is expressed by
the equation $d(gh) = gdh$, which implies that $(gh)^{-1}d(gh) = h^{-1}dh$.

The rather surprising fact is that this construction can be carried over to the
case of quantum groups. As an example we consider $SL_q(n, \mathbb{C})$. The differential
calculus $\Omega^*(SL_q(n, \mathbb{C}))$ is considerably restricted by the covariance condition and
$\Omega^1(SL_q(n, \mathbb{C}))$ is generated as a left (or right) module by the left-covariant 1-forms

$$\omega_j^i = S(a_k^i)da_j^k.$$

□

Example 4.75 A covariant derivative can be also constructed over the quantum
groups $SL_q(n, \mathbb{C})$. Using the differential calculus of the preceding example one can
construct the right- and left-invariant 1-form

$$\theta = \frac{q^{2n+1}}{q - q^{-1}} \sum_i q^{-2i}\omega_i^i.$$

Then the exterior derivative of a form α is given by $d\alpha = -[\theta, \alpha]$, with a graded
commutator. This is to be compared with (3.2.1).

If σ is any generalized flip then the covariant derivative defined by

$$D\alpha = -\theta \otimes \alpha + \sigma(\alpha \otimes \theta) \qquad (4.4.28)$$

defines a linear connection associated to σ. This is to be compared with (3.6.16) and
(3.6.22). It can be shown that, for each σ, Equation (4.4.28) is the only covariant
derivative for generic q. Further it can be shown that it has necessarily vanishing
torsion. This is in contrast to the commutative case ($q = 1$) where there are an
infinite number of linear connections not necessarily bicovariant and torsion-free and
where the generalized flip is constrained to be the ordinary flip. It is also in contrast
to the cases with q a root of unity. For generic q the arbitrariness in the deformed
case lies only in the generalized flip for which it can be shown that there is at least
a 2-parameter family. □

In Chapter 3 we constructed noncommutative analogues of partial derivatives as
derivations of the algebra, satisfying the Leibniz rule. Other definitions are possible.
As an example consider the quantum hyperplane and define e_i by

$$e_i x^j = \delta_i^j.$$

If we require that $d = \xi^i e_i$ be a graded derivation then it is easily seen from (4.4.26) that e_i must satisfy

$$e_i(x^j x^k) = \delta_i^j x^k + C^{jk}{}_{il} x^l.$$

There are basically two points of view. One can start with a set of derivations in the strict sense of the word, a set of linear maps of the algebra into itself which satisfy the Leibniz rule and use them as basis for the construction of the associated differential forms. Or one can start with a set of differential forms obtained for example from some covariance criterion and construct a set of possibly twisted derivations which are dual to the forms. By 'twisted' here we mean derivations which satisfy a modified form of the Leibniz rule. We shall compare the two points of view.

Let $\mathcal{A}$ be an algebra and λ_a, $1 \leq a \leq n$, a set of n elements of $\mathcal{A}$ which is such that only the identity commutes with it. This rule implies that only multiples of the identity will have a vanishing differential. We have obviously thereby excluded commutative algebras from consideration. In the example we consider this condition will not be satisfied, which explains why we can have a noncommutative geometry with only one dimension. We shall comment on this latter. We introduce a set of derivations e_a defined on an arbitrary element $f \in \mathcal{A}$ by $e_a f = [\lambda_a, f]$. We have here given the λ_a the physical dimensions of mass; we set this mass scale equal to one. Suppose that the algebra is generated formally by n elements x^i. If one defines the differential of $f \in \mathcal{A}$ by $df(e_a) = e_a f$ exactly as one does in ordinary geometry, or by any other method, then one finds that in general

$$dx^i(e_a) \neq \delta_a^i.$$

The 'natural' basis e_a of the derivations are almost never dual to the 'natural' basis dx^i of the 1-forms. There are basically two ways to remedy the above defect. One can try to construct a new basis θ^a which is dual to the basis of the derivations or one can introduce derivations ∂_i which satisfy a modified form of the Leibniz rule and which are dual to the dx^i. One has then either, or both, of the following equations:

$$\theta^a(e_b) = \delta_b^a, \qquad dx^i(\partial_j) = \delta_j^i.$$

In general these two points of view are equivalent. By construction the θ^a commute with all elements of the algebra. These commutation relations define the structure of the 1-forms as a bimodule over the algebra.

Example 4.76 Much of the analysis of Example 4.18 could be based on the twisted derivations ∂_1 and $\bar{\partial}_1$ of $\mathbb{R}_q^1$ defined by the commutation relations

$$\partial_1 x = 1 + qx\partial_1, \qquad \partial_1 \Lambda = q^{-1}\Lambda\partial_1,$$
$$\bar{\partial}_1 x = 1 + q^{-1}x\bar{\partial}_1, \qquad \bar{\partial}_1 \Lambda = q^{-1}\Lambda\bar{\partial}_1$$

analogous to Equations (4.1.52). We find the relations

$$\partial_1 = \theta_1^1 e_1 = \Lambda^{-1} x^{-1} e_1, \qquad \bar{\partial}_1 = \bar{\theta}_1^1 \bar{e}_1 = q^{-1} \Lambda x^{-1} \bar{e}_1 \qquad (4.4.29)$$

between the twisted derivations $(\partial_1, \bar{\partial}_1)$ and the derivations $(e_1, \bar{e}_1)$. We recall that the vector space $\mathrm{Der}(\mathbb{R}_q^1)$ is not a left module over the algebra $\mathbb{R}_q^1$. □

Example 4.77 The structure of a quantum group has even been found in the set of divergent Feynman diagrams of any renormalizable field theory. The coproduct is constructed from subgraphs and the counterterms are the antipodes. □

Notes

Anyone interested in Plato's inscription can read about it in the book by Riginos (1976). A good introduction to the algebraic theory of noncommutative rings is to be found in the book by Lam (1991). The Weyl algebra was introduced by Weyl (1950). The relation between quadratic algebras and curves in projective space has been extended by Artin *et al.* (1990) to the case of three generators. A succinct introduction to supersymmetry was given by Wess & Bagger (1983), a more mathematical treatise by DeWitt (1984). The mathematical structure of superalgebras is to be found, for example, in Cornwell (1992) or in Constantinescu & de Groote (1994). For an introduction to supergravity we refer to West (1990) or to Castellani *et al.* (1991). Several structures for the family of algebras of Example 4.7 have been considered in the past, by Snyder (1947b, 1947a), Madore (1989b) and Doplicher *et al.* (1995). Example 4.11 is due to Sophus Lie. For a more recent discussion we refer to the monograph by Vilenkin (1968). A review of the techniques involved in constructing and classifying Lie algebras of differential operators on $\mathbb{C}^n$ is to be found in González-López *et al.* (1994). The same techniques permitted Turbiner (1988, 1994) to introduce a new class of solvable Schrödinger operators, the quasi-exactly-solvable Schrödinger operators, so called because the finite numbers of eigenstates can be found algebraically. Example 4.12 is taken from Connes (1994). The quantum plane was introduced by Manin (1988). A differential calculus on it was proposed by Pusz & Woronowicz (1989) and by Wess & Zumino (1990); a differential calculus with two parameters by Maltsiniotis (1993). This latter result was to a certain extent anticipated in the analysis of the corresponding Heisenberg algebra by Fairlie and Zachos (1991, 1992). See also Wess (2000). Example 4.13 is taken from Dubois-Violette *et al.* (1995). See also Sitarz (1994b). Example 4.15 is based on the articles by Dimakis & Madore (1996) and by Fiore *et al.* (2000). In the first article an example with $n = 3$ is also worked out in some detail. An example with $n = 3$ had already been studied for completely different reasons by Schmüdgen & Schüler (1993). Example 4.16 is taken from Leitenberger (1996), Aghamohammadi *et al.* (1997) and

Cho *et al.* (1998). For a discussion of the Poincaré half-plane as an example of a classical or quantum phase space we refer to Emch *et al.* (1994). The quantum euclidean planes were introduced by Faddeev *et al.* (1990). The differential calculi on them were introduced by Carow-Watamura *et al.* (1991) and by Ogievetsky *et al.* (1992). The reality problem was first studied by Ogievetsky & Zumino (1992). For more details concerning Examples 4.18 and 4.19 we refer to Cerchiai *et al.* (1999b, 1999a). For more details of Example 4.21 we refer to Fiore (1994) and to Fiore & Madore (2000). The geometry of the higher-dimensional quantum planes has been also examined (Cerchiai et al. 2000).

The formal theory of deformations of algebraic structures is to be found in the article by Gerstenhaber (1964). A description of *-products and their possible relation to quantum mechanics is to be found for example in Weyl (1950) in Moyel (1949) or, more recently, in the article by Bayen *et al.* (1978). The construction and classification of *-products over arbitrary Poisson manifolds is due to Kontsevich (1997). For a introductory review we refer to Sternheimer (1999). For a recent discussion of the problems involved in defining *-products within the context of Kähler manifolds we refer to Schlichenmaier (1999). Example 4.22 is discussed in Chu *et al.* (1997). For more details on the idea expressed in Example 4.24 we refer to Madore (1999) and Bourgeois & Cahen (1999).

Historically the most important noncommutative algebra in physics was the quantized phase space of non-relativistic quantum mechanics. This algebraic approach to quantum mechanics was extended and developed by von Neumann (1955). Of importance in this respect in developing intuition about the commutative limit of a noncommutative geometry is the article by Ehrenfest (1927) on the classical limit of quantum-mechanical systems. For a more thorough analysis we refer to Ginibre & Velo (1980). The mathematics of the algebras of operators which arise in quantum field theory and in statistical mechanics is the subject of the books of Bratteli and Robinson (1987), Thirring (1979) and Haag (1992). An introduction to C^*-algebras is to be found in the monograph by Davidson (1996). The symplectic geometry of quantized phase space and its relation to noncommutative geometry have been discussed, for example, by Dubois-Violette (1991) and Dimakis & Müller-Hoissen (1993). For more details on Example 4.30 we refer to Dubois-Violette (1993). More details of Example 4.32 can be found in several texts on quantum field theory or, for example, in the articles by Ruijsenaars (1977), Carey & Ruijsenaars (1987) and Araki (1987). The discussion given here was taken from Grosse (1988). The quote at the end of Example 4.33 is by E. Nelson. Example 4.34 is taken from the collection of orig-inal articles edited by Stone (1992). We refer to Perelomov (1986) for a recent introduction to the coherent-state formalism. An exhaustive treatise on unbounded operators, including an introduction to the deficiency index of Example 4.36 is to be found in Dunford & Schwartz (1994). A standard reference on von Neumann algebras is the monograph by Kadison & Ringrose (1997). See also Sunder (1987) for an elementary introduction to the subject and Jones & Sunder (1997) for a detailed study of factors of type II_1. The presentation of Example 4.42 is based on unpublished lecture notes by K. Fredenhagen. The hyperfinite structure has been explained in detail (Landi et al. 1999a) in the case of the rotation algebra. The crossed-product construction is taken from the succinct summary by Jones (1995). Example 4.44 is due to Rieffel (1981, 1990). See also Connes (1994). Example 4.45 and Example 4.46 are standard results, to be found in Kadison & Ringrose (1997). More general results in the same direction are to be found in Woronowicz (1991). See also Renault (1980) and, for example, Vainerman (1995). More details of Examples 4.47, 4.48 and 4.49 are to be found in in Cerchiai *et al.* (1999b). We refer to Fiore (1996) for more details of Example 4.51. Further details of Example 4.52 are to be found in Schmüdgen (1998). A mathematically correct description of Example 4.53 can be found in the book by Connes (1994). For an introduction to foliations in general we refer to the book by Reinhart (1983). Example 4.54, the paradigm of noncommutative differential geometry, is due to Rieffel *(loc. cit.)* Example 4.55 is taken from Berezin (1975) and Klimek & Lesniewski (1992a, 1992b). A more thorough discussion of the relevance to physics of the Tomita-Takesaki theorem and the KMS condition can be found in Haag (1992) and in Bratteli & Robinson (1987). More details of Example 4.59 can be found in Haag (1992) and in Connes (1986).

The subject of quantum groups is at the present the object of active research much of which is based on the work of Jimbo (1985), Woronowicz (1987b, 1987a), Drinfeld (1988) and Manin (1988, 1989). For more details and applications we refer to the collections of articles edited by Doebner & Hennig (1991), Kulish (1991), Ge (1992) and to the monographs by Chari & Pressley (1994) and by Majid (1995). Mathematical articles are to be found in the collection edited by Boutet de Monvel *et al.* (1993). See also Kulish (1993), Klymik & Schmüdgen (1997) and Pittner (1998). For a recent discussion of the problems involved in making quantum groups into topological algebras we refer to Maes & Van Daele (1998). Todorov (1990) and Fuchs (1995) have given detailed accounts of the relation of quantum groups to conformal field theory. Fröhlich & Kerler (1993) have written a review of the relation between quantum groups and braid statistics. We refer to Wassermann (1995, 1998) for a description of conformal field theory from the point of view of operator algebras and which is close in spirit to noncommutative geometry. The book by Gómez *et al.* (1996) contains an introduction to 2-dimensional physics from the standpoint of quantum groups. Example 4.61 is taken from Kulish and Mudrov (1999). Example 4.62 is taken from an article by Dubois-Violette & Launer (1990). It was extended by Ewen & Ogievetsky (1994) and by Ohn (1998) to the case of three generators. For a recent discussion of differential calculi on quantum homogeneous spaces we refer to Schmüdgen (1999b). The q-deformed versions of the rotation groups and the associated euclidean spaces were introduced by Faddeev *et al.* (1990) and by Takeuchi (1990). Differential calculi were constructed on them by Carow-

Watamura *et al.* (1991). The braid matrix which we did not define in Example 4.66 can be found in Faddeev *et al.* (1990). The dilatator was added to construct inhomogeneous extensions of quantum groups by Ogievetsky *et al.* (1992), Schlieker *et al.* (1992) and by Majid (1993). Quantum deformations of the Lorentz group have been proposed by several authors, for example Podleś & Woronowicz (1990), Carow-Watamura *et al.* (1990), Schmidke *et al.* (1991), Ogievetsky *et al.* (1992) and by Majid (1993). Quantum deformations of the Poincaré group have been also proposed, for example by Podleś & Woronowicz (1990), Carow-Watamura *et al.* (1990), Ogievetsky *et al.* (1992), Lukierski *et al.* (1992, 1993), Kehagias *et al.* (1995), Aschieri & Castellani (1996) and Lukierski *et al.* (1996). Explicit representations were first found by Lorek *et al.* (1997) and by Cerchiai & Wess (1998). Example 4.69 is due to Dubois-Violette (unpublished). Example 4.70 is taken from Kulish & Reshetikhin (1983). On the comparison with other forms of the q-deformation we refer to Sudbery (1990) or Curtright & Zachos (1990); on the use of the Jackson derivative see, for example, Ruegg (1990) and Turbiner (2000). More details on Example 4.71 are to be found in Brzeziński & Majid (1993) and in Durđević (1996). A discussion of connections on them is to be found in Dabrowski *et al.* (1999). The h-deformation $SL_h(2)$ of $SL(2)$ was introduced by Demidov *et al.* (1990) and studied by Kupershmit (1992), Ohn (1992), Aghamohammadi (1993), Karimipour (1995), Aghamohammadi *et al.* (1995), Leitenberger (1996) and Abdesselam *et al.* (1998). Combined 'hybrid' deformations of $GL(2)$, with multiple parameters, have also been proposed, both of the structure algebra (Kupershmit 1992) as well as of the enveloping algebra (Dobrev 1992). We refer to Aneva *et al.* (2000) for a recent discussion of the extent to which they are all independent.

A differential calculus was first constructed over a quantum group by Woronowicz (1980, 1987b). For a detailed introduction to this subject as well as reference to the original literature of Woronowicz and others we refer to the book by Klymik & Schmüdgen (1997). More details of Example 4.75 can be found for example in the articles by Schupp *et al.* (1992), Aschieri & Schupp (1996), Georgelin *et al.* (1997) and Heckenberger & Schmüdgen (1997). For a taste of what can happen when one considers quantum groups with q a root of unity we refer to Steinacker (1998) or to Kastler (1998). Quantum deformations of Clifford algebras were introduced by Sylvester (1884) in the particular case when q is a root of unity. For a discussion of the general case we refer to Bautista *et al.* (1996). The discussion of twisted derivations is taken from Wess and Zumino (1990). In Example 4.56 we briefly mentioned a method of constructing a differential calculus using the notion of a spectral triple and in Section 4.4 we gave some examples of differential calculi constructed so as to be covariant under the coaction of a quantum group. The relation between these two constructions has yet to be completely understood. Although we have presented several formal Dirac operators it is not clear in which cases, if any, these can be actually realized as part of a spectral triple. Preliminary work in this direction has been done by Schmüdgen (1999b, 1999a). More details on Example 4.77 can be found, for example, in the review article by Connes (2000).

5 Vector Bundles

We gave the definition of a vector bundle in Section 2.1 and we mentioned that the set $\mathrm{Vect}(V)$ of all vector bundles has a sum and a product operation. In the first section of this chapter we examine a set of equivalence classes of bundles over a manifold V which we show to have the structure of a ring. This ring encodes some information on the topology of the manifold V, a result which we state without proof. The vector bundles over a manifold V can be described in terms of modules over the algebra $\mathcal{C}(V)$ of functions on V and they have natural noncommutative generalizations. The vector space $\mathcal{H}$ of smooth functions f defined on $\mathbb{R}^n$ with values in $\mathbb{C}^r$ can be considered as the space of sections of a trivial vector bundle H over $\mathbb{R}^n$ with fibre $\mathbb{C}^r$. Within the algebra of all operators on $\mathcal{H}$ the differential operators are of special interest. These are polynomials in the partial derivatives of f with respect to the coordinates of $\mathbb{R}^n$, with smooth functions as coefficients. Different copies of $\mathbb{R}^n$ can be patched together in different ways to form manifolds and at the same time the different copies of $\mathbb{C}^r$ can be patched to form non-trivial vector bundles. Consider a differential operator P defined on the corresponding vector space $\mathcal{H}$ of smooth sections of one of these bundles. In a long effort by many mathematicians, culminating in the work of Atiyah and Singer, an important and subtle general relation was found between the null space of the operator P and of its adjoint on the one hand and the topological structure of the vector bundle on the other. Of great mathematical interest, this relation has been used in physics to find the dimension of the solution space of field equations. For example, there is a possibility that the left-handed character of the neutrino and the breaking of parity by the weak interactions can be described, if not explained, by the fact that the Dirac operator must be defined on a non-trivial vector bundle. We shall introduce as quickly as possible the result of Atiyah and Singer so as to be able to formulate the equivalent statement in a noncommutative geometry. The algebra of pseudodifferential operators is defined and a few examples of differential operators used in physics are given. After a brief introduction to the theory of vector bundles over a compact manifold which we formulate in terms of projective modules, we give the corresponding theory for matrix geometry in Section 5.2. This furnishes simple examples of modules which are finite-dimensional vector spaces. In the final section we touch upon the more general infinite-dimensional case.

5.1 K-theory

Using connections, integral invariants can be constructed which aid in the classification of principal bundles and of their associated vector bundles. Let $I(\Omega)$ be a real-valued, symmetric, invariant polynomial of the curvature form Ω on a bundle P over V. A simple example of $I(\Omega) = I(F)$ appears in Equation (2.1.32). If the polynomial is homogeneous of order p then $I(\Omega)$ is a $2p$-form on V. Because of the

Bianchi identities, it is obviously closed:

$$dI(\Omega) = 0.$$

It is also independent of the connection ω to within an exact form. Let Ω_0 and Ω_1 be the curvature 2-forms of two connections ω_0 and ω_1 defined on the same bundle P. Then their difference $\omega_1 - \omega_0$ can be identified with a 1-form α on V. Consider $\omega_t = \omega_0 + t\alpha$. Then

$$\frac{d\Omega_t}{dt} = D_t\alpha$$

where D_t is the covariant derivative of α with respect to the connection ω_t. It is easy to verify that

$$I(\Omega_1) - I(\Omega_0) = d\Phi$$

where

$$\Phi = p\int_0^1 I(\alpha, \Omega_t, \cdots \Omega_t)dt.$$

The integral over V of $I(\Omega)$ defines therefore an element of the cohomology of V which is distinct for non-isomorphic principal bundles. The homomorphism which to each invariant polynomial associates an element in the ring $H^*(V, \mathbb{R})$ is known as the *Chern-Weil homomorphism*.

There are two invariant polynomials which are of particular importance in the classification of associated U_r-bundles over V. The *total Chern class* is defined to be

$$c(H) = \det(1 + \frac{i}{2\pi}\Omega). \qquad (5.1.1)$$

The unit is the unit in U_r. We have written $c(H)$ instead of $c(\Omega)$ because the right-hand side defines an element of $H^*(V, \mathbb{Z})$ which depends only on the vector bundle H associated to the principal bundle P on which the connection is defined. It depends in fact only on the class $[H]$ in the ring $K^0(V)$ which we introduce below. When expanded $c(H)$ produces a series of invariants:

$$c(H) = \sum_j c_j(H).$$

The j-th *Chern class* $c_j(H)$ is an element of $H^{2j}(V; \mathbb{Z})$. The normalization factor in the definition of $c(H)$ has been chosen so that the integral of $c_j(H)$ over a $2j$-cycle is an integer. Compare with Example 2.12. The total Chern class of the Whitney sum $H_1 \oplus H_2$ of two bundles is given as the product

$$c(H_1 \oplus H_2) = c(H_1)\, c(H_2).$$

Of more interest from the point of view of the noncommutative generalizations is the *Chern character*

$$\mathrm{ch}(H) = \mathrm{Tr}(\exp(\frac{i}{2\pi}\Omega)).$$

This is formally an infinite series but applied to any given manifold reduces to a polynomial. Again, we have written ch(H) instead of ch(Ω) because the right-hand side defines an element of $H^*(V, \mathbb{Q})$ which depends only on the vector bundle H associated to the principal bundle P on which the connection is defined. If we formally factorize $c(H)$,

$$c(H) = \prod_j (1 + x_j),$$

then the Chern character can be written as

$$\mathrm{ch}(H) = \sum_j e^{x_j}. \tag{5.1.2}$$

Again the sum reduces to a polynomial on any given bundle. The Chern character satisfies the identities

$$\mathrm{ch}(H_1 \oplus H_2) = \mathrm{ch}(H_1) + \mathrm{ch}(H_2),$$
$$\mathrm{ch}(H_1 \otimes H_2) = \mathrm{ch}(H_1)\,\mathrm{ch}(H_2). \tag{5.1.3}$$

For a U_r-bundle we have

$$\mathrm{ch}(H) = r + c_1(H) + \frac{1}{2}(c_1^2(H) - 2c_2(H)) + \cdots.$$

As one sees from the integers in the denominators of its formal series expansion, the Chern character takes its values in $H^*(V; \mathbb{Q})$, the cohomology ring with rational coefficients.

Consider now the set Vect(V) defined at the end of Section 2.1. It has almost the structure of a ring. A sum and a product can be defined but there is no notion of the negative of a given element. One can however define an equivalence relation on Vect(V) such that the equivalence classes admit a natural extension to a commutative ring which is designated $K^0(V)$. Let $H \mapsto [H]$ be the map which associates to each H in Vect(V) the corresponding class $[H]$ in $K^0(V)$. It is natural to require that $[H_1] = [H_2]$ if the bundles are isomorphic: $H_1 \simeq H_2$. That is, we should have the inclusion

$$[H] \supseteq \{H' : H' \simeq H\}.$$

We need to impose also the relation

$$[H] = [H_1] + [H_2]$$

if $H \simeq H_1 \oplus H_2$. We see then that two bundles can map to the same class only if the fibres have the same dimensions. We define $[l] \in K^0(V)$ to be the equivalence class of the trivial bundle with fibre $\mathbb{R}^l$. We introduce a new element $-[l]$ in the same way that the negative integers are constructed from the natural numbers. We have then $\mathbb{Z} \subseteq K^0(V)$. Now it can occur that, as a $GL(l_1 + l_2, \mathbb{R})$-bundle, the direct sum of a $GL(l_1, \mathbb{R})$-bundle H_1 and a $GL(l_2, \mathbb{R})$-bundle H_2 is trivial. In fact a theorem of Swan assures us that for any H_1 it is always possible to find an H_2 such that this is so. Therefore for some l we have $[H_1] + [H_2] = [l]$. One can define then the negative of $[H_1]$ as

$$-[H_1] = [H_2] - [l].$$

We shall see below in Example 5.6 that $[H_1] = [H_2]$ cannot imply that $H_1 \simeq H_2$. To see what is required we consider three bundles H_1, H_2 and H_3 and we suppose that $[H_1] + [H_3] = [H_2] + [H_3]$. There exists a bundle H_4 such that $H_3 \oplus H_4$ is trivial. We have then $[H_1] + [l] = [H_2] + [l]$ for some integer l and therefore $[H_1 \oplus V \times \mathbb{R}^l] = [H_2 \oplus V \times \mathbb{R}^l]$. We define accordingly the equivalence class of H to be the set of all bundles H' such that $H' \oplus V \times \mathbb{R}^l \simeq H \oplus V \times \mathbb{R}^l$ for some l. This is known as *stable equivalence*. Since $H_1 \otimes H_2 \simeq H_2 \otimes H_1$, the ring $K^0(V)$ is commutative.

Example 5.1 The same construction can be repeated for complex vector bundles. One of the simplest examples is furnished by the complex bundles over $V = S^2$. For each integer l there is exactly one nontrivial complex line bundle L_l over S^2, with $L_{-l} = L_l^*$, the complex conjugate bundle. Furthermore if $l \neq l'$ then $[L_l] \neq [L_{l'}]$. Any bundle whose fibre is of dimension greater than the dimension of the manifold splits as the sum of a bundle with dimension one less and a trivial line bundle. The tensor product $L_1 \otimes L_1$ is again a line bundle but the direct sum $L_1 \oplus L_1$ is a plane bundle. It can be shown that $2[L_1] = [L_1]^2 + [1]$. Therefore $K^0(S^2)$ is a ring with two generators $[L_1]$ and $[1]$ and one relation

$$([L_1] - [1])^2 = 0.$$

The element $[1]$ is the unit of the ring. Quite generally one can show that

$$K^0(S^{2n}) = \mathbb{Z} \oplus \mathbb{Z}$$

for all integers $n \geq 1$.   □

The 2-sphere plays a particularly important role because of the isomorphism

$$K^0(V) \otimes K^0(S^2) \simeq K^0(V \times S^2).$$

Odd spheres have no non-trivial complex bundles over them and

$$K^0(S^{2n+1}) = \mathbb{Z}$$

for all integers $n \geq 0$. Notice that in these examples $K^0(V)$ has the same number of generators as the even cohomology. This is a general fact; the ring $K^0(V)$ does not describe completely the topology of the manifold V. In the case of complex bundles one can show from (5.1.3) that the Chern character yields a ring isomorphism

$$\text{ch} : K^0(V) \otimes_{\mathbf{Z}} \mathbb{Q} \simeq H^{2*}(V; \mathbb{Q}). \tag{5.1.4}$$

In the case of line bundles over S^2 the isomorphism (5.1.4) is given by

$$[l] \mapsto l, \qquad [L_1] \mapsto 1 + c_1(L_1).$$

The integers on the right-hand side represent elements of $H^0(V; \mathbb{Q})$.

Example 5.2 There is an index on $K^0(V)$ to distinguish it from another group $K^1(V)$ which for complex bundles is defined in terms of equivalence classes of smooth maps of V into the group $GL(r, \mathbb{C})$, in the limit of large r. Since complex vector bundles over V are defined in terms of the set of groups $GL(r, \mathbb{C})$ it is to be expected that there is a close relation between $K^0(V)$ and $K^1(V)$. In fact together they satisfy most of the conditions of the cohomology groups $H^p(V, \mathbb{C})$ when the indices on the latter are taken modulo 2. The study of the relation of the ring $K^*(V)$ to the manifold V is called *K-theory*. □

Example 5.3 We have defined $K^0(V)$ over compact manifolds V without boundaries, but the cotangent bundle T^*V (as well as the tangent bundle) to such a manifold is not compact. We shall have occasion below to consider K-theory over this bundle and so we shall use it as an example to define *relative K-theory*. Recall from Section 2.2 the unit-ball bundle BV and the unit-sphere bundle SV over V. Consider two vector bundles over BV whose restrictions to SV are isomorphic. Let I be the unit interval. Over the boundary of $BV \times I$ one defines a new bundle by using the isomorphism to join together $SV \times [0]$ with $SV \times [1]$. The ring $K^0(T^*V)$ is constructed from equivalence classes of these bundles. From two bundles over BV one obtains a single element of $K^0(T^*V)$. This is known as the *difference construction*. □

Let $\mathcal{A} = \mathcal{C}(V)$ be the algebra of smooth, real-valued functions on V. We have already noticed in Section 2.2 that the space $\mathcal{H}$ of smooth sections of a vector bundle H is an $\mathcal{A}$-module. If the bundle H is trivial then the module $\mathcal{H}$ is a free module; if $H = V \times \mathbb{R}^l$ then $\mathcal{H}$ can be identified with the direct sum of l copies of $\mathcal{A}$. In general H is not trivial but since there always exists another bundle H' such that the direct sum is trivial, there always exists a second module $\mathcal{H}'$ such that for some integer m

$$\mathcal{H} \oplus \mathcal{H}' = \mathcal{A}^m \equiv \bigoplus_1^m \mathcal{A}.$$

This means that $\mathcal{H}$ is always a *projective module* of finite type. Conversely one can show that any projective $\mathcal{C}(V)$-module of finite type is isomorphic to the space of sections of some vector bundle over V.

The module $\mathcal{H}$ can be identified with an *idempotent* e (4.3.21),

$$e^* = e, \qquad e^2 = e,$$

in the algebra $M_m(\mathcal{A})$ of $m \times m$ matrices with values in $\mathcal{A}$. In fact if we interpret this algebra as the algebra of smooth functions on V with values in M_m then e is the projection which to each point $x \in V$ associates the fibre H_x. Conversely, from every idempotent $e \in M_m(\mathcal{A})$ of constant rank l one can construct a fibre bundle H with fibre $\mathbb{R}^l$ whose space of sections is the image of e. Let $M_\infty(\mathcal{A})$ be the algebra of sparse matrices, infinite matrices with zero in all but a finite number of entries. Then each $M_m(\mathcal{A})$ can be embedded in a natural way in $M_\infty(\mathcal{A})$, for example as

$$a \mapsto \begin{pmatrix} a & 0 \\ 0 & 0 \end{pmatrix},$$

and the idempotent e can be considered as an element of $M_\infty(\mathcal{A})$. Let H and H' be two vector bundles and e and e' the corresponding projections onto the modules $\mathcal{H}$ and $\mathcal{H}'$. Then

$$\mathcal{H} \oplus \operatorname{Ker} e = \mathcal{H}' \oplus \operatorname{Ker} e'. \tag{5.1.5}$$

If H and H' are isomorphic as vector bundles then there exists an $\mathcal{A}$-module isomorphism ϕ from $\mathcal{H}$ to $\mathcal{H}'$. Let u and v be the elements of $M_\infty(\mathcal{A})$ which in the decomposition (5.1.5) have the form

$$u = \begin{pmatrix} \phi & 0 \\ 0 & 0 \end{pmatrix}, \qquad v = \begin{pmatrix} \phi^{-1} & 0 \\ 0 & 0 \end{pmatrix}.$$

Then

$$e = vu, \qquad e' = uv. \tag{5.1.6}$$

Conversely if two idempotents e and e' can be written in this form then $\mathcal{H} = \operatorname{Im} e$ and $\mathcal{H}' = \operatorname{Im} e'$ are isomorphic, as are the corresponding vector bundles. If e and e' satisfy (5.1.6) they are said to be *equivalent*. This is to be compared with the notion of equivalence introduced in Equation (4.3.22).

Since $\mathcal{A}$ has a unit there is a natural embedding of $M_\infty(\mathbb{R})$ in $M_\infty(\mathcal{A})$. The trivial bundle with fibre $\mathbb{R}^l$ corresponds to the equivalence class $[l]$ of all idempotents e_l of rank l in the image of $M_\infty(\mathbb{R})$. The equivalence classe of idempotents which allows for the stable equivalence of vector bundles is obtained by requiring e and e' to be equivalent if and only if for some unit e_l,

$$e \oplus e_l = vu, \qquad e' \oplus e_l = uv. \tag{5.1.7}$$

Let e be an arbitrary idempotent and define

$$e' = \begin{pmatrix} 0 & 0 \\ 0 & e \end{pmatrix}.$$

Then e' and e are equivalent. So if two idempotents e_1 and e_2 are not orthogonal they are equivalent to two which are so. Let $[e]$ be the equivalence class of e. We can define then

$$[e_1] + [e_2] = [e_1 + e_2],$$

with $e_1 e_2 = 0$. For every e_1 there exists an e_2 such that $[e_1] + [e_2] = [l]$ for some l. So if we introduce the formal negatives $-[l]$ as before we can define the negative of $[e_1]$ as

$$-[e_1] = [e_2] - [l].$$

If e_1 and e_2 are idempotents in $M_m(\mathcal{A})$ then so is the tensor product $e_1 \otimes e_2$, which corresponds to the projection onto the module $\mathcal{H}_1 \otimes \mathcal{H}_2$, the space of sections of the bundle $H_1 \otimes H_2$. With this product the ring $K^0(V)$ can be identified then with a ring of equivalence classes of idempotents in $M_\infty(\mathcal{A})$, in which case it is written $K_0(\mathcal{A})$. The position of the index is governed by the same considerations as those which determine the position of the star on the tangent and cotangent bundles and on the homology and cohomology groups. A smooth map of V into a manifold V' defines, via the pull-back, a map of $K^0(V')$ into $K^0(V)$. As in (2.1.8), and for the same reasons, the induced arrow changes directions.

Example 5.4 Consider a principal U_r-bundle P over V and an associated bundle H which is a direct summand of a trivial bundle with fibre $\mathbb{C}^n$. Let e be the idempotent in $M_n(\mathcal{A})$ which projects onto the space $\mathcal{H}$ of sections of H. Then if ψ is a section of H we have $e\psi = \psi$. We can take the exterior derivative of ψ considered as an element of the trivial bundle. It will be a 1-form with values in $\mathcal{A}^n$. If we project back onto $\mathcal{H}$ with e we can define a covariant derivative $D\psi$ of ψ:

$$D\psi = ed\psi.$$

This is indeed a linear map of the form (2.2.28). From the fact that e is an idempotent follows the identity $e(de)e = 0$. Therefore

$$D^2\psi = eded\psi = eded(e\psi) = \Omega\psi$$

where

$$\Omega = edede.$$

Consider e as a function on V which takes its values in the elements of M_n of rank r. Then e can be written in the form $e = g^{-1}e_0 g$ for some $GL(n, \mathbb{C})$-valued function g and some constant matrix e_0 of rank r. From this follows the identity $de = -[\theta, e]$ where $\theta = g^{-1}dg$. The 1-form

$$\omega_0 = e_0 g d(g^{-1})e_0$$

is a U_r connection. Let Ω_0 be the corresponding curvature. Then $\Omega = g^{-1}\Omega_0 g$. The Chern classes can be defined using the above expression for Ω in Formula (5.1.1). For example if V is a 2-dimensional manifold then

$$\int_V c_1(H) = \frac{1}{2\pi i}\mathrm{Tr}\int_V edede \in \mathbb{Z}. \tag{5.1.8}$$

The trace is over M_n. We shall have occasion to refer to this equation when describing the Hall effect in Example 5.13.   □

One approach to the study of the structure of a differential manifold V is through the differential operators which act on its algebra of functions or, more generally, on sections of vector bundles defined over it. We saw, for example, in Equation (2.2.20) that the exterior derivative can be defined in terms of a differential operator. The dimension of the manifold is encoded in the asymptotic behaviour of the spectrum of this same operator. The space $\mathcal{H}$ of sections of an hermitian vector bundle H can be completed to form a Hilbert space such that the operator becomes a bounded operator. There is then a natural generalization to noncommutative geometry. The inverse of a differential operator, when it exists, is not a differential operator and it is useful therefore to consider the larger algebra $\mathcal{P}(\mathcal{H})$ of pseudodifferential operators on $\mathcal{H}$. These can be defined in terms of Fourier transformations. Let now $\mathcal{H}$ be the space of square-integrable sections of an hermitian vector bundle H over a compact manifold V. That is, $\mathcal{H}$ is the completion of the vector space of smooth sections of H with respect to the inner product

$$(\psi, \chi) = \int \psi^* \chi dx. \tag{5.1.9}$$

Here dx is the invariant volume element of V with respect to a metric. Choose a coordinate neighbourhood $\mathcal{O}$ and identify it with $\mathbb{R}^n$. This identification permits one also to identify the tangent space at each point of $\mathcal{O}$ with $\mathcal{O}$ itself. Let $\psi \in \mathcal{H}$ and consider its restriction to $\mathcal{O}$. We can write

$$\psi(x) = \int e^{ikx}\tilde{\psi}(k)dk. \tag{5.1.10}$$

The k is an element of *momentum space*, the cotangent space to V at the point x. The product kx is the ordinary pairing with the tangent space. A general element $P \in \mathcal{P}(\mathcal{H})$ is defined through the symbol $\sigma(P)$ by the equation

$$(P\psi)(x) = \int e^{ikx}\sigma(P)(x, k)\tilde{\psi}(k)dk. \tag{5.1.11}$$

The symbol can be an arbitrary smooth function of x and k but the condition that the operator P be well defined imposes restrictions on the transformation properties of $\sigma(P)$ under change of local coordinates. If $\sigma(P)$ is a polynomial of degree p

in k then P is a differential operator of order p. If P and Q are two differential operators of order p and q respectively then PQ is a differential operator of order pq but in general there is no simple relation between $\sigma(PQ)$ and $\sigma(P)\sigma(Q)$. These two expressions are equal only if $\sigma(Q)$ does not depend on x. If $\sigma(P)^{-1}$ exists then it can be used in (5.1.11) to define the inverse of P.

If the trace $\mathrm{Tr}(P)$ of P exists it can sometimes be expressed in a useful formula as an integral over the symbol of P. Suppose that the support of ψ lies within the coordinate neighbourhood $\mathcal{O}$ of V. Then the Fourier transform (5.1.10) can be inverted and we can write

$$P\psi(x) = \int e^{ik(x-y)}\sigma(P)(x,k)\psi(y)dydk.$$

This formula presupposes a correct normalization of dk, including a factor $(2\pi)^{-n}$. Suppose that P can be written in the form

$$P\psi(x) = \int K(x,y)\psi(y)dy.$$

Then since ψ is arbitrary we can conclude that

$$K(x,y) \simeq \int e^{ik(x-y)}\sigma(P)(x,k)dk.$$

The approximate equality is due to the fact that ψ is not quite arbitrary but has been supposed to have its support in one coordinate neighbourhood. Integration over V yields

$$\mathrm{Tr}(P) = \int K(x,x)dx \simeq \int \sigma(P)(x,k)dxdk. \qquad (5.1.12)$$

We shall use this formula in Example 5.12 in an asymptotic expression in which the cancellation of ψ is allowed.

For each real number $s \geq 0$, one can define the Sobolev norm $\|\psi\|_s$ of an element $\psi \in \mathcal{H}$. If the support of ψ lies within a coordinate neighbourhood $\mathcal{O}$ then we define

$$\|\psi\|_s^2 = \int (1+k^2)^s|\tilde{\psi}(k)|^2dk.$$

If the support of ψ does not lie within any given neighbourhood one must choose a finite covering $\mathcal{O}_i$ of V. There exists always a partition of unity, a set of smooth functions p_i whose support lies within $\mathcal{O}_i$ and such that $\sum_i p_i = 1$. We can define then

$$\|\psi\|_s^2 = \sum_i \|p_i\psi\|_s^2$$

The Sobolev space $\mathcal{H}_s$ is the completion of $\mathcal{H}$ with respect to this norm. It is clear that for $s' \geq s \geq 0$ we have

$$\mathcal{H}_{s'} \subseteq \mathcal{H}_s \subseteq \mathcal{H}.$$

Using the above norm one can also define $\mathcal{H}_{-s}$ for any $s \geq 0$ and show that the norm defined by the inner product (5.1.9) permits one to identify $\mathcal{H}_{-s}$ as the dual of $\mathcal{H}_s$. As a subalgebra of the algebra of all operators on $\mathcal{H}$, the algebra $\mathcal{P}(\mathcal{H})$ has a norm.

Differential operators are local operators; they depend only on the value of the function and a finite number of its derivatives. If $\sigma(P)$ is a polynomial of order p with $k = 0$ as the only real zero, then P is an *elliptic* differential operator of order p. One can show then that the spectra of P and of its adjoint P^* are discrete with finite multiplicities. In particular the null spaces $\text{Ker}(P)$ and $\text{Ker}(P^*)$ are of finite dimension. If P is self-adjoint as an operator on $\mathcal{H}$ then the spectrum is real. Let P be elliptic of order p and let P_s be the restriction of P to $\mathcal{H}_{s+p}$. By definition then it is a map of $\mathcal{H}_{s+p}$ into $\mathcal{H}_s$. It follows directly from the definitions that P_s is a bounded operator. In fact if $\psi \in \mathcal{H}_{s+p}$ then $\tilde{\psi}$ decreases faster than $|k|^{-s-p-n/2}$ as $k \to \infty$ and therefore $\tilde{P}\psi$ decreases faster than $|k|^{-s-n/2}$. It is therefore in $\mathcal{H}_s$ and for some constant c, we have $\|P\psi\|_s \leq c\|\psi\|_{s+p}$. The dual P_s^* of P_s maps $\mathcal{H}_{-s}$ into $\mathcal{H}_{-s-p}$. The restriction of P_s^* to $\mathcal{H}_{s+p} \subseteq \mathcal{H}_{-s}$ maps $\mathcal{H}_{s+p}$ into $\mathcal{H}_s$.

An *infinitely smoothing* operator takes $\mathcal{H}$ into $\mathcal{H}_s$ for all s. For example, if K is such that $\sigma(K) = e^{-k^2}$ then K is infinitely smoothing. Although in general an elliptic operator has no inverse, a *quasi-inverse* can be defined. This is an operator Q such that $PQ - 1$ and $QP - 1$ are infinitely smoothing operators. Consider a differential equation of the form

$$P\psi = J$$

where P is an elliptic differential operator on a vector bundle H. If $\text{Ker}(P)$ is not equal to zero then the solution will not be unique, if it exists. If $\text{Ker}(P^*)$ is not equal to zero then in general a solution will not exist. There will be a finite number of integrability conditions; the source J must be orthogonal to $\text{Ker}(P^*)$. If χ is an element of $\text{Ker}(P^*)$ then

$$(J, \chi) = (P\psi, \chi) = (\psi, P^*\chi) = 0.$$

For every source which satisfies these conditions a solution can be given in terms of Q:

$$\psi = QJ.$$

The spaces $\text{Ker}(Q)$, $\text{Ker}(P^*)$ and $\text{Coker}(P)$ all have the same dimension:

$$\dim \text{Ker}(Q) = \dim \text{Ker}(P^*) = \dim \text{Coker}(P). \tag{5.1.13}$$

The Laplace operator (2.2.10) is a self-adjoint, second-order, elliptic, differential operator on the algebra of forms $\Omega^*(V)$. If λ_l is its l-th eigenvalue, including multiplicities, then one can show that asymptotically for large l

$$\lambda_l \sim l^{2/n}$$

where n is the dimension of the space V.

Example 5.5 In particular consider the example $V = S^1$. The Laplace operator is given by

$$\Delta = -\frac{d^2}{dx^2}$$

for $0 \leq x \leq 2\pi$ and the eigenvalues are given by $\lambda_l = l^2$, each 2-fold degenerate except $l = 0$. The associated eigenvectors $e^{\pm ilx}$ are smooth functions and belong of course to $\mathcal{H}$. They belong in fact to $\mathcal{H}_s$ for all $s \geq 0$. This must be so since Δ takes $\mathcal{H}_s$ into $\mathcal{H}_{s-2}$ and so if $\psi \in \mathcal{H}$ is an eigenvector then $\Delta\psi \in \mathcal{H}$ also and therefore $\psi \in \mathcal{H}_2$. This is a general property of elliptic operators known as *elliptic regularity*. □

Let H and H' be two vector bundles over a smooth, compact manifold V and consider a (pseudo)differential operator

$$\mathcal{H} \xrightarrow{P} \mathcal{H}' \tag{5.1.14}$$

with a quasi-inverse Q. The *analytic index* $i_a(P)$ is defined to be the difference

$$i_a(P) = \dim \operatorname{Ker}(P) - \dim \operatorname{Ker}(Q). \tag{5.1.15}$$

From (5.1.13) it follows then that the index of a self-adjoint operator vanishes. It can be shown that for all $s \geq 0$

$$i_a(P) = i_a(P_s).$$

We can define then the index of an elliptic operator in terms of a family of bounded operators with a finite-dimensional kernel and cokernel.

At each point (x, k) of the unit-sphere bundle SV the leading term in the symbol $\sigma(P)$ defines a homomorphism

$$H_x \xrightarrow{\sigma(P)} H'_x. \tag{5.1.16}$$

The operator P is elliptic if and only if this homomorphism is an isomorphism. In particular the dimensions of the fibres of H and H' must be equal. Let π be the projection of the unit-ball bundle BV onto V and let π^*H be the pull-back of H over BV. The isomorphism (5.1.16) extends to an isomorphism between the restrictions to SV of π^*H and π^*H'. From our definition of relative K-theory, $\sigma(P)$ defines then an element

$$[\sigma(P)] \in K^0(T^*V).$$

One can show further that every element in $K^0(T^*V)$ is of the form $[\sigma(P)]$ for some elliptic P. The element $[\sigma(P)]$ depends only on the homotopy class of $\sigma(P)$. In particular if the isomorphism defined by (5.1.16) over SV can be extended to all of BV then $[\sigma(P)] = 0$.

It can be shown that if H and H' are fixed then the right-hand side of (5.1.15) does not depend on the operator P. In fact it depends only on the element of $K^0(T^*V)$ defined by H and H'. The analytic index can be considered then as a homomorphism

$$K^0(T^*V) \xrightarrow{i_a} \mathbb{Z}.$$

We have seen that there is an isomorphism (5.1.4) of $K^0(V)$ onto the even cohomology of V. A similar isomorphism

$$\text{ch} : K^0(T^*V) \otimes_{\mathbb{Z}} \mathbb{Q} \simeq H^{2*}(BV, SV; \mathbb{Q}) \qquad (5.1.17)$$

exists between $K^0(T^*V)$ and the relative cohomology of the unit-ball bundle with respect to the unit-sphere bundle. By the Thom isomorphism (2.3.8) however this cohomology can be identified with a cohomology of V. Integration over V yields therefore in particular a homomorphism

$$K^0(T^*V) \to \mathbb{Q}. \qquad (5.1.18)$$

Another index, the *topological index* $i_t(P)$ of P defines also a homomorphism of $K^0(T^*V)$ into the rational numbers:

$$K^0(T^*V) \xrightarrow{i_t} \mathbb{Q}.$$

In Example 5.10 we shall give the general form for $i_t(P)$. In particular cases it coincides with (5.1.18) but in general it does not. It is a fundamental result that the two indices are equal

$$i_a(P) = i_t(P) \equiv i(P).$$

In particular then the image of $K^0(T^*V)$ under the topological index lies in $\mathbb{Z} \subset \mathbb{Q}$. The most interesting proof of this result for a physicist is based on the so-called 'heat-kernel' method since it uses methods which resemble closely those which are used to calculate regularized versions of infinite determinants which appear in the 1-loop quantum corrections to classical actions. We shall briefly mention this method in Example 5.12.

Example 5.6 For $n = 1, 3, 7$ the tangent bundle TS^n is trivial. Let N be the normal bundle to S^n for general n. By definition the direct sum $TS^n \oplus N$ is isomorphic to $S^n \times \mathbb{R}^{n+1}$. Therefore $[TS^n] + [N] = [n + 1]$ as elements of $K^0(S^n)$. But N is itself trivial: $[N] = [1]$. Therefore $[TS^n] = [n]$. However, TS^n and $S^n \times \mathbb{R}^n$ are not isomorphic as vector bundles for $n \neq 1, 3, 7$. We see then that the condition that $[H_1]$ and $[H_2]$ be equal as elements of $K^0(V)$ does not in general imply that the corresponding bundles H_1 and H_2 are isomorphic. Now one can show that TS^2 is isomorphic to L_{-2} as a complex line bundle and we saw that each complex line bundle L_l gives rise to a nontrivial element $[L_l]$ in the ring $K^0(S^2)$ of complex vector bundles. That is, TS^2 is stably trivial as a real bundle but not stably trivial as a complex bundle. This is so since the normal bundle is not a complex line bundle. $\square$

Example 5.7 The Kähler-Dirac operator $d + \delta$ is a self-adjoint map of $\Omega^*(V)$ into itself. It takes $\Omega^{*\pm}(V)$ into $\Omega^{*\mp}(V)$. The restriction to the even forms $\Omega^{*+}(V)$ is no longer self-adjoint and has a non-vanishing topological index, given by

$$i_t(d + \delta) = e(V) = \int_V \chi(V).$$

Here $\chi(V)$ is the Euler class, the element of $H^n(V; \mathbb{Z})$ whose integral yields the Euler number

$$e(V) = \sum_{i=0}^{n} (-1)^i \dim H^i(V; \mathbb{Z}).$$

It vanishes if the dimension n of V is odd. □

Example 5.8 Let A and F be as in Section 2.2 and consider a principal $(SU_2 \times \mathrm{Spin}(4))$-bundle over S^4. Since a sphere can be covered by two coordinate charts a principal bundle can be completely characterized by one map of the form (2.1.16), that is, by a map of S^3 into SU_2. These maps are classified by a *winding number l*. The Chern class is given by $c = 1 + c_2$ with

$$c_2(H) = -\frac{1}{8\pi^2} \mathrm{Tr}(\Omega^2), \qquad \int_V c_2(H) = l. \tag{5.1.19}$$

Let $\not{D}$ be the Dirac operator formed using the derivative (2.2.31) on an associated bundle with fibre $\mathbb{C}^2 \otimes \mathbb{C}^4$ and let D^+ be its restriction to positive-helicity spinors. The helicity decomposition was given in (2.2.18). The topological index of D^+ is given by

$$i_t(D^+) = -\int_V c_2(H) = -l.$$

If $l \geq 0$, a more detailed study shows that $\dim \mathrm{Ker}(D^+) = 0$ and that $\dim \mathrm{Ker}(D^-) = l$. There are then l negative-helicity solutions to the Dirac equation and no positive-helicity solutions. The isomorphism (5.1.4) classifies all vector bundles over S^4 in terms of the cohomology of S^4. In this case the index is equivalent to the isomorphism. □

Example 5.9 The topological index of D^+ on a $\mathrm{Spin}(n)$-bundle which includes a factor H can be given as an integral

$$i_t(D^+) = \int_V \mathrm{ch}(H)\hat{A}(V).$$

The integrand on the right-hand side is to be considered as an exterior product of two forms of varying degrees of which only the terms of total degree $\dim(V)$ are to be retained. The $\hat{A}$-*genus*

$$\hat{A}(V) = \prod_j \frac{x_j/2}{\sinh(x_j/2)}$$

is an even polynomial in the x_j defined in the factorization of $c(H)$ and so can be expressed in terms of the Pontrjagin classes

$$\hat{A}(V) = 1 - \frac{1}{24}p_1(V) + \frac{1}{5760}(7p_1(V)^2 - 4p_2(V)) + \cdots.$$

The Pontrjagin classes $p_j(V)$ of a manifold are defined in terms of the Chern classes of the complexified tangent space. Like the Chern class, the $\hat{A}$-genus is multiplicative with respect to Whitney sums. The Dirac operator D^+ interchanges the two $\mathrm{Spin}(n)$-bundles $H^{\pm}$ defined in Section 2.2. The total bundle structure over T^*V factorizes into H which lifts from V and which gives rise to $\mathrm{ch}(H)$ and the spin-structure which generates the $\hat{A}$-genus. $\square$

Example 5.10 The index of any elliptic (pseudo)differential operator P can be expressed quite generally as an integral over V. Let P be as in (5.1.14) and let $[\sigma(P)]$ be the equivalence class in $K^0(T^*V)$ of its symbol. Then the topological index of P is given by

$$i_t(P) = (-1)^{n(n+1)/2} \int_V \mathrm{ch}([\sigma(P)])\mathrm{td}(T^*V \otimes_{\mathbb{R}} \mathbb{C}). \qquad (5.1.20)$$

The integer n is the dimension of V and ch is the isomorphism (5.1.17). We have here identified the relative cohomology $H^*(BV, SV; \mathbb{Q})$ with $H^*(V; \mathbb{Q})$ using the Thom isomorphism (2.3.8). The complexified cotangent space $T^*V \otimes_{\mathbb{R}} \mathbb{C}$ is a U_n-bundle over V. The *Todd genus*

$$\mathrm{td}(H) = \prod_j \frac{x_j}{1 - e^{-x_j}}$$

of an arbitrary U_r-bundle H can be expressed in terms of the Chern classes:

$$\mathrm{td}(H) = 1 + \frac{1}{2}c_1(H) + \frac{1}{12}(c_1(H)^2 + c_2(H)) + \cdots.$$

It is also multiplicative with respect to Whitney sums and, like the Chern character, takes its values in $H^*(V; \mathbb{Q})$.

The index vanishes in general if P is a differential operator and the dimension of V is odd. It simplifies considerably if the bundles H and H' are trivial. If they are $\mathbb{C}^r$-bundles then the leading term of the symbol is a map of SV into $GL(r, \mathbb{C})$ and the factor $\mathrm{ch}([\sigma(P)])$ in (5.1.20) reduces to an expression which depends only on the corresponding image of $H^*(GL(r, \mathbb{C}); \mathbb{Q})$ in $H^*(SV; \mathbb{Q})$. In Example 5.7 $H = \Omega^{*+}(V)$ and $H' = \Omega^{*-}(V)$. The integrand reduces to the Euler class $\chi(V)$. In the previous example $H = H^+$ and $H' = H^-$ are also closely related. The Chern character of the spin factor combines with the Todd genus to yield the $\hat{A}$-genus. $\square$

Example 5.11 Let V and G be as in the Example 5.8. The map

$$A \mapsto *D * F$$

is given by a second-order nonlinear differential operator. If J is a given 1-form with values in the Lie algebra of SU_2 then the equation

$$*D * F = J$$

is the Yang-Mills field equation. The map

$$A \mapsto *F - F$$

is given by a first-order nonlinear differential operator. If F is *self-dual*, if the equation

$$*F = F \qquad\qquad (5.1.21)$$

holds then the equation $*D * F = 0$ follows immediately from the Bianchi identities. Let $\Omega^2_-(V) \subset \Omega^2(V)$ be the anti-self-dual 2-forms, solutions to the equation $*\alpha = -\alpha$. Let A_0 be a particular solution to Equation (5.1.21). A solution A which is near to A_0 can be written in the form $A = A_0 + \alpha$, where α is a section of the associated bundle H with fibre $\underline{su}_2 \otimes \Omega^1(V)$. Let H' be the associated bundle with fibre $\underline{su}_2 \otimes (\Omega^2_-(V) \oplus \Omega^0(V))$. Write $F = F_0 + \beta$ and let $f \in \underline{su}_2 \otimes \Omega^0(V)$. Then the linearized form of Equation (5.1.21) can be written in the form $P\alpha = 0$, where P is an elliptic differential operator from H to H' given by

$$*(D\alpha) - D\alpha = \beta, \qquad \delta\alpha = f. \qquad\qquad (5.1.22)$$

The second equation is a gauge condition. The adjoint P^* of P can be written in the form

$$*D * \beta + df = \alpha.$$

When the integer l of (5.1.19) is equal to zero then one can see that $\mathrm{Ker}(P) = 0$ and that $\dim \mathrm{Ker}(P^*) = 3$ with the 3 solutions given by $\beta = 0$ and f a constant element of $\underline{su}_2$. The index in this case is therefore given by $i(P) = -3$. The general case is most conveniently treated using the 'heat-kernel' method. The final result is

$$i(P) = 8l - 3.$$

In this case then, although the index $i(P)$ is equal to the image of the element $[\sigma(P)] \in K^0(T^*V)$ under the map (5.1.18), we have expressed it in terms of the image of $[H]$ under the isomorphism (5.1.4). $\square$

Example 5.12 Let P be a first-order elliptic differential operator between H and H' and let P^* be its adjoint. Suppose that the leading term of the symbol of $\Delta = P^*P$ and of that of $\Delta' = PP^*$ are both equal to that of the ordinary Laplace operator. Then the index theorem can be proved using the 'heat-kernel' method. The two operators Δ and Δ' can only differ in their zero eigenvalues and it is easy to see that for all $t > 0$ the index $i(P)$ of P can be written

$$i(P) = \mathrm{Tr}(e^{-t\Delta}) - \mathrm{Tr}(e^{-t\Delta'}).$$

Using (5.1.12) we can write

$$\mathrm{Tr}(e^{-t\Delta}) \simeq \mathrm{Tr} \int_{T^*V} \sigma(e^{-t\Delta})dxdk$$

where the trace on the right-hand side is over the symbol matrix. The equality is valid asymptotically in the limit when $t \to 0$. The symbol $\sigma(e^{-t\Delta})$ can be expressed in terms of $\sigma(\Delta)$ and its derivatives. Using this fact one can show that in the limit $t \to 0$ the right-hand side has an asymptotic expansion in powers of t and therefore that

$$\mathrm{Tr}(e^{-t\Delta}) \simeq \sum_0^\infty t^{(i-n)/2} B_i[\Delta]. \tag{5.1.23}$$

The $B_i[\Delta]$ are integrals over V of invariants formed from $\sigma(\Delta)$ and its derivatives and n is the dimension of V. Since V has no boundary the coefficients B_i vanish for odd values of i. The index is given by the t-independent term:

$$i(P) = B_n[\Delta] - B_n[\Delta'].$$

$\square$

5.2   A matrix analogue

We saw that a vector bundle in the commutative case was equivalent to a projective $\mathcal{A}$-module. The analogue in the matrix case is therefore an M_n-module. (All M_n-modules are projective.) Let $\mathcal{H}$ be an M_n-module. The simplest M_n-module is the vector space $\mathcal{H} = \mathbb{C}^n$. This is the matrix analogue of a non-trivial line bundle. It would be possible to consider $\mathbb{C}^n$ as the trivial bundle over a point, in which case it would more properly be considered as a free $\mathbb{C}$-module of rank n. What corresponds to the trivial line bundle is the algebra M_n itself. Consider the idempotents e_1 and e_2 which project onto respectively the first l and the last $n - l$ columns of M_n. Then $e_1 + e_2 = 1$. Define the two left M_n-modules $\mathcal{H}_l = M_n e_1$ and $\mathcal{H}_{n-l} = M_n e_2$. Then $\mathcal{H}_l$ is projective and the direct sum $\mathcal{H}_l \oplus \mathcal{H}_{n-l} = M_n$ is the free M_n-module of rank 1.

The algebra $M_\infty(M_n)$ is isomorphic to the algebra of sparse matrices over the complex numbers:

$$M_\infty(M_n) \simeq M_\infty(\mathbb{C}).$$

The analogue of a general vector bundle is an idempotent in this algebra. Since all idempotents are classified by their rank we have

$$K_0(M_n) = \mathbb{Z}.$$

To underline the analogy with the calculations of Section 5.1 in the commutative case we define $[l]$ to be the equivalence class of idempotents which correspond to free

M_n-modules of rank l and we set $-[l]$ to be its formal negative. For every idempotent e_1 there is an e_2 such that $e_1 + e_2$ is equivalent to $[l]$ for some l and so the negative of e_1 is

$$-[e_1] = [e_2] - [l].$$

The trace of the idempotent $[l]$ is equal to nl. We divide by n and define a rescaled trace $\tau([e]) = \text{Tr}(e)/n$. The decomposition of a module into a free part and a projective part corresponds then to the decomposition $\tau([e]) = l + j/n$ of $\tau([e])$ into an integer plus an integral multiple of $1/n$. In Example 5.19 we shall mention a similar but non-trivial situation with $1/n$ replaced by an irrational number $\alpha < 1$.

We saw in the previous section that in the formulation of the topological index an important role was played by the Chern character and the isomorphism (5.1.4). We would like therefore to have a natural object to associate to M_n whose even cohomology is also $\mathbb{Z}$. In Chapter 7 we shall discuss a sense in which M_n tends to the algebra of functions on the 2-sphere and the geometry of M_n tends to that of S^3. In this respect we can say that the matrix analogue of the isomorphism (5.1.4) is the trivial identity map

$$K_0(M_n) \to H^0(S^3; \mathbb{Z}) = \mathbb{Z}.$$

In the next section we shall introduce a dual to the Chern character, of which the matrix algebras will furnish also a trivial example.

The analogue of a differential operator is an element of the matrix algebra M_{n^2}. We shall study one such operator in Chapter 7. An elliptic operator is a matrix which has no negative eigenvalues. This case is too simple to be of help in understanding the topological significance of the analytic index. The index vanishes for every operator of finite dimension.

Example 5.13 We can now return again to the *quantum Hall effect* of Example 4.34. The wave function ψ of the electron is a section of a vector bundle over the torus. From the definition it is clear then that the condition (4.3.18) follows from the fact that the integral of the first Chern class over a 2-cycle must be an integer. The quantization condition (4.3.19) on the Hall conductivity can be derived in a more general context, in the presence, for example, of an external potential. Consider an infinite plane with again a constant perpendicular magnetic field but with the hamiltonian H given by

$$H = -\frac{1}{2\mu}D_\lambda D^\lambda + U,$$

where U is an arbitrary periodic external potential:

$$U(x^1 + a^1, x^2) = U(x^1, x^2 + a^2) = U(x^1, x^2).$$

The restriction of the wave function ψ to the unit cell in the plane defined by the periodicity of the potential is not a section of a U_1-bundle. Consider the case where the parameter α, which measures the magnetic flux through the unit cell, is rational, a ratio of two positive, mutually prime integers l and n:

$$\alpha = \frac{1}{2\pi} e B a^1 a^2 = l/n.$$

Then the restriction of ψ to n contiguous cells is a section of a U_1-bundle L whose first Chern number is equal to l. To see this it is most convenient to first fix a gauge. For example we set $A_\lambda = (0, ieBx^1)$. The generators of the magnetic translations are given then by

$$t_1 = -i\partial_1 + eBx^2, \qquad t_2 = -i\partial_2$$

and we can impose the boundary conditions

$$\psi(x^1 + a^1, x^2) = e^{-ieBa^1x^2}\psi(x^1, x^2), \qquad \psi(x^1, x^2 + a^2) = \psi(x^1, x^2).$$

From the condition on α we find then that $u^n v\psi = v u^n \psi$. Let $\mathcal{A}$ be the algebra generated by u and v. We see that u^n and v^n belong to the center of $\mathcal{A}$ and, as in Example 3.4 and Example 4.44, the algebra $\mathcal{A}$ is a matrix algebra over a commutative algebra. The commutative algebra is generated by translations of n units along the two axes. It can therefore be identified with the algebra of functions on the 'magnetic Brillouin zone', the region V of the space of p_λ defined by the inequalities $0 \leq p_1 \leq 2\pi/na^1$ and $0 \leq p_2 \leq 2\pi/na^2$. That is, we have

$$\mathcal{A} = \mathcal{C}(V) \otimes M_n.$$

If $\mathcal{A}$ is completed to a C^*-algebra then $\mathcal{C}(V)$ is replaced by $C^0(V)$. With this identification the wave function ψ becomes a vector in $\mathbb{C}^n$ with components in $\mathcal{C}(V)$. The hamiltonian can be considered as an element of $\mathcal{A}$ and the n eigenvalues are the equivalent of the Landau levels. An interesting point of this example is the fact that the (noncommutative) algebra of observables $\mathcal{A}$ is not equal to the (commutative) algebra of functions.

Consider the trivial bundle over V with fibre $\mathbb{C}^n$. Then ψ is a section of a line bundle which is a direct summand of this bundle and which can be characterized by an idempotent $e \in \mathcal{A}$. Under certain conditions there is a formula, the Kubo formula, which gives the conductivity σ_H as an integral over V. There are several forms for this formula which are discussed in the literature. We shall take the one which is most convenient for our purposes:

$$\sigma_H = \alpha_0 \int_V c_1(L) = \frac{\alpha_0}{2\pi i} \operatorname{Tr} \int_V edede.$$

Here $\alpha_0 = e^2/(2\pi)$ with e^2 the square of the electric charge. The quantization condition follows from (5.1.8). The real physics lies in the discussion of the conditions under which the Kubo formula is valid.   □

5.3   Fredholm modules

Let $\mathcal{A}$ be an algebra with a representation as a closed subalgebra of the algebra $\mathcal{B}(\mathcal{H})$ of all bounded operators on some Hilbert space $\mathcal{H}$. In the particular case where $\mathcal{H}$ comes from the space of smooth sections of a vector bundle H we saw that the index of an elliptic operator can be defined in terms of the index of bounded operators with finite kernel and cokernel. The elements $P \in \mathcal{B}(\mathcal{H})$ which are such that $\mathrm{Ker}(P)$ and $\mathrm{Coker}(P)$ are of finite dimension constitute the subalgebra $\mathcal{F}$ of *Fredholm* operators.

An operator whose range is of finite dimension is said to be of finite rank. In an appropriate basis then it is a finite matrix. The norm closure of all such operators is the algebra $\mathcal{K}$ of *compact* operators on $\mathcal{H}$. It is a closed 2-sided ideal in $\mathcal{B}(\mathcal{H})$. The smoothing operators which we mentioned in Section 5.1 are compact. An element $a \in \mathcal{K}$ has a discrete spectrum with zero as unique possible accumulation point.

An operator P is Fredholm if and only if it has a *quasi-inverse*, an operator Q such that $PQ - 1$ and $QP - 1$ are compact. The analytic index (5.1.15) can be defined on the elements of $\mathcal{F}$. One can show that if P is Fredholm and K is compact then $P + K$ is a Fredholm operator with the same index as P. In particular the index of a compact operator is zero.

In Section 5.1 we introduced the homomorphism ch from $K^0(V)$ into $H^{2*}(V; \mathbb{Q})$. If we use cohomology with complex coefficients it can be written in the form of a map

$$K_0(\mathcal{C}(V)) \xrightarrow{\text{ch}^\bullet} H^{2*}(V; \mathbb{C}). \tag{5.3.1}$$

This is the (cohomology) Chern character. There is a dual homomorphism ch$_*$ called the *homology Chern character* which has a noncommutative extension.

Suppose $\mathcal{A}$ has a unit. Then we define $K_0(\mathcal{A})$, exactly as in the case $\mathcal{A} = \mathcal{C}(V)$, as the equivalence classes of idempotents in $M_\infty(\mathcal{A})$ under the equivalence relation defined in (5.1.7). The elements $[l]$ are still defined and every idempotent e is less than e_l for some l: $ee_l = e_l e = e$. Therefore $e_l - e$ is an idempotent and we have $[e] + [e_l - e] = [l]$. One can identify the matrix algebras as operators of finite rank on $\mathcal{H}$, and since $\mathcal{A}$ is a normed algebra it is sometimes convenient to replace $M_\infty(\mathcal{A})$ by its closure, the tensor product $\mathcal{A} \otimes \mathcal{K}$.

Let $e \in M_m(\mathcal{A})$ represent an element of $K_0(\mathcal{A})$ and let P be Fredholm. Let f be an $n \times n$ complex matrix. Then a general element of $M_m(\mathcal{A})$ is a linear combination of elements of the form $f \otimes a$. Define P_e to be the Fredholm operator eP acting on the Hilbert space $e(\mathbb{C}^m \otimes \mathcal{H})$:

$$P_e = ePe.$$

On the right-hand side we have identified P with $1 \otimes P$ on the tensor product $\mathbb{C}^m \otimes \mathcal{H}$. For every P we have then a map

$$K_0(\mathcal{A}) \to \mathbb{Z},$$

given by the analytic index

$$e \mapsto i_a(P_e).$$

Since we shall be concerned with the index of P it is convenient to consider it together with its quasi-inverse Q as a pair acting on an enlarged Hilbert space which is the direct sum of two copies $\mathcal{H}^{\pm}$ of $\mathcal{H}$:

$$\mathcal{H} = \mathcal{H}^+ \oplus \mathcal{H}^-. \tag{5.3.2}$$

Introduce ϵ as in (2.2.16) but where now the units are the units on $\mathcal{H}^{\pm}$. If $\psi = \psi^+ + \psi^-$ then $\epsilon\psi = \psi^+ - \psi^-$. The algebra $\mathcal{B}(\mathcal{H})$ has a $\mathbb{Z}_2$ grading,

$$\mathcal{B}(\mathcal{H}) = \mathcal{B}^+(\mathcal{H}) + \mathcal{B}^-(\mathcal{H}),$$

where an element of $\mathcal{B}^+(\mathcal{H})$ commutes with ϵ and an element of $\mathcal{B}^+(\mathcal{H})$ anticommutes with it. We choose the elements of $\mathcal{A}$ to be even:

$$\mathcal{A} \subset \mathcal{B}^+(\mathcal{H}).$$

We write P and Q together as an odd element

$$F = \begin{pmatrix} 0 & Q \\ P & 0 \end{pmatrix}.$$

The couple $(\mathcal{H}, F)$ is called a *Fredholm module*. Let λ_l be the spectrum of an element of $\mathcal{K}$. For each $p \geq 1$ define the ideals $\mathcal{L}^p \subset \mathcal{K}$ as the set of elements of $\mathcal{K}$ for which

$$\sum_l |\lambda_l|^p < \infty.$$

For example, $\mathcal{L}^1$ are trace-class operators and $\mathcal{L}^2$ are Hilbert-Schmidt. If for every element $a \in \mathcal{A}$, the commutator $[F, a]$ lies in $\mathcal{L}^p$ then $(\mathcal{H}, F)$ is a *p-summable* Fredholm module (over $\mathcal{A}$).

The motivation of the above definition comes from the commutative case. Let $\mathcal{H}$ be the space of sections of some vector bundle H over V and let $\hat{f}$ be left multiplication by an element $f \in \mathcal{C}(V)$. Then if P is a differential operator on $\mathcal{H}$ of order p the commutator $[P, \hat{f}]$ is of order $p-1$. In particular if P is a pseudodifferential operator of order 0 then $[P, \hat{f}]$ is a compact operator. Concerning such properties as the index of an operator, one need only consider operators of order 0. For example if Δ is the Laplace operator on V then P of order p has the same index as $P(\Delta+1)^{-p/2}$ of order 0.

Suppose that $(\mathcal{H}, F)$ is 1-summable and that P is invertible; therefore $F^2 = 1$. Suppose further that $Q = P^*$; therefore $F = F^*$. Define a *character* τ_0 as a linear map of $\mathcal{A}$ into the complex numbers given by

$$\tau_0(a) = \frac{1}{2}\mathrm{Tr}(\epsilon F[F, a]).$$

The form of this expression is suggested by the fact that if P is not invertible its index $i(P)$ can be written in the form of a trace:

$$i(P) = \text{Tr}(1 - QP) - \text{Tr}(1 - PQ) = \text{Tr}([P, Q]).$$

In Section 6.2 we shall introduce the *cyclic cohomology* groups $HC^*(\mathcal{A})$ and show that τ_0 is an element of $HC^0(\mathcal{A})$. A τ_0 can be defined for every element of $\mathcal{L}^1$. We have therefore a map

$$\mathcal{L}^1 \xrightarrow{\text{ch.}} HC^0(\mathcal{A}). \tag{5.3.3}$$

of the $1-$summable modules into the zeroth cyclic cohomology group.

Let f be a $m \times m$ complex matrix. Then a general element of $M_m(\mathcal{A})$ is a linear combination of elements of the form $f \otimes a$. We shall see in Section 6.3 that

$$HC^*(M_m(\mathcal{A})) \simeq HC^*(\mathcal{A})$$

for any integer m. The map τ_0 has a natural extension to $M_m(\mathcal{A})$ given by

$$\tau_0(f \otimes a) = \frac{1}{2}\text{Tr}(f)\text{Tr}(\epsilon F[F, a])$$

and we find that

$$\tau_0(e) = \text{Tr}([P_e, Q_e]).$$

We see then that $\tau_0(e) = i(P_e)$ and in particular that

$$\tau_0(e) \in \mathbb{Z}.$$

This can be written as a pairing

$$K_0(\mathcal{A}) \times \mathcal{L}^1 \to \mathbb{Z}$$

of $K_0(\mathcal{A})$ with the $1-$summable modules to yield an integer.

The construction has a generalization to p-summable modules. In Chapter 3 we introduced a differential algebra Ω_η^* with a $\mathbb{Z}_2$ grading and a differential based on (3.2.1). In Example 2.10 we showed that the ordinary exterior derivative on a manifold can be written as a commutator with the Dirac operator. These definitions can be extended to more general situations of algebras of operators on an arbitrary Hilbert space. On the elements $a \in \mathcal{B}(\mathcal{H})$ define a differential d by

$$da = i[F, a], \tag{5.3.4}$$

where the bracket is a graded commutator as in Section 3.2. Then by assumption d maps $\mathcal{A}$ into the odd elements of $\mathcal{L}^p$. By construction d is a graded derivation of $\mathcal{B}(\mathcal{H})$ and if we suppose that $F^2 = 1$ we have $d^2 = 0$. Since $F = F^*$ we have $d(a^*) = (da)^*$. In the noncommutative calculus the 'infinitesimals' are the compact operators and the 'infinitesimals of order $1/p$' are the elements of the ideal $\mathcal{L}^p$. An infinitesimal of order one can be integrated; the integral is the trace.

The construction of Section 3.2 is purely algebraic and independent of the dimension of $\mathcal{H}$. Moreover it simplifies in the present case since $d^2 = 0$. It yields a differential algebra Ω_F^*, which depends on the operator F. Define $\Omega_F^0 = \mathcal{A}$ and define $\Omega_F^1 = \overline{d\Omega_F^0} \subset \mathcal{L}^p$ to be the $\mathcal{A}$-module generated by the image of Ω_F^0 in $\mathcal{L}^p$ by d. As in (3.2.5) define

$$\Omega_F^0 \xrightarrow{d_F} \Omega_F^1$$

by $d_F = d$. With the appropriate identifications we have for each $k \geq 1$

$$\Omega_F^k = \{\alpha = a_0 da_1 \cdots da_k : a_i \in \mathcal{A}\}.$$

The differential d_F extends to a map

$$\Omega_F^k \xrightarrow{d_F} \Omega_F^{k+1}$$

by

$$d_F \alpha = da_0 \cdots da_k.$$

We shall give another definition of Ω_F^* in Section 6.1.

One can show that $\Omega_F^k \subset \mathcal{L}^{p/k}$ for every $k \geq 1$. Let k be an even integer and let $\alpha = a_0 da_1 \cdots da_k \in \Omega_F^k$. Then $[F, \alpha] \in \mathcal{L}^{p/(k+1)}$ is of trace class for $k \geq p - 1$. We can define therefore a *character* τ_k by

$$\tau_k(a_0, \cdots, a_k) = \frac{1}{2}\mathrm{Tr}(\epsilon F[F, \alpha]).$$

The τ_k are multilinear maps of $\mathcal{A}^{\otimes k}$ into the complex numbers which we shall discuss in more generality in Section 6.2. We shall see that τ_k can be identified with an element of the cyclic cohomology group $HC^k(\mathcal{A})$. A τ_k can be defined for every element in $\mathcal{L}^p$ with $p \leq k + 1$. We have extended then (5.3.3) to a map

$$\mathcal{L}^p \xrightarrow{\mathrm{ch}.} HC^{2*}(\mathcal{A}). \tag{5.3.5}$$

of the p-summable modules into the cyclic cohomology.

If we extend τ_k as we did τ_0 to the algebra $M_m(\mathcal{A})$ we find also that

$$i(P_e) = \tau_k(e, \cdots, e) \tag{5.3.6}$$

for any $k \geq p - 1$. This can again be written as a pairing

$$K_0(\mathcal{A}) \times \mathcal{L}^p \to \mathbb{Z} \tag{5.3.7}$$

of $K_0(\mathcal{A})$ with the p-summable modules to yield an integer. We shall illustrate this in Example 5.14.

The index of P_e is given by (5.3.6) for any sufficiently large even k. There is in fact a map

$$HC^{2*}(\mathcal{A}) \xrightarrow{S} HC^{2*}(\mathcal{A})$$

such that $\tau_{k+2} = S\tau_k$. There is also an extension of ch$_*$ to odd values of k.

Since a character is analogous to an integral it is suggestive to introduce the corresponding notation and write

$$\tau_k(a_0, \cdots, a_k) = \frac{1}{(2\pi i)^{k/2}} \int \alpha.$$

From the definitions it follows that this integral satisfies, *mutatis mutandis*, the conditions (3.4.14). As a particular case we find a condition analogous to (5.1.8),

$$\tau_2(e, e, e) = \frac{1}{2\pi i} \int edede \in \mathbb{Z},$$

where however the trace over the matrices and the integral over the manifold have been replaced by a trace over the algebra $M_m(\mathcal{A})$.

Example 5.14 Consider the Dirac operator $\displaystyle{\not}D$ on an associated Spin(n)-bundle over V of dimension n. Then

$$i(D^+) = \int_V \hat{A}(V).$$

If H is a vector bundle over V and e the corresponding element in $K_0(\mathcal{C}(V))$ then the index of D_e^+ is given by

$$i(D_e^+) = \int_V \mathrm{ch}(H)\hat{A}(V). \tag{5.3.8}$$

Although the integral of the Chern character alone would yield a rational number, with inclusion of the $\hat{A}$-genus one obtains an integer on the right-hand side. Let $\mathcal{H}$ be the sections of the trivial Spin(n)-bundle and set $P = D^+$. Then $(\mathcal{H}, F)$ is a p-module for $p > n$. The index of D_e^+ is also given by (5.3.6) for any $k > n$. Since e is an element of $K_0(\mathcal{C}(V))$ and the Dirac operator is an element of $\mathcal{L}^p$ for $p > n$, Equation (5.3.8) is an example of the pairing (5.3.7).   □

Example 5.15 Suppose that $\mathcal{H}$ is a vector space of even dimension n which is split as in (5.3.2) into the direct sum of two equal subspaces. Then $\mathcal{L}^p = \mathcal{K} = \mathcal{B}(\mathcal{H}) = M_n = M_n^+ \oplus M_n^-$ where the even elements M_n^+ leave invariant the two subspaces and the odd elements M_n^- interchange them. The algebra $\mathcal{A}$ can be chosen equal to M_n^+ and $\Omega_F^k \subseteq M_n^+(M_n^-)$ for even (odd) k. If we set $F = i\eta$ then (5.3.4) is the same as (3.2.1) of Section 3.2 with $\eta^2 = -1$ and we find $\Omega_F^* = \Omega_\eta^*$.   □

Example 5.16 Let $\mathcal{H}$ come from the space of smooth sections of some vector bundle H over V. The operator $\hat{f}$ defined as left multiplication by the function f is not of trace class. It would be useful to have a regularized trace on the elements of $\mathcal{A} \subset \mathcal{B}(\mathcal{H})$ such that the trace of $\hat{f}$ is equal to the integral of f. The first coefficient B_0 in the expansion (5.1.23) was shown by Weyl to depend only on the volume of V:

$$B_0 = \frac{\mathrm{Vol}(V)}{(4\pi)^{n/2}}.$$

Therefore for any second-order elliptic operator Δ we have

$$\mathrm{Vol}(V) = \lim_{t \to 0}(4\pi t)^{n/2}\mathrm{Tr}(e^{-t\Delta}).$$

For example, if Δ is the ordinary Laplace operator on the flat n-torus this follows from the identity

$$\theta(t) = (1/\sqrt{t})\theta(1/t)$$

satisfied by the classical θ-function. In this case $B_i = 0$ for $i > 0$. The expression for the volume can be modified to yield

$$\int_V f\,dx = \lim_{t \to 0}(4\pi t)^{n/2}\mathrm{Tr}(\hat{f}e^{-t\Delta}).$$

One could choose therefore the left-hand side as a general regularization procedure for an arbitrary element a of the algebra and set

$$\mathrm{Tr}_R(a) = \lim_{t \to 0}(4\pi t)^{n/2}\mathrm{Tr}(ae^{-t\Delta})$$

provided the limit exists. Another expression for the same limit can be obtained using the Mellin transform. We can write

$$(\Delta + \mu^2)^{-s} = \frac{1}{\Gamma(s)}\int_0^\infty e^{-t\Delta}e^{-\mu^2 t}t^{s-1}dt.$$

Using the expansion (5.1.23) we see that the trace of the left-hand side can be written as

$$\mathrm{Tr}((\Delta + \mu^2)^{-s}) = \frac{1}{\Gamma(s)}B_0\mu^{n-2s}\Gamma\left(s - \frac{n}{2}\right) + F(s)$$

with $F(s)$ analytic at $s = n/2$. The right-hand side has a simple pole at $s = n/2$. The residue Res of the pole is given by

$$\mathrm{Res} = \frac{B_0}{\Gamma(\frac{n}{2})}.$$

It is independent of the mass parameter μ as it should be. We can write then

$$\mathrm{Tr}_R(a) = \mathrm{Res}(a)$$

where

$$\text{Res}(a) = (4\pi)^{n/2}\Gamma(\frac{n}{2}) \lim_{s\to n/2} (s - \frac{n}{2})\text{Tr}(a(\Delta + \mu^2)^{-s}).$$

Yet a third way of writing the regularized trace is in terms of the eigenvalues. Since a is bounded, $a\Delta^{-n/2}$ is compact and has a discrete spectrum λ_l. Using the expression (5.1.12) for the trace one can see that the limit

$$\text{Tr}_\omega(a\Delta^{-n/2}) = \frac{1}{\log \Lambda} \sum_{l \le \Lambda} \lambda_l$$

exists. It is known as the *Dixmier trace* and is proportional to the regularized trace:

$$\text{Tr}_\omega(a\Delta^{-n/2}) = \frac{1}{(4\pi)^{n/2}\Gamma(\frac{n}{2})} \text{Tr}_R(a).$$

To define the action of a gauge theory defined on a noncommutative geometry one needs a generalized integral or trace which reduces in the commutative case to the ordinary integral. For example, let $\mathcal{A}$ be the algebra (3.3.4) with V a manifold of dimension 4. The trace $\text{Tr}_R(\hat{f})$ for any element $f \in \mathcal{A}$ is given by

$$\text{Tr}_R(\hat{f}) = \text{Tr} \int_V f dx,$$

where the trace on the right-hand side is the ordinary matrix trace. □

Example 5.17 The Dirac operator $\not{D}$ on the manifold V is a Fredholm operator if we consider as Hilbert space a completion of the space (2.2.17). So we can set $F = -\not{D}$. Then $F^2 \ne 1$. Suppose on the other hand that $\not{D}$ has an inverse and define $F = \Delta^{-1/2}\not{D}$ where $\Delta = \not{D}^2$. Then $F^2 = 1$. Let $V = S^4$ considered as compactified $\mathbb{R}^4$ with coordinates x^λ and let $\mathcal{H}$ be the Hilbert space completion of $\mathcal{C}(S^4) \otimes \mathbb{C}^4$. Let $\mathcal{A}$ be the closure of $\mathcal{C}(S^4)$ considered as an algebra of operators acting on $\mathcal{H}$ by left multiplication. Let λ_l be the spectrum of Δ. For large l we have $\lambda_l \sim \sqrt{l}$. Therefore for all $p > 4$ we have $d\hat{f} \in \mathcal{L}^p$ for every $\hat{f} \in \mathcal{A}$ and $\mathcal{H}$ is a Fredholm module. The cocycles τ_k exist for all $k \ge p - 1 > 3$. In the limiting case $k = 3$ the trace no longer exists. A useful positive regularized trace τ_{4R}^+ exists however as in the previous example. We define in general

$$\tau_k^+(a_0, \cdots, a_k) = \text{Tr}(\epsilon\alpha).$$

Consider the 4-form

$$\alpha = \hat{f}d\hat{x}^1 d\hat{x}^2 d\hat{x}^3 d\hat{x}^4 = \hat{f}\Delta^{-2}[\not{D}, \hat{x}^1][\not{D}, \hat{x}^2][\not{D}, \hat{x}^3][\not{D}, \hat{x}^4].$$

From (2.2.19) we have $\alpha = -\epsilon\hat{f}\Delta^{-2}$ and therefore we have formally

$$\tau_4^+(\hat{f}, \hat{x}^1, \hat{x}^2, \hat{x}^3, \hat{x}^4) = -\text{Tr}(\hat{f}\Delta^{-2}).$$

We find that

$$\tau^+_{4R}(\hat{f}, \hat{x}^1, \hat{x}^2, \hat{x}^3, \hat{x}^4) \propto \int_{S^4} f \, dx.$$

The limiting value of k at which τ^+_k must be regularized coincides with the dimension of the manifold.   □

Example 5.18 We mentioned that K-theory was in some respects a cohomology theory. There is a dual theory which is analogous to an homology theory and which is called K-*homology*. One defines $K^0(\mathcal{A})$ to be a set of equivalence classes $[P]$ of Fredholm operators P which have the property that for every $a \in \mathcal{A}$ the commutator $[P, a]$ is in $\mathcal{K}$. Then one can introduce a natural pairing

$$K_0(\mathcal{A}) \times K^0(\mathcal{A}) \rightarrow K^0(\mathcal{A}) \tag{5.3.9}$$

analogous to the cap product (2.3.7). Let $e \in M_m(\mathcal{A})$ represent an element of $K_0(\mathcal{A})$ and P an element of $K^0(\mathcal{A})$. To define the pairing we construct the element $[P \cap e] \in K^0(\mathcal{A})$ by setting

$$P \cap e = P_e.$$

The Fredholm modules $\mathcal{L}^p$ can be naturally embedded in $K^0(\mathcal{A})$:

$$\mathcal{L}^p \hookrightarrow K^0(\mathcal{A}).$$

They are the 'smooth' elements of $K^0(\mathcal{A})$. The triple $(\mathcal{A}, \mathcal{H}, F)$ is also called a K-*cycle*. In general ch$_*$ cannot be extended to a map of $K^0(\mathcal{A})$ into the cyclic cohomology.

If $\mathcal{A} = \mathcal{C}(V)$ no extension is needed since in this case all elements are 'smooth'. It follows on the other hand from the results of Section 6.2 that the homology groups $H_{2*}(V; \mathbb{C})$ are contained in the cyclic cohomology groups $HC^{2*}(\mathcal{A}(V))$. The maps (5.3.5) define in fact a map

$$K^0(\mathcal{C}(V)) \xrightarrow{\text{ch}_*} H_{2*}(V; \mathbb{C})$$

dual to (5.3.1). If we apply ch$_*$ and ch* to both sides of (5.3.9) we obtain the pairing (2.3.7), with complex coefficients. Just as (2.3.6) is a particular case of (2.3.7) the index pairings (5.3.7) follow as particular cases from the character maps.   □

Example 5.19 We can now return yet again to the *quantum Hall effect* of Example 4.34 and Example 5.13. We would like to consider the case of α not rational. This can be considered as a sort of limit where $n \rightarrow \infty$. There is no longer a matrix algebra and there is no V. However, one can still define a C^*-algebra $\mathcal{A}$ generated by u and v and one can still define maps τ_k from the idempotents of $\mathcal{A}$ into the integers. The Fredholm operator F used to define the differential (5.3.4) is constructed from the Fourier transform of the Dirac operator. We assume that $(\mathcal{H}, F)$ is p-summable for $p > 2$ as it is in the absence of a magnetic field. For all $k > 1$ then τ_k exists. It can be shown that

$$\sigma_H = \alpha_0\, \tau_2(e, e, e). \qquad (5.3.10)$$

Here $\alpha_0 = e^2/(2\pi)$ with e^2 the square of the electric charge. The simplest way to prove (5.3.10) is to show that $\tau_2(e, e, e)$ coincides for rational α with the expression given in (5.1.8):

$$\tau_2(e, e, e) = \int_V c_1(L).$$

Our notation suggests this. Here L is the line bundle which corresponds to the idempotent e.

Consider the idempotent e which projects onto the wave function of the electrons with energy less than or equal to the energy μ_F. Then e belongs to $\mathcal{A}$ if and only if μ_F belongs to a gap in the spectrum of the hamiltonian. The integral values of the conductivity appear then at the plateaux.

As a von Neumann algebra $\mathcal{A}$ is the hyperfinite II_1 factor of Example 4.44. The trace defines a map

$$K_0(\mathcal{A}) \xrightarrow{\tau} \mathbb{Z} \oplus \alpha\mathbb{Z}.$$

In the same sense that $\mathcal{A}$ is the limit of the algebras M_n of Example 5.13, this is the limit of the expression for the trace defined in Section 5.2. In Example 4.54 we partially indicated how one shows this result.   □

Notes

Further developments on the K-theory of spaces can be found in the article by Atiyah (1969) and in the books by Atiyah (1967) and Karoubi (1978). There is a detailed introduction to the K-theory of C^*-algebras in Blackadar (1998), Murphy (1990) and Wegge-Olsen (1993). We refer for example to Gilkey (1984) for more details on the definition of the symbol of a differential operator. For a proof of the Atiyah-Singer index theorem we refer to Shanahan (1978), Booss & Bleecker (1985) or to Berline et al. (1991).

The proposal to use a generalized Dirac operator to define the exterior derivative was first made by Connes (1986). The residue Res is discussed in Wodzicki (1984) and in Connes (1988). The use of the Dixmier trace as a generalized integral in noncommutative geometry, in particular to define the classical action in applications to field theory, was first advocated by Connes (1988). Connes & Rieffel (1987) have also developed and extended the notion of a Dixmier trace on certain types of algebras as a possible generalization of the notion of an integral. Topological invariants were introduced into noncommutative geometry by Connes (1986), Cunz

& Quillen (1995), Moscovici (1997). Example 5.13 is taken from the collection of original articles edited by Stone (1992). More details on K-homology, mentioned in Example 5.18, and its importance in the study of index theorems is to be found in the article by Baum & Douglas (1982). The noncommutative theory is developed by Connes (1986, 1994). The presentation given here is also based partly on private conversations with A. Čap. Example 5.19 is taken from Bellissard *et al.* (1994).

6   Cyclic Homology

We saw in Section 5.1 that the number of solutions to a differential equation on a manifold V can depend in an essential way on the global structure of V, as is for example encoded in the cohomology ring. It is important then to find an appropriate noncommutative generalization of cohomology. This means in particular that we must be able to express the de Rham cohomology of a manifold V in terms of the algebra of functions $\mathcal{C}(V)$. Recall from Section 2.1 that if ϕ is a map from one manifold into another then the induced map ϕ^* on the algebra of functions goes in the opposite direction. When we speak of chains in this chapter then we mean objects which resemble more the cochains of Chapter 2. We shall see in Example 6.9 below how the 1-chains defined over an algebra of functions are related to 1-cochains on the corresponding manifold. In Section 6.1 we shall give a general construction of a differential calculus which is purely algebraic and which can be used on any arbitrary associative algebra with a unit element including, for example, the algebra of continuous functions on a manifold. It is related to the finite-difference calculus which is used in numerical calculations. A generalization of the de Rham cohomology to noncommutative geometry is cyclic homology. It is defined in Section 6.2. In the last two sections Morita equivalence and the theorem of Loday and Quillen are stated but not proven.

6.1   The universal calculus

We have seen that it is possible to construct more than one differential calculus over a given algebra. Over the algebra M_n we constructed in Section 3.1 a 'spheroidal' calculus as well as a 'toroidal' one and in Example 3.12 we compared two different differential calculi over an ordinary smooth manifold. It is possible to construct at least one differential calculus over an arbitrary associative algebra. Consider an algebra $\mathcal{A}$ with unit element and define the set C_p of p-*chains* by

$$C_p = C_p(\mathcal{A}) = \mathcal{A}^{\otimes(p+1)}. \tag{6.1.1}$$

The tensor product is over the complex numbers. Each C_p is obviously an $\mathcal{A}$-bimodule. To simplify the notation we write $(a_0, \cdots, a_p)$ instead of $a_0 \otimes \cdots \otimes a_p$. We define a *coboundary operator* or *differential* d_u of C_p into C_{p+1} by the formula

$$\begin{aligned}
d_u(a_0, \cdots, a_p) = &(1, a_0, \cdots, a_p) \\
&+ \sum_{i=1}^{p} (-1)^i (a_0, \cdots, a_{i-1}, 1, a_i, \cdots, a_p) \\
&+ (-1)^{p+1}(a_0, \cdots, a_p, 1).
\end{aligned}$$

For example $d_u a_0 = (1, a_0) - (a_0, 1)$ lies in the kernel of the product of $\mathcal{A}$. We have written this before in Equation (3.2.14). It is easy to verify that

$$d_u^2 = 0.$$

Define $\alpha = (a_0, \cdots, a_p)$ and $\beta = (b_0, \cdots, b_q)$. Then the product

$$\alpha\beta = (a_0, \cdots, a_p b_0, \cdots, b_q)$$

makes the union C_* of all C_p into an associative algebra with

$$C_p C_q \subset C_{p+q}.$$

From the definition of d_u it follows that

$$d_u(\alpha\beta) = (d_u\alpha)\beta + (-1)^p \alpha d_u \beta$$

and so C_* is a graded differential algebra.

Let χ be any linear form on $\mathcal{A}$ with values in $\mathbb{C}$ normalized such that $\chi(1) = 1$. Then for each $p \geq 1$ the map

$$h_\chi(a_0, a_1, \cdots, a_p) = \chi(a_0)(a_1, \cdots, a_p).$$

of C_p into C_{p-1} is a *contracting homotopy*: $d_u h_\chi + h_\chi d_u = 1$. It follows that the sequence

$$0 \to \mathbb{C} \xrightarrow{\epsilon} C_0 \xrightarrow{d_u} C_1 \xrightarrow{d_u} \cdots$$
$$\xrightarrow{d_u} C_p \xrightarrow{d_u} C_{p+1} \xrightarrow{d_u} \cdots$$

is an exact sequence. The map ϵ associates to a complex number the corresponding multiple of the identity in $C_0 = \mathcal{A}$.

We can define also for each $p \geq 1$ a *boundary operator* b' of C_p into C_{p-1} by the formula

$$b'(a_0, \cdots, a_p) = \sum_{i=0}^{p-1} (-1)^i (a_0, \cdots, a_i a_{i+1}, \cdots, a_p).$$

For example $b'(a_0, a_1) = a_0 a_1$ is the product in $\mathcal{A}$. It is easy to see that

$$b'^2 = 0,$$

and that

$$d_u b' + b' d_u = 0.$$

For each $p \geq 0$, define the map s of C_p into C_{p+1} given by

$$s(a_0, \cdots, a_p) = (1, a_0, \cdots, a_p).$$

Then s is a contracting homotopy: $b's + sb' = 1$. It follows that the sequence

$$\cdots \xrightarrow{b'} C_{p+1} \xrightarrow{b'} C_p \xrightarrow{b'} \cdots$$
$$\xrightarrow{b'} C_2 \xrightarrow{b'} C_1 \xrightarrow{b'} C_0 \rightarrow 0. \tag{6.1.2}$$

is an exact sequence.

Using C_* we can construct the *universal calculus* $\Omega_u^*(\mathcal{A})$ over $\mathcal{A}$. Let $\Omega_u^0(\mathcal{A}) = \mathcal{A}$ and $\Omega_u^1(\mathcal{A}) \subset \mathcal{A} \otimes \mathcal{A}$ be the $\mathcal{A}$-bimodule generated by all elements of the form $d_u a$. For $p \geq 2$ define $\Omega_u^p(\mathcal{A})$ to be the $\mathcal{A}$-bimodule generated by all elements of the form $d_u a_1 \cdots d_u a_p$. The product here is the product in C_*. It can also be defined as the tensor product in $\Omega_u^1(\mathcal{A})$ taken over $\mathcal{A}$. A general element of $\Omega_u^p(\mathcal{A})$ can be written as a sum of elements of the form $a_0 d_u a_1 \cdots d_u a_p$ and this in turn can be expanded in terms of tensor products. By construction

$$\Omega_u^p(\mathcal{A}) \subset C_p.$$

For example

$$d_u a_0 d_u a_1 = (1, a_0, a_1) - (1, a_0 a_1, 1) - (a_0, 1, a_1) + (a_0, a_1, 1)$$

is an element of C_2. From the right-hand side of this equation one notices that

$$d_u a_0 d_u a_1 \neq -d_u a_1 d_u a_0.$$

From the definition of the product in C_* we have

$$\Omega_u^p(\mathcal{A})\Omega_u^q(\mathcal{A}) = \Omega_u^{p+q}(\mathcal{A}).$$

The contracting homotopy h_χ can be restricted to $\Omega_u^p(\mathcal{A})$. It follows therefore that the cohomology of $\Omega_u^*(\mathcal{A})$ is also trivial,

$$H^0(\Omega_u^*(\mathcal{A})) = \mathbb{C}, \quad H^p(\Omega_u^*(\mathcal{A})) = 0, \ p \geq 1,$$

and that the sequence

$$0 \longrightarrow \mathbb{C} \xrightarrow{\epsilon} \Omega_u^0(\mathcal{A}) \xrightarrow{d_u} \Omega_u^1(\mathcal{A}) \xrightarrow{d_u} \cdots$$
$$\xrightarrow{d_u} \Omega_u^p(\mathcal{A}) \xrightarrow{d_u} \Omega_u^{p+1}(\mathcal{A}) \xrightarrow{d_u} \cdots$$

is an exact sequence.

The universal calculus is distinguished by a *universal property*. Let $(\Omega^*(\mathcal{A}), d)$ be some other differential calculus over $\mathcal{A}$. Then there is a unique d_u-homomorphism $\hat{\phi}$,

$$\Omega_u^*(\mathcal{A}) \xrightarrow{\hat{\phi}} \Omega^*(\mathcal{A}),$$

of $\Omega_u^*(\mathcal{A})$ onto $\Omega^*(\mathcal{A})$. It is given by

$$\hat{\phi}(a) = a, \qquad \hat{\phi}(d_u a) = da. \tag{6.1.3}$$

The restriction $\hat{\phi}_p$ of $\hat{\phi}$ to each $\Omega_u^p(\mathcal{A})$ is defined by

$$\hat{\phi}_p(a_0 d_u a_1 \cdots d_u a_p) = a_0 da_1 \cdots da_p.$$

Example 6.1 Let $\mathcal{A}$ and $\mathcal{A}'$ be two algebras. Then the tensor product $\Omega_u^*(\mathcal{A}) \otimes \Omega_u^*(\mathcal{A}')$ is a differential calculus over $\mathcal{A} \otimes \mathcal{A}'$. From the universal property there is a projection of $\Omega_u^*(\mathcal{A} \otimes \mathcal{A}')$ onto $\Omega_u^*(\mathcal{A}) \otimes \Omega_u^*(\mathcal{A}')$ but in general the two are not isomorphic. By definition

$$\Omega^0(\mathcal{A} \otimes \mathcal{A}') = \Omega^0(\mathcal{A}) \otimes \Omega^0(\mathcal{A}')$$

and one can easily see that

$$\Omega^1(\mathcal{A} \otimes \mathcal{A}') = \Omega^1(\mathcal{A}) \otimes \mathcal{A}' + \mathcal{A} \otimes \Omega^1(\mathcal{A}').$$

However in higher orders $\hat{\phi}$ is in general a proper projection. To see this we choose the element

$$\alpha = (a \otimes a') d_u (b \otimes b') d_u (c \otimes c')$$

in $\Omega^2(\mathcal{A} \otimes \mathcal{A}')$. Since

$$\hat{\phi}_1(d_u(b \otimes b')) = d_u b \otimes b' + b \otimes d_u b',$$
$$\hat{\phi}_1(d_u(c \otimes c')) = d_u c \otimes c' + c \otimes d_u c'$$

we see that

$$\hat{\phi}_2(\alpha) = a d_u b d_u c \otimes a'b'c' + (a d_u b c \otimes a'b'd_u c' + ab d_u c \otimes a'd_u b'c') + abc \otimes a'd_u b'd_u c'.$$

It is clear that if we choose $ab = 0$ and $a'b' = 0$ then $\hat{\phi}_2(\alpha) = 0$. In this case α can be written as

$$\alpha = (a \otimes a', b \otimes b', c \otimes c') - (a \otimes a', bc \otimes b'c', 1).$$

$\square$

Suppose that $\mathcal{A}$ is a subalgebra of the even elements $\mathcal{B}^+(\mathcal{H})$ of the algebra of bounded operators $\mathcal{B}(\mathcal{H})$ on a graded Hilbert space $\mathcal{H}$ and let F be a self-adjoint Fredholm operator on $\mathcal{H}$. Define on $\mathcal{A}$ a graded derivation $\hat{d}$ by $\hat{d}a = i[F, a]$. In general $\hat{d}^2 \neq 0$. Let $\hat{\phi}$ be the graded map of $\Omega_u^*(\mathcal{A})$ into $\mathcal{B}(\mathcal{H})$ given by (6.1.3). That is, $\hat{\phi}(a) = a \in \mathcal{A} \subset \mathcal{B}(\mathcal{H})$ and $\hat{\phi}(d_u a) = \hat{d}a \in \mathcal{B}^-(\mathcal{H})$. Then we can construct from $\hat{\phi}$ a *differential algebra* Ω_F^*. The kernel $\text{Ker}\,\hat{\phi}_p$ of $\hat{\phi}_p$ is a submodule of $\Omega_u^p(\mathcal{A})$. Define for each $p \geq 1$

$$\Omega_F^p = \Omega_u^p(\mathcal{A})/(\text{Ker}\,\hat{\phi}_p + d_u \text{Ker}\,\hat{\phi}_{p-1}). \tag{6.1.4}$$

Let $[\alpha]$ be the equivalence class in Ω_F^p of a cochain $\alpha \in \Omega_u^p(\mathcal{A})$. Define

$$\Omega_F^p \xrightarrow{d_F} \Omega_F^{p+1}$$

by $d_F[\alpha] = [d_u\alpha]$. This is well defined since if $[\alpha] = 0$ then $\alpha \in \mathrm{Ker}\,\hat{\phi}_p + d_u\mathrm{Ker}\,\hat{\phi}_{p-1}$ and thus $d_u\alpha \in d_u\mathrm{Ker}\,\hat{\phi}_p$, which implies that $d_F[\alpha] = [d_u\alpha] = 0$. Also by construction $d_F^2 = 0$. Therefore (Ω_F^*, d_F) is a differential calculus over $\mathcal{A}$.

Example 6.2 Let M_n be the graded matrix algebra of Section 3.2 and $\mathcal{A} = M_n^+$. Set $F = i\eta$. We have by definition $\Omega_F^0 = \Omega_\eta^0$. By construction $\Omega_F^1 = \Omega_u^1/\mathrm{Ker}\,\hat{\phi}_1$ is equal to the $\mathcal{A}$-module generated by $\hat{d}\Omega_F^0$. On the other hand Ω_η^1 is the M_n^+-module generated by $\hat{d}\Omega_\eta^0$. Therefore $\Omega_F^1 = \Omega_\eta^1$. By construction, for every $p \geq 1$ a general element of Ω_η^p can be written as a sum of elements of the form $\alpha = a_0\hat{d}a_1\cdots\hat{d}a_p$. Let $\beta = a_0 d_u a_1 \cdots d_u a_p \in \Omega_u^p$. Then $\hat{\phi}_p(\beta) = \alpha$ and α vanishes in Ω_η^p if and only if $\beta \in (\mathrm{Ker}\,\hat{\phi}_p + d_u\mathrm{Ker}\,\hat{\phi}_{p-1})$. Therefore $\Omega_F^p = \Omega_\eta^p$.   □

Example 6.3 The Dirac operator $\not{D}$ on the manifold V is a Fredholm operator if we consider as Hilbert space a completion of the space (2.2.17). So we can set $F = -\not{D}$. Then $F^2 \neq 1$. Consider the algebra $\mathcal{A} = \mathcal{C}(V)$ as embedded by left multiplication in the algebra of bounded operators. Then $\hat{d}$ was defined already in Equation (2.2.20). On functions it coincides with the ordinary differential. So in this case the forms (6.1.4) can be identified with the module of de Rham p-forms:

$$\Omega_F^* = \Omega^*(V).$$

□

The diagram (3.2.16) can be continued to yield a differential calculus which completes the module of 1-forms. The construction is essentially the same as the previous one. The calculus so constructed is the largest which is consistent with the constraints which are a consequence of the module structure of the 1-forms but it is always possible to construct others which are smaller. Suppose that there are given an $\mathcal{A}$-bimodule $\Omega^1(\mathcal{A})$ and a map

$$\mathcal{A} \xrightarrow{d} \Omega^1(\mathcal{A})$$

which satisfies the Leibniz rule $d(fg) = f(dg) + (df)g$. Then one can define by induction a series of bimodules $\Omega^p(\mathcal{A})$ for $p \geq 2$ and construct an extension

$$
\begin{array}{ccccccc}
\mathcal{A} & \xrightarrow{d_u} & \Omega_u^1(\mathcal{A}) & \xrightarrow{d_u} & \Omega_u^2(\mathcal{A}) & \xrightarrow{d_u} & \cdots \\
\parallel & & \phi_1 \downarrow & & \phi_2 \downarrow & & \\
\mathcal{A} & \xrightarrow{d} & \Omega^1(\mathcal{A}) & \xrightarrow{d} & \Omega^2(\mathcal{A}) & \xrightarrow{d} & \cdots
\end{array}
$$

of the diagram (3.2.16).

Let ξ be the image in $\Omega^1(\mathcal{A})$ of some element ξ_u in $\Omega^1_u(\mathcal{A})$. We must require that $d\xi = 0$ whenever $\xi = 0$. That is, the image of $d_u\xi_u$ should vanish in $\Omega^2(\mathcal{A})$. The largest consistent choice would be to set then

$$\Omega^2(\mathcal{A}) = \Omega^2_u(\mathcal{A})/\overline{d_u\mathrm{Ker}\,\phi_1} \qquad (6.1.5)$$

where $\overline{d_u\mathrm{Ker}\,\phi_1}$ is the bimodule generated by $d_u\mathrm{Ker}\,\phi_1$. The map ϕ_2 is defined to be the projection of $\Omega^2_u(\mathcal{A})$ onto $\Omega^2(\mathcal{A})$. An element $\alpha \in \overline{d_u\mathrm{Ker}\,\phi_1}$ can be written as a sum

$$\alpha = \sum_i f_i d_u\xi_i g_i,$$

with $f_i, g_i \in \mathcal{A}$ and $\xi_i \in \mathrm{Ker}\,\phi_1$. This can be rewritten in the form

$$\alpha = \sum_i d_u(f_i\xi_i g_i) - \sum_i d_u f_i \otimes \xi_i g_i - \sum_i f_i\xi_i \otimes d_u g_i$$

using the properties of the universal calculus. We see then that one can write

$$\overline{d_u\mathrm{Ker}\,\phi_1} = d_u\mathrm{Ker}\,\phi_1 + \Omega^1_u(\mathcal{A}) \otimes \mathrm{Ker}\,\phi_1 + \mathrm{Ker}\,\phi_1 \otimes \Omega^1_u(\mathcal{A}).$$

From the definition of the universal calculus we have

$$\Omega^2_u(\mathcal{A}) = \Omega^1_u(\mathcal{A}) \otimes_{\mathcal{A}} \Omega^1_u(\mathcal{A})$$

and hence a map $\phi_1 \otimes \phi_1$ of $\Omega^2_u(\mathcal{A})$ onto $\Omega^1(\mathcal{A}) \otimes_{\mathcal{A}} \Omega^1(\mathcal{A})$. Therefore $\Omega^2(\mathcal{A})$ can be written also as

$$\Omega^2(\mathcal{A}) = (\Omega^1(\mathcal{A}) \otimes_{\mathcal{A}} \Omega^1(\mathcal{A}))/\mathcal{K}$$

with

$$\mathcal{K} = (\phi_1 \otimes \phi_1)(\overline{d_u\mathrm{Ker}\,\phi_1}) = (\phi_1 \otimes \phi_1)(d_u\mathrm{Ker}\,\phi_1).$$

It can happen that $\mathcal{K} = \Omega^1(\mathcal{A}) \otimes_{\mathcal{A}} \Omega^1(\mathcal{A})$, in which case $\Omega^2(\mathcal{A}) = 0$.

Let π be the projection (3.1.39). Then one sees that ϕ_2 is given by

$$\phi_2 = \pi \circ (\phi_1 \otimes \phi_1).$$

We have supposed that π has a right inverse (3.1.40) so we can identify $\Omega^2(\mathcal{A})$ as a submodule of $\Omega^1(\mathcal{A}) \otimes_{\mathcal{A}} \Omega^1(\mathcal{A})$.

Let ξ_u be an inverse image in $\Omega^1_u(\mathcal{A})$ of ξ in $\Omega^1(\mathcal{A})$. Then the map d from $\Omega^1(\mathcal{A})$ to $\Omega^2(\mathcal{A})$ can be written in terms of d_u as

$$d\xi = \phi_2(d_u\xi_u).$$

We have defined $\Omega^2(\mathcal{A})$ so that this map is independent of the choice of ξ_u. Equation (6.1.5) defines the largest set of 2-forms consistent with the constraints on the module structure of $\Omega^1(\mathcal{A})$.

The above procedure can be continued by iteration to arbitrary order in $p \geq 2$. Introduce for this the map

$$\Omega_u^p(\mathcal{A}) \xrightarrow{\hat{\phi}_p} \bigotimes_1^p \Omega^1(\mathcal{A})$$

defined by $\hat{\phi}_p = \phi_1 \otimes \cdots \otimes \phi_1$ and set

$$\Omega^p(\mathcal{A}) = \Omega_u^p(\mathcal{A})/(\text{Ker}\,\hat{\phi}_p + d_u\text{Ker}\,\hat{\phi}_{p-1}). \tag{6.1.6}$$

This is the same as (6.1.4) in a more formal context. We have then by definition

$$\text{Ker}\,\phi_p = \text{Ker}\,\hat{\phi}_p + d_u\text{Ker}\,\hat{\phi}_{p-1}. \tag{6.1.7}$$

For example

$$\text{Ker}\,\hat{\phi}_2 = \Omega_u^1(\mathcal{A}) \otimes \text{Ker}\,\phi_1 + \text{Ker}\,\phi_1 \otimes \Omega_u^1(\mathcal{A})$$

and using (6.1.7) we find (6.1.5) when $p = 2$. The choice (6.1.6) for $\Omega^p(\mathcal{A})$ is made so that the differential d will be well defined. Let $\alpha = \phi_p(\alpha_u)$ be the image in $\Omega^p(\mathcal{A})$ of $\alpha_u \in \Omega_u^p(\mathcal{A})$ and define

$$\Omega^p(\mathcal{A}) \xrightarrow{d} \Omega^{p+1}(\mathcal{A})$$

by $d\alpha = \phi_{p+1}(d_u\alpha_u)$. This is well defined since if $\alpha = 0$ then

$$\alpha_u \in \text{Ker}\,\phi_p = \text{Ker}\,\hat{\phi}_p + d_u\text{Ker}\,\hat{\phi}_{p-1}$$

and thus $d_u\alpha_u \in d_u\text{Ker}\,\hat{\phi}_p$, which implies that $d\alpha = \phi_{p+1}(d_u\alpha_u) = 0$.

Equation (6.1.6) can be rewritten as

$$\Omega^p(\mathcal{A}) = \bigotimes_1^p \Omega^1(\mathcal{A})/\hat{\phi}_p(d_u\text{Ker}\,\hat{\phi}_{p-1})$$

or in the form

$$\Omega^p(\mathcal{A}) = \bigotimes_1^p \Omega^1(\mathcal{A})/\mathcal{K}_p$$

with the $\mathcal{K}_p$ defined by the recurrence relations

$$\mathcal{K}_p = \mathcal{K}_{p-1} \otimes \Omega^1(\mathcal{A}) + \Omega^1(\mathcal{A}) \otimes \mathcal{K}_{p-1}, \qquad \mathcal{K}_2 = \mathcal{K}.$$

In particular we find the expression

$$\Omega^3(\mathcal{A}) = \frac{\Omega^1(\mathcal{A}) \otimes \Omega^1(\mathcal{A}) \otimes \Omega^1(\mathcal{A})}{\mathcal{K} \otimes \Omega^1(\mathcal{A}) + \Omega^1(\mathcal{A}) \otimes \mathcal{K}}$$

for the module of 3-forms.

If a frame exists then from the associativity rule for the product in $\Omega^3(\mathcal{A})$ one finds that the $C^{ij}{}_{kl}$ defined in Equation (3.1.51) must satisfy a weak form of the *braid relation*. Consider first the case when the elements of the frame anticommute and $C^{ij}{}_{kl}$ is therefore given by Equation (3.1.54). One can interchange the second two factors then the first two and then again the second two. This yields a trivial sequence of identities: $\theta^i\theta^j\theta^k = -\theta^i\theta^k\theta^j = \theta^k\theta^i\theta^j = -\theta^k\theta^j\theta^i$. One can also interchange the first two factors then the second two and then again the first two. This yields another trivial sequence of identities: $\theta^i\theta^j\theta^k = -\theta^j\theta^i\theta^k = \theta^j\theta^k\theta^i = -\theta^k\theta^j\theta^i$. The consistency condition for the existence of an associative product within the algebra of forms is the condition that these two procedures yield the same result, as indeed they do. This must also be true in general when the product satisfies the condition $\theta^i\theta^j = -C^{ij}{}_{kl}\theta^k\theta^l$. One finds that the consistency conditions are equivalent to a weak form

$$C^{ijk}{}_{lmn}\theta^l\theta^m\theta^n = 0 \tag{6.1.8}$$

of the braid relation. We have here defined

$$C^{ijk}{}_{lmn} = C^{ji}{}_{pq}C^{kp}{}_{lr}C^{rq}{}_{mn} - C^{kj}{}_{pq}C^{qi}{}_{rn}C^{pr}{}_{lm}.$$

The braid relation (4.4.13) written out with indices is the condition that $C^{ijk}{}_{lmn} = 0$. If one writes

$$C^{ij}{}_{kl} = \delta^j_k\delta^i_l + C^{ij}_{(1)kl}$$

as a perturbation of (3.1.54) then from (3.1.52) one finds that to first order

$$C^{ij}_{(1)lk} + C^{ji}_{(1)kl} = 0.$$

To first order in $C^{ij}_{(1)kl}$ the braid relation is identically satisfied. However to second order the condition (6.1.8) is non-trivial. To this order one can suppose that the forms on the left-hand side of (6.1.8) anticommute.

If a frame exists then the structure of $\Omega^*(\mathcal{A})$ is easy to calculate. Define $\bigwedge^*_C$ to be the algebra generated by the θ^i with the product given by Equation (3.1.43). Then because of the trivial commutation relations (3.1.34) we have obviously

$$\Omega^*(\mathcal{A}) = \mathcal{A} \otimes \bigwedge^*_C. \tag{6.1.9}$$

This identification is an isomorphism of algebras. The differential will not necessarily respect it except if the structure elements $C^i{}_{jk}$ lie in the center of the algebra.

Example 6.4 Consider again the calculus of $\Omega^*(M_n)$ of Section 3.1 constructed using the complete Lie algebra of all derivations. We saw in Example 3.13 that in this case the map ϕ_1 is an isomorphism and therefore the construction we have just given yields simply the universal calculus, which is larger than $\Omega^*(M_n)$. It is easy to compare the two. Define $\bigwedge^*$ and T^* to be respectively the exterior algebra and the tensor algebra over the vector space spanned by the θ^a. Then we have

$$\Omega^*(M_n) = M_n \otimes \textstyle\bigwedge^*, \qquad \Omega_u^*(M_n) = M_n \otimes T^*.$$

The latter has a natural projection onto the former.   □

Example 6.5 Let $\mathbb{R}^{2m}$ be the phase space of a hamiltonian system and suppose that there are $m - n$ constraints $\phi_a = 0$ which reduce the possible configurations to a smooth manifold Σ of dimension $m + n$. Let ω be the symplectic form on $\mathbb{R}^{2m}$ and let X_a be the hamiltonian flow associated to the function ϕ_a. Suppose that the X_a close under the Lie bracket to form a Lie algebra $\underline{g}$ and that they are regular in some open neighbourhood $\mathcal{O}$ of Σ. Then the construction of Example 4.12 yields a differential calculus $(\Omega'^*(\mathcal{O}), d')$ over $\mathcal{C}(\mathcal{O})$. Let $H^*(d')$ be the associated cohomology. It follows from the definitions that $H^0(d')$ can be identified with the algebra of functions on $\mathcal{O}$ which are invariant under the action of $\underline{g}$. It is an algebra of functions on a space of dimension $2m - (m - n) = m + n$. The assumptions we have made exclude most hamiltonian systems of interest but the construction yields a finite-dimensional analogue of the hamiltonian formulation of gauge theories. In this context the differential forms are called *ghosts* and they are often represented by the letter c.   □

Example 6.6 With the notation of the previous example let $\bigwedge^* \underline{g}$ be the exterior algebra over $\underline{g}$. Consider the complex

$$\Omega''^{-*} = \mathcal{C}(\mathcal{O}) \otimes \textstyle\bigwedge^* \underline{g}.$$

We shall construct a differential d'' on this complex to yield a second differential calculus over (or rather under) $\mathcal{C}(\mathcal{O})$. We define

$$\Omega''^{-p} \xrightarrow{\ d''\ } \Omega''^{-p-1}$$

by replacing in one occurrence the vector X_a by the function ϕ_a and summing over all possibilities, with alternating sign. For example $d''(f^a X_a) = f^a \phi_a$ and $d''(f^{ab} X_a \wedge X_b) = 2 f^{ab} \phi_a X_b$. Let $H^*(d'')$ be the associated cohomology. It follows from the definitions that $H^0(d'')$ can be identified with the space of functions on Σ. It is an algebra of functions on a space of dimension $2m - (m - n) = m + n$. In this context the elements of the Lie algebra are called *anti-ghosts* and they are often represented by the symbol $\bar{c}$.   □

Example 6.7 The differential calculi of the two previous examples can be used to construct a third differential calculus over $\mathcal{C}(\mathcal{O})$. Let $\bigwedge^* \underline{g}^*$ be the exterior algebra over $\underline{g}^*$, the dual of $\underline{g}$, generated by the ghosts. Consider the double complex

$$\Omega^{*,-*} = \mathcal{C}(\mathcal{O}) \otimes \textstyle\bigwedge^* \underline{g}^* \otimes \textstyle\bigwedge^* \underline{g}.$$

There is a unique extension of d' and d'' to all of $\Omega^{*,-*}$ so that the sum $d = d' + d''$ satisfies

$$d^2 = 0.$$

Define now the complex Ω^* by

$$\Omega^r = \bigoplus_{k-l=r} \Omega^{k,-l}.$$

Then (Ω^*, d) is a differential calculus over $\mathcal{C}(\mathcal{O})$. It follows from the definitions that $H^0(d)$ can be identified with the space of gauge-invariant functions on Σ. It is an algebra of functions on a space of dimension $2m - 2(m-n) = 2n$. The operator d creates a ghost through d' and destroys an anti-ghost through d''. One can include the original coordinates as well as the ghosts and anti-ghosts as $2(m+n)$ generators of a graded algebra. The latter correspond to negative degrees of freedom. The ghosts subtract $m - n$ gauge degrees of freedom and the anti-ghosts subtract $m - n$ constraints. The system is reduced then to one of dimension $2n$. The differential d acting on an element α can be written as the commutator of the operator Q defined by

$$Q = \phi_a c^a + \frac{1}{2} C^a{}_{bc} c^b c^c \bar{c}_a.$$

This is to be compared with the operator η defined in Equation (3.2.12).   □

6.2   Cyclic homology

Let $\mathcal{M}$ be an $\mathcal{A}$-bimodule and consider the sequence

$$\cdots \xrightarrow{b} \mathcal{M} \otimes C_{p+1} \xrightarrow{b} \mathcal{M} \otimes C_p \cdots$$
$$\xrightarrow{b} \mathcal{M} \otimes C_1 \xrightarrow{b} \mathcal{M} \otimes C_0 \xrightarrow{b} \mathcal{M} \longrightarrow 0$$

with *boundary operator* b given by

$$b(m, a_1, \cdots, a_p) = (ma_1, \cdots, a_p)$$
$$+ \sum_{i=1}^{p-1} (-1)^i (m, a_1, \cdots, a_i a_{i+1}, \cdots, a_p)$$
$$+ (-1)^p (a_p m, a_1, \cdots, a_{p-1}).$$

It is easy to see that

$$b^2 = 0.$$

The homology of $\mathcal{A}$ with coefficients in $\mathcal{M}$ is by definition the homology of $\mathcal{M} \otimes C_{*-1}$:

$$H_*(\mathcal{A}; \mathcal{M}) = H_*(\mathcal{M} \otimes C_{*-1}).$$

The p-th homology group is the homology of the sequence at $\mathcal{M} \otimes C_{p-1}$. Consider the special case $\mathcal{M} = \mathcal{A}$. We can replace then m by a_0 and we obtain

$$b(a_0, \cdots, a_p) = b'(a_0, \cdots, a_p) + (-1)^p (a_p a_0, \cdots, a_{p-1}).$$

For example $b(a_0, a_1) = a_0 a_1 - a_1 a_0$. The result has been to obtain the same chains (6.1.1) as previously but with a different boundary operator and the sequence C_* is no longer in general exact. The homology

$$H_* = H_*(\mathcal{A}) \equiv H_*(\mathcal{A}; \mathcal{A}) \tag{6.2.1}$$

is the *Hochschild homology* of $\mathcal{A}$.

Dual to this complex of chains there is a complex of cochains. A p-*cochain* is an element of the vector space C^p of multilinear maps ϕ of C_p into the complex numbers: $\phi(a_0, \cdots, a_p) \in \mathbb{C}$. Dual to b is the coboundary operator δ defined by

$$\delta\phi = \phi b. \tag{6.2.2}$$

Therefore

$$\delta\phi(a_0, \cdots, a_p) = \sum_{i=0}^{p-1} (-1)^i \phi(a_0, \cdots, a_i a_{i+1}, \cdots, a_p) + (-1)^p \phi(a_p a_0, \cdots, a_{p-1}).$$

For example, if $\phi \in C^0$,

$$\delta\phi(a_0, a_1) = \phi b(a_0, a_1) = \phi(a_0 a_1) - \phi(a_1 a_0).$$

From the definition we see that
$$\delta^2 = 0.$$

The homology of the complex C^* is the *Hochschild cohomology* of the algebra $\mathcal{A}$.

The homology of the complex C_* would be a natural candidate for the generalization of the de Rham cohomology to an arbitrary algebra $\mathcal{A}$. It is, however, much too big. We shall see below in Example 6.9 that when $\mathcal{A}$ is the algebra of smooth functions on V the Hochschild homology is equal to $\Omega^*(V)$. Also, since differential forms are completely antisymmetric, it is natural to impose a symmetry condition also on the elements of C_* which are to be considered as their generalizations. To consider only those elements of C_* which are completely antisymmetric products would be too strong a condition. The boundary operator b would not respect it. One considers therefore a weaker *cyclic condition*. Define the map $t : C_p \to C_p$ by

$$t(a_0, \cdots, a_p) = (-1)^p (a_p, a_0, \cdots, a_{p-1})$$

The map $1 - t$ can be used to intertwine the two boundary operators b and b':

$$b(1 - t) = (1 - t)b'.$$

We shall construct a complex whose elements are those of C_* modulo elements which are invariant under t.

If we set

$$C_p^0 = \mathrm{Im}\,(1-t), \qquad CC_p = \mathrm{Coker}\,(1-t),$$

then by the definition of Im and Coker,

$$0 \to C_*^0 \to C_* \to CC_* \to 0$$

is an exact sequence of complexes. We have already introduced in (6.2.1) the homology H_* of C_*. Let H_*^0 and HC_* be the homology respectively of C_*^0 and CC_*. The homology

$$HC_* = HC_*(\mathcal{A})$$

is the *cyclic homology* of the algebra $\mathcal{A}$.

Let $\alpha \in HC_p$. Then $b\alpha = 0$ and there exists $\alpha' \in C_p$ such that $b\alpha'$ projects onto zero in CC_{p-1}. There is therefore an element $\beta \in C_{p-1}^0$ which has $b\alpha'$ as image in C_{p-1}. We define $\partial\alpha = \beta$. Then $b\beta = 0$ and ∂ defines a map $HC_p \to H_{p-1}^0$. There is therefore a sequence

$$\cdots \xrightarrow{\partial} H_p^0 \longrightarrow H_p \longrightarrow HC_p \xrightarrow{\partial} H_{p-1}^0 \longrightarrow \cdots$$
$$\xrightarrow{\partial} H_0^0 \longrightarrow H_0 \longrightarrow HC_0 \longrightarrow 0 \qquad (6.2.3)$$

which can be shown to be an exact sequence. In particular $H_0^0 = 0$ and $HC_0 = H_0$.

Define a map B of C_p into C_{p+1} by

$$B = (1-t)s(1 + t + t^2 + \cdots + t^p)$$

where s is the contracting homotopy of (6.1.2). Then

$$B^2 = 0, \qquad bB + Bb = 0$$

and B induces an isomorphism $HC_p \simeq H_{p+1}^0$. For example if $p = 0$ and $a \in HC_0$, then $Ba = (1, a) + (a, 1) \in H_1^0$; if $p = 1$ and $\alpha = (a_0, a_1) - (a_1, a_0)$, with $a_0 a_1 = a_1 a_0$, a cycle representing an element of HC_1 then

$$B\alpha = (1, a_0, a_1) - (a_1, 1, a_0) - (1, a_1, a_0) + (a_0, 1, a_1)$$

is a cycle and represents the image element in H_2^0. The homology groups H_* and HC_* are related therefore through the exact sequence

$$\cdots \xrightarrow{\partial} HC_{p-1} \xrightarrow{B} H_p \longrightarrow HC_p \xrightarrow{\partial} HC_{p-2} \xrightarrow{B} \cdots$$
$$\xrightarrow{\partial} HC_0 \xrightarrow{B} H_1 \longrightarrow HC_1 \longrightarrow 0. \qquad (6.2.4)$$

Dual to CC_* one can define the cyclic cochains CC^*. The *cyclic cohomology* groups are isomorphic to the corresponding cyclic homology groups.

Example 6.8 Consider the case $\mathcal{A} = \mathbb{C}$. Then all of the C_p reduce to $\mathbb{C}$ and the boundary operator is alternately the zero map and the identity map. We have then

$$H_0(\mathbb{C}) = \mathbb{C}, \qquad H_p(\mathbb{C}) = 0, \quad p \geq 0.$$

It follows from the exact sequence (6.2.4) that for all p

$$HC_{2p}(\mathbb{C}) = \mathbb{C}, \qquad HC_{2p+1}(\mathbb{C}) = 0. \tag{6.2.5}$$

$\square$

Example 6.9 Let $\mathcal{A} = C(V)$ be the algebra of smooth functions on a compact manifold and consider the algebra $\Omega^*(V)$ of differential forms on V. Since by definition $\Omega^0(V) = \mathcal{A} = H_0(\mathcal{A})$ there is an obvious isomorphism $\gamma_0 : \Omega^0(V) \to H_0(\mathcal{A})$. Consider the map $\gamma_1 : \Omega^1(V) \to H_1(\mathcal{A})$ given by $f_0 df_1 \mapsto (f_0, f_1)$. If $df_1 = 0$ then f_1 is proportional to the unit function. But $(1, 1) = b(1, 1, 1)$ which vanishes in $H_1(\mathcal{A})$. So γ_1 is well defined. It is obviously onto. On the other hand we have

$$b(f_0, f_1, f_2) = (f_0 f_1, f_2) - (f_0, f_1 f_2) + (f_2 f_0, f_1) = f_0 \gamma_1 (f_1 df_2 + f_2 df_1 - d(f_1 f_2)).$$

So γ_1 is an isomorphism. We have used here the fact that $f_2 df_1 = (df_1) f_2$. Compare with Example 3.12.

The exterior derivative maps $\Omega^1(V)$ into $\Omega^2(V)$ and B maps $H_1(\mathcal{A})$ into $H_2(\mathcal{A})$. So it is natural to define $\gamma_2 : \Omega^2(V) \to H_2(\mathcal{A})$ by the condition that the diagram

$$
\begin{array}{ccc}
\Omega^1(V) & \xrightarrow{\ d\ } & \Omega^2(V) \\
\gamma_1 \downarrow & & \gamma_2 \downarrow \\
H_1(\mathcal{A}) & \xrightarrow{\ B\ } & H_2(\mathcal{A})
\end{array}
$$

commute. That is, by setting

$$\gamma_2 d = B\gamma_1.$$

We define γ_2 on $\Omega^2(V)$ by $d(f_1 df_2) \mapsto B(f_1, f_2)$. It is easy to show that it is again an isomorphism. The exterior derivative is a graded derivation for the algebraic structure of $\Omega^1(V)$ and it is possible to construct a product in $H_*(\mathcal{A})$ with respect to which B is also a graded derivation. So the above diagram can be continued to yield an algebra isomorphism

$$\gamma : \Omega^*(V) \simeq H_*(\mathcal{A}).$$

From the long exact sequence (6.2.4) we find

$$0 \to H_p / \text{Im}\, B \to HC_p \to \text{Ker}\, B \to 0,$$

and therefore

$$HC_p = (H_p/\mathrm{Im}\, B) \oplus \mathrm{Ker}\, B,$$

where $\mathrm{Ker}\, B \subset HC_{p-2}$ for $p \geq 2$. Using this equality and γ one can deduce a relation between the cyclic homology groups and the de Rham cohomology groups $H^*(V;\mathbb{C})$. One finds

$$HC_0(\mathcal{A}) = H_0(\mathcal{A}) = \Omega^0(V), \qquad HC_1 = \Omega^1(V)/d\Omega^0(V),$$

and by induction

$$\begin{aligned}
HC_2(\mathcal{A}) &= \Omega^2(V)/d\Omega^1(V) \oplus H^0(V;\mathbb{C}), \\
HC_3(\mathcal{A}) &= \Omega^3(V)/d\Omega^2(V) \oplus H^1(V;\mathbb{C}), \qquad (6.2.6)\\
HC_4(\mathcal{A}) &= \Omega^4(V)/d\Omega^3(V) \oplus H^2(V;\mathbb{C}) \oplus H^0(V;\mathbb{C})
\end{aligned}$$

and so forth. The de Rham cohomology groups are therefore contained as direct summands in the cyclic homology groups. They can be identified as the kernel of B in the sequence (6.2.4). The singular homology groups are contained as direct summands of the cyclic cohomology groups.

There is a natural map of the p-chains in V into the p-cochains, defined by the integration. Let f_i for $0 \leq i \leq p$ be a set of $p+1$ elements of $\mathcal{C}(V)$ and define

$$\phi(f_0 \cdots f_p) = \int f_0 df_1 \cdots df_p,$$

where the integral is taken over some closed p-chain. Then ϕ is a cyclic p-cochain. If p is equal to n, the dimension of V, then $\delta\phi = 0$ and the cochain is a cyclic n-cocycle.
$\square$

Example 6.10 In the language of this section the multilinear maps τ_k introduced in Section 5.3 are cochains. They are in fact cyclic cocycles. They satisfy the cyclic condition

$$\tau_k(a_0, \cdots, a_k) = (-1)^k \tau_k(a_k, a_0, \cdots, a_{k-1}),$$

as well as the cocycle condition

$$\delta\tau_k = 0.$$

These identities follow directly from the definition when $a_0 \in \mathcal{L}^1$. A p-summable Fredholm module defines therefore an element of $HC^k(\mathcal{A})$ for each even $k \geq p-1$.
$\square$

Example 6.11 Refer back to Example 2.1. The de Rham cohomology of the 2-sphere is given by $H^0(S^2;\mathbb{C}) = \mathbb{C}$, $H^1(S^2;\mathbb{C}) = 0$, $H^2(S^2;\mathbb{C}) = \mathbb{C}$. As in the preceding example from (6.2.6) we see that $HC_{2p+1} = 0$ for $p \geq 1$, $HC_{2p} = \mathbb{C} \oplus \mathbb{C}$

for $p \geq 2$ and that $\dim(HC_p) = \infty$ for $p = 0, 1, 2$. In Chapter 7 we shall discuss a series of approximations to the 2-sphere by noncommutative geometries with the matrix algebras M_n as structure algebras. Their cyclic homology groups are given in Example 6.14 below. This approximation is singular from the point of view of cyclic homology.   □

Example 6.12 Refer back to Example 2.2. The de Rham cohomology of the 2-torus is given by $H^0(\mathbb{T}^2; \mathbb{C}) = \mathbb{C}$, $H^1(\mathbb{T}^2; \mathbb{C}) = \mathbb{C} \oplus \mathbb{C}$, $H^2(\mathbb{T}^2; \mathbb{C}) = \mathbb{C}$. From (6.2.6) we see that $HC_p = \mathbb{C} \oplus \mathbb{C}$ for $p \geq 3$ and that $\dim(HC_p) = \infty$ for $p = 0, 1, 2$. The lattice approximation to the torus has as structure algebra the product of n^2 copies of $\mathbb{C}$. Its cyclic homology groups are given then by (6.2.5). Therefore $\dim(HC_p) = n^2$ for even values of p and $HC_p = 0$ for odd values. Except for $p = 0, 2$ the limit of the homology is not equal to the homology of the limit. The lattice approximation also is rather singular from the point of view of cyclic homology.   □

Example 6.13 We can now return yet again to the Schwinger term s of Examples 2.24 and 4.32 and express once more the Jacobi identity as a cocycle condition, but this time within the context of cyclic cohomology. The simple fact that $\delta s = 0$ follows immediately from the definition (6.2.2) of the coboundary. If the expression (4.3.16) for s is well defined then obviously s is a coboundary; $s = \delta t$ with

$$t(a) = \mathrm{Tr}(P_- a P_-).$$

The Schwinger term is interesting when the expression (4.3.16) is ill defined but the expression (4.3.17) can be defined using a limiting procedure. It is possible then to show that the Schwinger term is actually a character τ_1 if the algebra of observables $\mathcal{A}$ is endowed with a differential calculus defined by a p-summable Fredholm module with $p \leq 2$. We can not do this here since we have defined characters τ_k only for k even. Introduce the Fredholm operator $F = P_- - P_+$. Then if $[F, a]$ and $[F, b]$ both belong to $\mathcal{L}^2$ (Hilbert-Schmidt) one can show that the expression (4.3.17) is well defined. Let $[a, b]_+$ be the anticommutator of a and b. If also $[F, a]_+$ and $[F, b]_+$ both belong to $\mathcal{L}^1$ (trace class) one can show that the expression (4.3.16) is also well defined as is the boundary $t(a)$. Consider a Bogoliubov transformation to a new basis and let P'_- be the projector onto the negative-energy states of the new Hilbert space $\mathcal{H}'$. Set $F' = P'_- - P'_+$. If $F' - F$ belongs to $\mathcal{L}^1$ one can show that the cohomology class $[s] \in HC^1(\mathcal{A})$ is invariant; one can write $s'(a, b) - s(a, b) = \delta t(a, b)$ with t explicitly given in terms of $F' - F$.   □

6.3 Morita equivalence

Let $\mathcal{A}$ and $\mathcal{B}$ be two algebras and denote by $_\mathcal{A}\mathcal{M}_\mathcal{B}$ a left module with respect to $\mathcal{A}$ and a right module with respect to $\mathcal{B}$. If there exist $_\mathcal{A}\mathcal{M}_\mathcal{B}$ and $_\mathcal{B}\mathcal{N}_\mathcal{A}$ and bimodule isomorphisms such that

$$\mathcal{M} \otimes_{\mathcal{B}} \mathcal{N} \simeq \mathcal{A}, \qquad \mathcal{N} \otimes_{\mathcal{A}} \mathcal{M} \simeq \mathcal{B}$$

then $\mathcal{A}$ and $\mathcal{B}$ are said to be *Morita equivalent*. It is obvious that M_n and M_m are Morita equivalent for all m, n and that the appropriate modules are the $n \times m$ and the $m \times n$ matrices. The algebra M_∞ of sparse matrices is also Morita equivalent to M_n for all n.

Using the fact that $M_n(\mathcal{A}) \simeq M_n(\mathbb{C}) \otimes \mathcal{A}$ we can define a map

$$C_*(M_n(\mathcal{A})) \to C_*(\mathcal{A})$$

by taking the trace over the product of the matrix factors:

$$(f_0 \otimes a_0, \cdots, f_p \otimes a_p) \mapsto \mathrm{Tr}(f_0 \cdots f_p)(a_0, \cdots, a_p).$$

This map commutes with the boundary operator b and so it induces a homomorphism of homology groups:

$$H_*(M_n(\mathcal{A})) \to H_*(\mathcal{A}).$$

One can show that this homomorphism is in fact an isomorphism. We have therefore also, using (6.2.3), the isomorphisms

$$HC_*(M_\infty(\mathcal{A})) \simeq HC_*(M_n(\mathcal{A})) \simeq HC_*(\mathcal{A}).$$

More generally, any two algebras which are Morita equivalent have the same cyclic and Hochschild homology groups. The same result is valid for the respective cohomology groups:

$$HC^*(M_\infty(\mathcal{A})) \simeq HC^*(M_n(\mathcal{A})) \simeq HC^*(\mathcal{A}).$$

Example 6.14 Using Morita equivalence and the result of Example 6.2.1 we find that the Hochschild homology of $M_n(\mathbb{C})$ is trivial. One can give a direct proof of this fact using only the following property of matrices. Any complex matrix g can be written in the form $g = k + [p, q]$ where $k \in \mathbb{C}$ and p, q are matrices. Further, if f commutes with g then p and q can be chosen to commute also with f. Suppose that $\rho = (f, g)$ with $b\rho = 0$. Define $\sigma = (kf, 1, 1) + (f, q, p) - (f, p, q)$. Then a straightforward calculation shows that $\rho = b\sigma$. Therefore $H_1(M_n(\mathbb{C})) = 0$. The general proof consists of an explicit construction of a boundary for each cycle. It follows that

$$HC_{2p}(M_n(\mathbb{C})) = \mathbb{C}, \qquad HC_{2p+1}(M_n(\mathbb{C})) = 0.$$

Cyclic cohomology differs therefore from the cohomology which was defined in Section 3.6.   □

6.4   The Loday-Quillen theorem

Let $\underline{g}$ be a Lie algebra and $\bigwedge^* \underline{g}$ the associated exterior algebra. Let $x_i \in \underline{g}$ and define d^* by

$$d^*(x_0 \wedge \cdots \wedge x_p) = \sum_{i<j} (-1)^{i+j}[x_i, x_j] \wedge x_0 \wedge \cdots \hat{x}_i \cdots \hat{x}_j \cdots \wedge x_p.$$

Then $\bigwedge^* \underline{g}$ is a complex. It is a graded coalgebra with a coproduct similar to the one introduced in Section 4.2 for the enveloping algebra $\mathcal{U}$ of a Lie algebra. There is also a product but the differential is not compatible with it. For example $d^*(x_0 \wedge x_1) = -[x_0, x_1]$ but $d^*x_0 \wedge x_1 - x_0 \wedge d^*x_1 = 0$. We have used the symbol d^* for the boundary operator since it is similar to a dual of the operator d defined by Equation (2.1.2). The associative algebra $M_n(\mathcal{A})$ is also a Lie algebra which we write as $\underline{gl}_n$. To distinguish the Lie-algebra homology from the Hochschild homology we shall write $H_*(\underline{gl}_n)$ for the former and $H_*(M_n)$ for the latter. Similarly we write $H_*(\underline{gl}_\infty)$ and $H_*(M_\infty)$.

Let $C_p(M_\infty(\mathcal{A}))$ be the complex (6.1.1) for the algebra $M_\infty(\mathcal{A})$ and consider the map

$$\bigwedge^{p+1}\underline{gl}_\infty \to C_p(M_\infty(\mathcal{A}))$$

defined by

$$x_0 \wedge \cdots \wedge x_p \mapsto \sum_\sigma (-1)^{|\sigma|}(x_0, x_{\sigma(1)}, \cdots, x_{\sigma(p)}).$$

The sum is taken over all permutations σ, with parity $|\sigma|$. The map (6.4) is not well defined since, for example, $x_0 \wedge x_1 \wedge x_2$ and $x_2 \wedge x_0 \wedge x_1$ go onto different elements. However, if we use the projection of $C_p(M_\infty(\mathcal{A}))$ onto $CC_p(M_\infty(\mathcal{A}))$ defined in Section 6.2 we find that the composite map

$$\bigwedge^{p+1}\underline{gl}_\infty \xrightarrow{\gamma} CC_p(M_\infty(\mathcal{A}))$$

is well defined. This map is similar to the one mentioned in Example 6.9 and we use the same symbol. From the definitions of d^* and b one sees that the diagram

$$
\begin{array}{ccc}
\bigwedge^{p+1}(\underline{gl}_\infty) & \xrightarrow{d^*} & \bigwedge^p(\underline{gl}_\infty) \\
\gamma \downarrow & & \gamma \downarrow \\
CC_p(M_\infty) & \xrightarrow{b} & CC_{p-1}(M_\infty)
\end{array}
$$

commutes. Therefore γ is a map of complexes. It induces for all p a homomorphism

$$H_{p+1}(\underline{gl}_\infty) \xrightarrow{\gamma} HC_p(M_\infty) \simeq HC_p(\mathcal{A}).$$

The isomorphism is due to Morita equivalence. The $HC_*(M_\infty)$ is a set of vector spaces with no algebraic structure. The $H_*(\underline{gl}_\infty)$ on the other hand has the structure of a differential coalgebra. It also has a compatible product which gives it the structure of a Hopf algebra but that fact is not relevant here. The coproduct Δ of $H_*(\underline{gl}_\infty)$ comes from the coproduct of $\bigwedge^* \underline{g}$.

In general let $\mathcal{A}$ be a coalgebra with coproduct Δ. The set of *primitive elements* of $\mathcal{A}$ is the set $\mathcal{P}$ of elements which satisfy the condition $\Delta(a) = 1 \otimes a + a \otimes 1$. If $\mathcal{A}$ is a Hopf algebra we can think of $\mathcal{P}$ as a set of generators of $\mathcal{A}$ as an algebra. If $\mathcal{A} = \mathcal{U}$, the enveloping algebra of a Lie algebra $\underline{g}$, then $\mathcal{P} = \underline{g}$. The set $\mathcal{P}_*$ of primitive elements of $H_*(\underline{gl}_\infty)$ is a graded vector space. The theorem of Loday and Quillen states that γ is an isomorphism on the primitive elements of $H_{p+1}(\underline{gl}_\infty)$ and therefore that

$$\mathcal{P}_{p+1} \simeq HC_p(\mathcal{A}).$$

We outline the proof of this theorem in two simple cases. First suppose that $\mathcal{A} = \mathbb{C}$. We have shown in (6.2.5) that $HC_*(\mathbb{C})$ has one generator in each even group. It is a standard result of Lie algebra theory that the homology algebra of $\underline{gl}_n$ is an exterior algebra generated by one element in each of the first n odd degrees. The homology algebra of $\underline{gl}_\infty$ has one primitive element in each odd degree. Consider a general algebra $\mathcal{A}$ but with $p = 0$. The map γ is the identity and it induces an isomorphism

$$H_1(\underline{gl}_\infty) \simeq H_0(M_\infty) \simeq HC_0(M_\infty) \simeq HC_0(\mathcal{A}).$$

Every element in $H_1(\underline{gl}_\infty)$ is necessarily primitive.

Notes

For a general introduction to the algebraic aspects of (co)homology we refer to the book by Mac Lane (1963). The construction of a differential algebra from a graded derivation is due to Connes & Lott (1992). The general construction of a differential calculus using a Fredholm operator is due to Connes (1994). The version we have presented here is taken from Dimakis & Madore (1996) and Madore & Mourad (1998). Some details of the relation between $\Omega_F^*(\mathcal{A} \otimes \mathcal{A}')$ and $\Omega_F^*(\mathcal{A}) \otimes \Omega_F^*(\mathcal{A}')$ have been given by Kalau *et al.* (1995). Ghosts were introduced into physics by Feynman, DeWitt, Faddeev and Popov to treat constraints and gauge freedom in field theory using the language of Feynman diagrams. We refer to Henneaux (1985) for more details of the differential calculus they form as well as for references to the original literature.

Cyclic homology was introduced by Tsygan (1983). For a short historical review one can consult Cartier (1984). For more details on cyclic (co)homology we refer to Connes (1986), Seibt (1987) or Loday (1992). The presentation given here is partly based on notes taken at a seminar organized by S. Halpern, A. Nicas and P. Selick at the University of Toronto in 1983/84. The idea of Example 6.13 is due to Araki (1987). It was developed by Grosse (1988), Grosse *et al.* (1993),

Langmann (1994) and by Langmann & Mickelsson (1994), to which we refer for more details. We refer to Seibt (1987), Loday (1992) or Connes (1994) for more details on Sections 6.3 and 6.4 as well as for references to the original literature.

7 Modifications of Space-Time

We now return to the suggestion made in the Introduction that at sufficiently small length scales the geometry of space-time might be better described by a noncommutative algebra. The physical hypothesis is that geometry based on a set of commuting coordinates is only valid at length scales greater than some fundamental length. At smaller scales it is impossible to localize a point and a new geometry must be used. We can use a solid-state analogy and think of the ordinary Minkowski coordinates as macroscopic order parameters obtained by 'coarse-graining' over regions whose size is determined by the fundamental length. They break down and must be replaced by elements of the noncommutative algebra $\mathcal{A}$ when one considers phenomena on smaller length scales. If a coherent description could be found for the structure of space-time which were pointless on small length scales, then the ultraviolet divergences of quantum field theory could be eliminated. In fact the elimination of these divergences is equivalent to coarse-graining the structure of space-time over small length scales; if an ultraviolet cut-off Λ is used then the theory does not see length scales smaller than Λ^{-1}. When a physicist calculates a Feynman diagram he is forced to place a cut-off Λ on the momentum variables in the integrands. This means that he renounces any interest in regions of space-time of volume less than Λ^{-4}. As Λ becomes larger and larger the forbidden region becomes smaller and smaller but it can never be made to vanish. There is a fundamental length scale, much larger than the Planck length, below which the notion of a point is of no practical importance. The simplest and most elegant, if certainly not the only, way of introducing such a scale in a Lorentz-invariant way is through the introduction of noncommuting space-time 'coordinates' x^μ.

In the first section we shall outline some of the features a noncommutative space-time might be expected to have and we shall present in detail two particular models which have been proposed. In Example 7.9 we present a model which is described by an algebra which admits a natural action of the Poincaré group and in Example 7.10 we briefly describe a q-deformed version of Minkowski space which is invariant under the coaction of a q-deformed version of the Lorentz group in the sense we described in Example 4.67. In the second section we shall present a finite model of space-time which lacks points but which does have well defined directions. In the solid-state analogy this would correspond to a crystal with dislocations, defects which shift the lattice but which do not change the local orientation, but no disclinations or shifts in orientation. The model is a noncommutative modification of the 2-sphere; not only is the dimension wrong but, more important, also the signature of the metric. Even on a compact 2-dimensional manifold, however, quantum field theory is still impossible to solve in general, and in order to do explicit calculations beyond the perturbative approximation, further modifications are necessary. One possibility is to modify the geometry at small length scales so that points disappear in favour of a cellular structure.

In the last section we present in some detail the description of a quantized scalar field on the noncommutative model of Section 7.2 to illustrate how quantum field theory might be modified by the cellular structure. There is a convenient description of quantum physics as a path integral over all possible classical configurations. With the euclidean metric this integral is identical to the integral which defines the partition function of statistical mechanics. It becomes an integral over the components of the elements of a matrix algebra. In Example 7.29, we examine a model in which topology also has an ambiguous meaning at small length scales. In Example 7.30, we give a brief discussion of a lattice model. We close the section with a discussion of some problems which one might like to be able to solve on a noncommutative version of Minkowski space-time and which are based on an infinite-dimensional algebra.

7.1 Noncommutative space-time

LE MAÎTRE DE PHILOSOPHIE

On les peut mettre premièrement comme vous avez dit: *Belle marquise, vos beaux yeux me font mourir d'amour.* Ou bien : *D'amour mourir me font, belle marquise, vos beaux yeux.* Ou bien : *Vos beaux yeux d'amour me font, belle marquise, mourir.* Ou bien : *Mourir vos beaux yeux, belle marquise, d'amour me font.* Ou bien : *Me font vos yeux beaux mourir, belle marquise, d'amour.*

M. JOURDAIN

Mais de toutes ces facons-là, laquelle est la meilleure? *

For a physicist noncommutative geometry is geometry in which the 'coordinates' do not commute; order is important. To cope with the divergences in the newly discovered quantum field theory, throughout the 1930's Heisenberg and others had flirted with the idea of replacing space-time with a fundamental lattice. The lattice spacing would serve as cut-off. The difficulty here was that the lattice destroyed Lorentz invariance. A solution to this problem was proposed in 1947 by Snyder who showed that by using noncommutative 'coordinates' it was possible to have a version of 'space-time' which had some of the desirable features of a lattice but which was nevertheless Lorentz invariant. His suggestion came however just at the time when the renormalization program finally successfully became an effective if rather *ad hoc* prescription for predicting numbers from the theory of quantum electrodynamics and it was for the most part ignored. Some time later von Neumann introduced the term 'noncommutative geometry' to refer in general to a geometry in which an algebra of functions is replaced by a noncommutative algebra. As in the phase-space example

---

*Molière

coordinates are replaced by generators of the algebra. Since these do not commute they cannot be simultaneously diagonalized and the space disappears. We shall argue with an example below that, just as Bohr cells replace classical-phase-space points, the appropriate intuitive notion to replace a 'point' is a Planck cell of dimension given by the Planck area (4.1.5).

Since points are therefore ill-defined we shall use the expression *fuzzy space* to designate what would have been the space of ordinary geometry. When describing a fuzzy version of space-time one typically replaces the four Minkowski coordinates $\tilde{x}^\mu$ by four generators x^μ of a noncommutative algebra which satisfy commutation relations of the form

$$[x^\mu, x^\nu] = i\hbar J^{\mu\nu} \tag{7.1.1}$$

discussed already in Example 4.7. The analogues of the Heisenberg uncertainty relations imply then that

$$\Lambda^2 \hbar \lesssim 1.$$

The existence of a forbidden region around each point in space-time means that the standard description of Minkowski space as a 4-dimensional continuum is redundant. By analogy with quantum mechanics we shall suppose that the generators x^μ can be represented as hermitian operators on some Hilbert space. The presence of the factor i in (7.1.1) implies that the $J^{\mu\nu}$ are also hermitian operators. The x^μ have real eigenvalues but because of the relations (7.1.1) they cannot be simultaneously diagonalized. Points are therefore ill-defined and *fuzzy space-time* consists of elementary cells of volume of the order of $(2\pi\hbar)^2$. A more accurate description would depend also on the eigenvalues of the operator $J^{\mu\nu}$. Equation (7.1.1) contains in fact little information about the algebra. If the right-hand side does not vanish it states that at least some of the x^μ do not commute. It states also that it is possible to identify the original coordinates with the generators x^μ in the limit where the Planck mass μ_P tends to infinity:

$$\tilde{x}^\mu = \lim_{\hbar \to 0} x^\mu. \tag{7.1.2}$$

For mathematical simplicity we shall suppose this to be the case although we might reasonably have included a singular 'renormalization constant' Z and replaced (7.1.2) by an equation of the form

$$Z\,\tilde{x}^\mu = \lim_{\hbar \to 0} x^\mu.$$

If, as we shall argue, gravity acts as a universal regulator for ultraviolet divergences then one could reasonably expect the limit $\hbar \to 0$ to be a singular limit.

Example 7.1 Certainly the simplest example of a fuzzy space is the quantized version of a 2-dimensional phase space, described by the 'coordinates' q and p. This example has the advantage of illustrating what is for us the essential interest of the relation of the form (7.1.1) as expressed in the Heisenberg uncertainty relations.

Since one cannot measure simultaneously q and p to arbitrary precision quantum phase space has no longer a notion of a point. It can however be thought of as divided into *Bohr cells* of volume $2\pi\hbar$. If the classical phase space is of finite total volume there will be a finite number of cells and the quantum system will have a finite number of possible states. A 'function' then on quantum phase space will be defined by a finite number of values and can be represented by a matrix.   □

Example 7.2 As a second example consider the phase space of a classical particle moving in a plane, described by two position coordinates (x^1, x^2) and two momentum coordinates (p_1, p_2). In the language of quantum mechanics these four coordinates are commuting operators. When the system is quantized they no longer commute. They satisfy the canonical commutation relations:

$$[x^1, p_1] = i\hbar, \qquad [x^2, p_2] = i\hbar. \tag{7.1.3}$$

This introduces a cellular structure in phase space. In the presence of a magnetic field B normal to the plane the momentum operators are further modified and they also cease to commute:

$$[p_1, p_2] = i\hbar eB. \tag{7.1.4}$$

This introduces a cellular structure in the momentum plane. It becomes divided into *Landau cells* of area proportional to $\hbar eB$. More details are given in Example 4.34. Consider in this case the divergent integral

$$I = \int \frac{dp_1 dp_2}{p^2}.$$

The commutation relation (7.1.4) does not permit p_1 and p_2 simultaneously to take the eigenvalue zero and the operator $p^2 = p_1^2 + p_2^2$ is bounded below by $\hbar eB$. The magnetic field acts as an *infrared cut-off*. If one adds an *ultraviolet cut-off* Λ then p^2 is bounded also from above and the integral becomes finite:

$$I \sim \log(\frac{\Lambda^2}{\hbar eB}). \tag{7.1.5}$$

If the position space were curved, with constant Gaussian curvature K, one would have (7.1.5) with $\hbar eB$ replaced by $\hbar^2 K$ and one would obtain again an infrared regularization for I. If instead of quantizing the system or introducing a magnetic field, we suppose the analogue of (7.1.1) then we must replace the coordinates of position space by two operators which do not commute:

$$[x^1, x^2] = i k J^{12}. \tag{7.1.6}$$

By the new uncertainty relation there is no longer a notion of a point in position space since one cannot measure both coordinates simultaneously but as before, position space can be thought of as divided into *Planck cells*. It has become fuzzy. This cellular structure would serve as an ultraviolet cut-off similar to a lattice structure. If we consider for example the divergent integral I and use the same logic that led to (7.1.5) we find that the commutation relations (7.1.6) introduce an ultraviolet cut-off. If we introduce also a Gaussian curvature and use the equivalent of (7.1.4) we have

$$I \sim \log(\hbar K). \tag{7.1.7}$$

The integral becomes completely regularized.

There is however now a new complication; the right-hand side of (7.1.7) seems not to depend on the operator J^{12}. We shall argue below that, endowed with an appropriate differential structure, each fuzzy space-time supports a uniquely determined gravitational field and that the latter is a classical manifestation of the commutation relations (7.1.1) plus a differential structure. From this point of view what we put on the right-hand side of (7.1.6) will depend on which gravitational field we wish to regularize the integral with. That is, in fact K does depend on J^{12}. We studied in Example 4.16 the noncommutative version of the Poincaré half-plane, the surface of constant negative Gaussian curvature. It was obtained by setting $i\hbar J^{12} = h(x^2)^2$. Similarly in higher dimensions we shall argue that the right-hand side of (7.1.1) is what determines what we call 'gravity'.   □

Example 7.3 An immediate consequence of (7.1.6) is that in general the commutation relations (7.1.3) will have to be modified. In fact from the Jacobi identities one finds

$$[x^1, [x^2, p_1]] + [x^2, [p_1, x^1]] = i\hbar[J^{12}, p_1].$$

Therefore unless J^{12} commutes with p_1 both $[x^2, p_1]$ and $[x^1, p_1]$ cannot lie in the center of the algebra $\mathcal{A}$ generated by the position and the momenta. One has

$$[x^j, p_k] = i\hbar(\delta^j_k + \hbar A^j_k)$$

with $A^i_j \notin \mathcal{Z}(\mathcal{A})$. In the Example 7.9 below no such modification is necessary.   □

Example 7.4 As a simple illustration of how a 'space' can be 'discrete' in some sense and still covariant under the action of a continuous symmetry group one can consider the ordinary round 2-sphere, which has acting on it the rotational group SO_3. As a simple example of a lattice structure one can consider two points on the sphere, for example the north and south poles. One immediately notices of course that by choosing the two points one has broken the rotational invariance. It can be restored at the expense of commutativity. The set of functions on the two points can be identified with the algebra of diagonal 2×2 matrices, each of the two entries on the diagonal corresponding to a possible value of a function at one of the two points. Now an action of a group on the lattice is equivalent to an action of the group on

the matrices and there can obviously be no (non-trivial) action of the group SO_3 on the algebra of diagonal 2×2 matrices. However, if one extends the algebra to the noncommutative algebra of all 2×2 matrices one recovers the invariance. The two points, so to speak, have been smeared out over the surface of a sphere; they are replaced by two cells. An 'observable' is an hermitian 2×2 matrix and has therefore two real eigenvalues, which are its values on the two cells. Although what we have just done has nothing to do with Planck's constant it is similar to the procedure of replacing a classical spin which can take two values by a quantum spin of total spin $1/2$. Only the quantum spin is invariant under the rotation group. By replacing the spin $1/2$ by arbitrary spin s one can describe a 'lattice structure' of $n = 2s + 1$ points in an SO_3-invariant manner. The algebra becomes then the algebra M_n of $n \times n$ complex matrices. We shall discuss this example in more detail in the next section. In general, a static closed surface in a fuzzy space-time as we define it can only have a finite number of modes and will be described by some finite-dimensional algebra.
□

Example 7.5 It is worth mentioning something which we do *not* mean by the expression 'fuzzy space-time'. To explain the Zitterbewegung of an electron Schrödinger and others considered center-of-mass position operators of the form

$$\mathbf{q} = \mathbf{x} + m^{-2}\mathbf{S} \times \mathbf{p}.$$

Here $\mathbf{S}$ is the spin vector. Because of the canonical commutation relations the operators $\mathbf{q}$ do not commute with each other. One could introduce an algebra generated by them, for each value of the time coordinate, and even a differential calculus to construct a noncommutative geometry. The algebra of the commutative limit would be, however, simply the real (or complex) numbers, the possible values of a function at the position of the particle; it would not be an algebra of functions on space-time. An associated noncommutative geometry would be simply a noncommutative generalization of a point. This example has a 'string' and 'membrane' generalization. The fact that the center-of-mass coordinates of an extended object do not commute does not mean necessarily that the noncommutative algebra which they generate is a noncommutative version of space-time. It could however be considered as a noncommutative Kaluza-Klein extension of a point of ordinary space-time. For example if one truncates the modes of a closed membrane and modifies the product to construct a matrix algebra the resulting geometry is a matrix extension of space-time. See in this respect Example 8.10 below.   □

But 'geometry' is more than a set of points and so more is needed that just an algebra. This problem was solved by Connes who introduced the notion of 'noncommutative differential geometry' to refer in general to a noncommutative geometry with an associated differential calculus. Just as it is possible to give many differential structures to a given topological space it is possible to define many differential calculi over a given algebra. We shall use the term 'noncommutative geometry' to mean 'noncommutative differential geometry' in the sense of Connes.

Let $\mathcal{A}_k$ be the algebra generated in some sense by the elements x^μ. We shall be here working on a formal level so that one can think of $\mathcal{A}_k$ as an algebra of polynomials in the x^μ although we shall implicitly suppose that there are enough elements to generate smooth functions on space-time in the commutative limit. If one represents the generators as hermitian operators on some Hilbert space $\mathcal{H}$ one can identify $\mathcal{A}_k$ as a subalgebra of the algebra of all operators on $\mathcal{H}$. If $\mathcal{A}_k$ were defined in terms of generators and relations as was described in Section 4.1, if the corresponding limit manifold V_0 of Section 4.2 were compact and if the metric were of euclidean signature then the most obvious structure would be that of a von Neumann algebra. We have added the subscript k to underline the dependence on this parameter but of course it alone does not determine the structure of $\mathcal{A}_k$. We in fact conjecture that every possible gravitational field can be considered as the commutative limit of a noncommutative equivalent which in turn is strongly restricted if not determined by the structure of the algebra $\mathcal{A}_k$. We need then a large number of algebras $\mathcal{A}_k$ for each value of k.

We argued above that the noncommutative structure gives rise to an ultraviolet cut-off. This idea has been developed by several authors since the original work of Snyder. It is the left-hand arrow of the diagram

$$
\begin{array}{ccc}
\mathcal{A}_k & \Longleftarrow & \Omega^*(\mathcal{A}_k) \\[4pt]
\Downarrow & & \Uparrow \\[4pt]
\text{Cut-off} & & \text{Gravity}
\end{array}
\qquad (7.1.8)
$$

The top arrow is a mathematical triviality; the $\Omega^*(\mathcal{A}_k)$ is what gives a differential structure to the algebra. We devoted much of Chapter 3 to it. We attempted there also, not completely successfully, to argue with models that each metric is the unique 'shadow' in the limit $k \to 0$ of some differential structure over some noncommutative algebra. If we identify the metric with the gravitational field this would define the right-hand arrow of the diagram. The uniqueness claim will be based on an assumption we shall make that the center of the algebra is trivial. If this be not the case then one cannot claim at all that a differential structure determines a metric; the larger the center the larger the ambiguity. In the limit when the algebra is commutative, in ordinary differential geometry for example, the differential structure in no way determines a metric. We would like to be able to prove that, when the center is trivial, the geometry is encoded in the $\mathcal{A}_k$-module structure of $\Omega^*(\mathcal{A}_k)$ just as the Poisson structure is encoded in the commutation relations of $\mathcal{A}_k$ itself.

The composition of the three arrows in (7.1.8) is an expression of an old idea, due to Pauli (1956), that perturbative ultraviolet divergences will somehow be regularized

by the gravitational field. The possibility which we shall consider here is that
the mechanism by which this works is through the introduction of noncommuting
'coordinates' such as the x^μ. A hand-waving argument can be given which allows
one to think of the noncommutative structure of space-time as being due to quantum
fluctuations of the light-cone in ordinary 4-dimensional space-time. This relies
actually on the existence of quantum gravitational fluctuations but we shall suppose
that the classical gravitational field is to be considered as regularizing the ultraviolet
divergences through the introduction of the noncommutative structure of space-time.
This can be strengthened as the conjecture that the classical gravitational field and
the noncommutative nature of space-time are two aspects of the same thing. It is
our main purpose here to explore this relation. If the gravitational field is quantized
then presumably the light-cone will fluctuate and the two points could have a time-
like separation on a time scale of the order of the Planck time, in which case the
operators will no longer commute. So even in flat space-time quantum fluctuations
of the gravitational field could be expected to introduce a non-locality in the theory.
This is one possible source of noncommutative geometry on the order of the Planck
scale.

   The fundamental open problem of the noncommutative theory of gravity concerns
of course the relation it might have to a future quantum theory of gravity either
directly or via the theory of 'strings' and 'membranes'. But there are more immediate
technical problems which have not received a satisfactory answer. We have already
discussed in Section 3.7 the problem of the definition of the curvature. It is not
certain that the ordinary definition of curvature taken directly from differential
geometry is the quantity which is most useful in the noncommutative theory. Cyclic
homology groups have been proposed by Connes as the appropriate generalization
to noncommutative geometry of topological invariants; these are discussed briefly in
the Section 6.2. The appropriate definition of other, non-topological, invariants in
not clear. It is not in fact even obvious that one should attempt to define curvature
invariants. On this point we refer to Example 7.28 below.

Example 7.6 There is an interesting theory of gravity, due to Sakharov and
popularized by Wheeler, called induced gravity, in which the gravitational field is a
phenomenological coarse-graining of more fundamental fields. Flat Minkowski space-
time is to be considered as a sort of perfect crystal and curvature is to be thought of as
a manifestation of elastic tension, or possibly of defects, in this structure. Quantum
fluctuations in flat space-time produce the vacuum energy which we discussed in
Example 4.29. A deformation in the crystal produces a variation in the vacuum
energy which we perceive as gravitational energy. 'Gravitation is to particle physics
as elasticity is to chemical physics: merely a statistical measure of residual energies.'
The description of the gravitational field which we are attempting to formulate using
noncommutative geometry is not far from this. We noticed in the Introduction

that the use of noncommuting coordinates is a convenient way of making a discrete structure like a lattice invariant under the action of a continuous group. In this sense what we would like to propose is a Lorentz-invariant version of Sakharov's crystal. Each coordinate can be separately measured and found to have a distribution of eigenvalues similar to the distribution of atoms in a crystal. The gravitational field is to be considered as a measure of the variation of this distribution just as elastic energy is a measure of the variation in the density of atoms in a crystal. We shall return to the mathematics of this idea that the graviton is a sort of space-time phonon in Example 7.27. Compare also Example 7.31.   □

Example 7.7 We have argued that the differential structure of a noncommutative algebra determines the metric. One could also argue the inverse. Suppose that a metric $g^{ij} = g(dx^i \otimes dx^j)$ exists at least in some quasi-commutative approximation. From the relations between fg^{ij} and $g^{ij}f$ one finds a severe restriction on the module structure of the 1-forms.   □

Example 7.8 An axiomatic approach to noncommutative geometry has been proposed by Connes. The principal hypothesis is that the algebra $\mathcal{A}$ which describes a noncommutative geometry must be a subalgebra of operators on a Hilbert space $\mathcal{H}$ and the differential calculus must be defined in terms of a generalized 'Dirac' operator F as described in Section 5.3. Supplementary assumptions are added to insure that the 'space' is in some sense real and that it has at least some of the properties of a 'manifold', for example a noncommutative version of *Poincaré duality*. As well as the ordinary geometry of smooth manifolds, it has been found that the Weyl algebra of Example 4.2 and a set of geometries based on matrix algebras satisfy the axioms.
□

Example 7.9 We return now again to the Example 4.7 and we shall suppose that the $J^{\mu\nu}$ lie in the center of the algebra. This permits us to suppose further that the matrix $J^{\mu\nu}$ has an inverse $J^{-1}_{\lambda\mu}$:

$$J^{-1}_{\lambda\mu} J^{\mu\nu} = \delta^\nu_\lambda.$$

We shall use this inverse to lower the indices of the generators x^μ:

$$\lambda_\mu = -i\mu_P^2 J^{-1}_{\mu\nu} x^\nu.$$

Recall the general formalism at the end of Section 3.1. A natural choice of n is $n = 4$. The associated derivations satisfy then

$$e_\mu x^\lambda = \delta^\lambda_\mu \qquad\qquad (7.1.9)$$

and it follows that

$$[e_\mu, e_\nu] = 0.$$

From (7.1.9) it follows that

$$\theta^\lambda = dx^\lambda, \qquad \theta = -\lambda_\mu dx^\mu$$

from which we deduce the values

$$P^{\mu\nu}{}_{\rho\sigma} = \frac{1}{2}(\delta^\mu_\rho \delta^\nu_\sigma - \delta^\nu_\rho \delta^\mu_\sigma), \qquad F^\lambda{}_{\mu\nu} = 0, \qquad K_{\mu\nu} = i\mu_P^2 J_{\mu\nu}^{-1}.$$

for the coefficients of Equation (3.1.50). One can interpret θ as a connection on a trivial bundle with the unitary elements of the algebra as structural group. We see from the above formula for $K_{\mu\nu}$ that $J^{\mu\nu}$ is related to the corresponding curvature. This is the noncommutative analogue of the classical result of mechanics which interprets the symplectic 2-form as the curvature of a line bundle.

From the commutation relations (3.1.43) one finds that the θ^λ anticommute. A possible form for σ is given therefore by

$$S^{\mu\nu}{}_{\rho\sigma} = \delta^\nu_\rho \delta^\mu_\sigma. \tag{7.1.10}$$

From (3.6.23) and (3.6.24) we see that in this case the coefficients of the connection necessarily lie in the center of the algebra. From (3.6.5) we see that the most general $S^{\mu\nu}{}_{\rho\sigma}$ must satisfy the constraint

$$S^{\mu\nu}{}_{[\rho\sigma]} + \delta^\mu_{[\rho} \delta^\nu_{\sigma]} = 0. \tag{7.1.11}$$

The most general σ is defined by a solution to the Equations (3.6.27) and (7.1.11). If we restrict the $g_{\lambda\mu}$ to be the components of the Minkowski metric then the unique solution is given by (7.1.10). The Minkowski metric is then symmetric also with respect to σ. From (3.6.25) and (3.6.26) we see that if we require that the torsion vanish then we have

$$\omega^\lambda{}_{\mu\nu} = 0.$$

The algebra describes therefore a noncommutative version of Minkowski space.

It is of interest to notice that Equation (7.1.9) defines a derivation of the algebra whatever the form of the matrix $J^{\mu\nu}$. The derivation is inner if the matrix is invertible; otherwise it is outer. Let θ^λ be a set of Grassmann variables. Define

$$x^\lambda = \tilde{x}^\lambda + \mu_P^{-1}\theta^\lambda, \qquad J^{\mu\nu} = -2i\theta^\mu\theta^\nu.$$

Then (7.1.1) is satisfied and $J^{\mu\nu} \in \mathcal{Z}(\mathcal{A})$. In this case the matrix $J^{\mu\nu}$ is not invertible; it is in fact nilpotent.

Notice also that in this example the center $\mathcal{Z}(\mathcal{A})$ is nontrivial and in fact it is possible to impose that it be a smooth 6-dimensional manifold with the $J^{\mu\nu}$ as coordinates. One can impose the conditions

$$J_{\mu\nu}J^{\mu\nu} = 0,$$

as well as

$$\epsilon_{\mu\nu\rho\sigma} J^{\mu\nu} J^{\rho\sigma} = 12,$$

which is equivalent to

$$J_{\mu\nu}^{-1} = -\frac{1}{3}\epsilon_{\mu\nu\rho\sigma} J^{\rho\sigma}.$$

The normalization is arbitrary. The manifold can then be reduced to 4 dimensions. In the limit $\mu_P \to 0$ we have in fact a structure which can be regarded as a 4-dimensional manifold with a noncommutative extension à la Kaluza-Klein similar to the structures to be studied in Chapter 8. In the limit $\mu_P \to \infty$ the structure can be considered to be that of an ordinary space-time with an extra 4- or 6-dimensional factor in which the Poisson structure defined by the commutator (7.1.1) takes its values. An element of $\mathcal{Z}(\mathcal{A})$ can in no way correspond to a function on space-time in the commutative limit. Also if f is an element of $\mathcal{A}$ such that $e_\mu f = 0$ then we can only conclude that f is an arbitrary function of the $J^{\mu\nu}$; we cannot conclude that it is proportional to the identity. We regard the non-trivial center as a defect which must eventually be eliminated.   □

Example 7.10 A q-deformed version of Minkowski space also exists. Consider the braid matrix of Example 4.66 with $n = 4$. The group $SO(4)$ is exceptional in that it is not simple; to within a factor $\mathbb{Z}_2$ it is the product of two copies of $SO(3)$. This has for consequence that it has not only the deformation mentioned in Example 4.66 which makes the corresponding space $\mathbb{C}_q^4$ look like two distinct commuting copies of $\mathbb{C}_q^2$, but (at least) another which makes $\mathbb{C}_q^4$ contain a $\mathbb{C}_q^1$ and $\mathbb{C}_q^3$ which commute. It is obviously this second deformation which is the better adapted to serve as a deformation of Minkowski space. We can think then of q-Minkowski space as an algebra with four generators x^μ and a dilatator Λ such that x^0 and Λ generate the algebra we introduced in Example 4.18 and the remaining three x^i and the dilatator generate the algebra introduced in Example 4.21. There are again two complex differential calculi. A q-deformed version of the corresponding phase space can be constructed by introducing a second copy of the algebra, identified as momentum space, and by introducing commutation relations between the two copies which depend on the q-deformed version of the Lorentz group.   □

7.2   A finite model

We would like now to show that the geometry of the smooth 2-sphere S^2 embedded in $\mathbb{R}^3$ with the standard euclidean metric can be obtained as the limit of a sequence of matrix geometries M_n when $n \to \infty$. We consider these geometries, one for each n, as finite approximations to the smooth differential geometry of the 2-sphere S^2. For each n then, the set of all derivations of M_n is to be considered as a finite approximation of the set of all smooth vector fields on S^2. There is a problem in the approximation because of the ambiguity in the ordering when passing from commutative to noncommutative. Associated with this ambiguity is the fact that the 1-forms on M_n are approximations, not of the 1-forms on S^2, but of 1-forms on SU_2 considered as a U_1-bundle over S^2. Under the map which defines the approximation, the image of a general 1-form on M_n is a U_1-connection on SU_2. Connections are 1-forms which possess an invariance property under the action of the group U_1. The image of M_n can also be considered as an algebra of functions on SU_2 invariant under the action of U_1. The calculations which we shall perform are formal algebraic manipulations and we make little attempt to justify the convergence of the limit $n \to \infty$ with appropriate topologies.

Recall the definition of the 2-sphere S^2 which was given in Example 2.1 and consider in particular the expansion (2.1.21):

$$\tilde{f}(\tilde{x}^a) = f_0 + f_a \tilde{x}^a + \frac{1}{2} f_{ab} \tilde{x}^a \tilde{x}^b + \cdots.$$

We mentioned that if we truncate the expansion to order n we obtain a vector space of dimension n^2. If we truncate all functions to the constant term we reduce the algebra $\mathcal{C}(S^2)$ to the algebra $\mathcal{A}_1 = \mathbb{C}$ of complex numbers and the geometry of S^2 is reduced to that of a point. Suppose that we keep the term linear in the $\tilde{x}^a$. That is, we consider the set $\mathcal{A}_2$ of functions of the form (2.1.21) with all coefficients other than f_0 and f_a put equal to zero. This set is a 4-dimensional vector space. What we wish to do is introduce a new product in the $\tilde{x}^a$ which will make $\mathcal{A}_2$ into an algebra. There are several ways of doing this. If we require that the radical of $\mathcal{A}_2$ be equal to zero then there are two ways. We can define the product so that $\mathcal{A}_2$ becomes equal to the product of 4 copies of $\mathbb{C}$. This algebra is commutative and the sphere looks like a set of 4 points. This would be a lattice approximation. We shall choose the second possibility. We shall define the product so that $\mathcal{A}_2$ becomes equal to the algebra M_2 of complex 2×2 matrices. That is, we replace

$$\tilde{x}^a \mapsto x^a = \hbar r^{-1}(\sigma^a/2).$$

The σ^a are the Pauli matrices and the parameter $\hbar$ must be related to r by the equation $4r^4 = 3\hbar^2$ in order to have the Casimir relation $g_{ab} x^a x^b - r^2 = 0$ for the x^a. The algebra M_2 describes the sphere very poorly and it is fuzzy. Only the north and south poles can be distinguished.

Suppose next that we keep the term quadratic in the $\tilde{x}^a$. That is we consider the set $\mathcal{A}_3$ of expansions of the form (2.1.21) with all coefficients other than f_0, f_a and f_{ab} put equal to zero. This set is a 9-dimensional vector space because of the constraint (2.1.20). We wish to introduce a new product in the $\tilde{x}^a$ which will make $\mathcal{A}_3$ into an algebra. There are again several ways of doing this. We shall define the product so that $\mathcal{A}_3$ becomes equal to the algebra M_3 of complex 3×3 matrices. That is, we replace

$$\tilde{x}^a \mapsto x^a \tag{7.2.1}$$

using the matrices x^a defined in Equation (3.1.16).

In general suppose that we suppress the terms of n-th order in the $\tilde{x}^a$. The resulting set is a vector space $\mathcal{A}_n$ of dimension n^2 and we can introduce a new product in the $\tilde{x}^a$ which will make it into the algebra M_n of complex $n \times n$ matrices. We make the replacement (7.2.1) but using J^a which generate the n-dimensional irreducible representation of $\underline{su}_2$. Because of the commutation relations (3.1.18) in the limit $\bar{k} \to 0$ the x^a commute and all of the points of the sphere can be distinguished.

Consider now the map ϕ_n of M_n into $\mathcal{C}(S^2)$ given by the inverse of (7.2.1) on the generators x^a. Every element $f \in M_n$ has a unique expansion in the x^a of the form (2.1.21). Let $\tilde{f}$ be the element of $\mathcal{C}(S^2)$ obtained from f by replacing x^a by $\tilde{x}^a$. Then $f \mapsto \tilde{f}$ defines a linear map ϕ_n of M_n into $\mathcal{C}(S^2)$. The image of ϕ_n is the set of functions on S^2 which are polynomials in the $\tilde{x}^a$ of degree up to $n-1$. Let $l \le n-1$. If we consider the vector space V_l of elements in M_n which possess an expansion of order at most l then we have for $f, g \in V_l$

$$\phi_n(fg) - \phi_n(f)\phi_n(g) = o(R(l)/n) \qquad (7.2.2)$$

where $R(l)$ is a polynomial in l. Consider first the case $l = 1$. Then $f = f_0 + f_a x^a$ and $g = g_0 + g_a x^a$ and

$$\phi_n(fg) - \phi_n(f)\phi_n(g) = \frac{1}{2}i\bar{k}\, C_{abc} f^a g^b \tilde{x}^c.$$

The right-hand side is of order $1/n$. In general there are l^2 terms in the product, each of which is of the form $f_{a_1 \cdots a_i} g_{b_1 \cdots b_j} x^{a_1} \cdots x^{a_i} x^{b_1} \cdots x^{b_j}$ with $1 \le i, j \le l$, and all of them have to be symmetrized. Each transposition in the corresponding symmetrized product contributes a term $1/n$ on the right-hand side of (7.2.2) and $R(l)$ is the total number of all transpositions. If $R(l) \sim n$ then (7.2.2) is an empty assertion. As the order of the polynomials approaches n, the error involved in considering ϕ_n an algebra morphism becomes more and more important.

We can consider then the algebra M_n as a finite-dimensional approximation to $\mathcal{C}(S^2)$. It plays a role similar to the algebra of functions on a lattice but it has the advantage of respecting all of the symmetries of the algebra of functions to which it is an approximation. In a formal sense the limit of the sequence of ϕ_n exists as an algebra morphism. It can be considered as the composition of the limit defined in Example 3.3 and the limit $\hbar \to 0$. A set of smooth functions on S^2 must be the image of a set of matrices which commute to within order $\bar{k}$. Most sequences of matrices would tend to singular functions. Let $\mathcal{C}_l$ be the image in the limit $n \to \infty$ of the matrix polynomials of order l. Then the set $\{\mathcal{C}_l\}$ is a filtration of $\mathcal{C}(S^2)$ which we shall refer to as the *fuzzy sphere*. The fact that the sequence of matrix algebras tends to $\mathcal{C}(S^2)$ and not to the algebra of functions on some other manifold is based on the choice of $\mathcal{C}_1$ and the corresponding derivations.

To see better in what sense we can approximate a commutative algebra $\mathcal{C}(S^2)$ by a noncommutative sequence M_n we must define a norm on M_n and show that in the limit $n \to \infty$ the algebra $\mathcal{C}(S^2)$ can be considered as the image of the 'almost diagonal' matrices in M_n. For each element $f \in M_n$ we set

$$\|f\|_n^2 = \frac{1}{n}\mathrm{Tr}(f^*f).$$

The generators x^a have a norm given by

$$\|x^a\|_n^2 = \frac{1}{3}r^2.$$

This is independent of n. We define the norm of an element $\tilde{f} \in \mathcal{C}(S^2)$ by

$$\|\tilde{f}\|^2 = \frac{1}{4\pi r^2}\int |\tilde{f}|^2 d\tilde{x}.$$

Here $d\tilde{x}$ is the invariant volume element of S^2. We have $\|x^a\|_n = \|\tilde{x}^a\|$. Let $f \in M_n$ and set $\tilde{f} = \phi_n(f)$. Then we have

$$\frac{1}{n}\mathrm{Tr}(f) = f_0 = \frac{1}{4\pi r^2}\int \tilde{f}d\tilde{x}.$$

At this point it is convenient to introduce the parameter $k = 2\pi\bar{k}$. From (3.1.17) we can write then

$$4\pi r^2 \simeq kn$$

and we have

$$k\mathrm{Tr}(f) \to \int \tilde{f}d\tilde{x}. \tag{7.2.3}$$

Suppose that f is a polynomial of degree l. Then from (7.2.2) we deduce that

$$\|f\|_n - \|\tilde{f}\| = o(R(l)/n).$$

A general element $f \in M_n$ with entries of order $o(1)$ will have a norm $\|f\|_n^2 = o(n)$. A diagonal matrix with entries of order $o(1)$ or a matrix with only a number of order $o(1)$ of off-diagonal terms will on the other hand have a norm $\|f\|_n^2 = o(1)$. For large n then bounded functions will be the image of almost-diagonal matrices, that is of matrices which commute to within order $\bar{k}$.

In Section 3.1 we introduced the matrices λ^a which generated an n-dimensional representation of $\underline{su}_2$. They were chosen anti-hermitian and normalized by the condition that the metric g_{ab} in (3.1.1) be the ordinary euclidean metric. At this point we must introduce a parameter μ with the dimensions of mass. The equations of Chapter 3 should have been more correctly written with the replacement

$$\lambda_a \to \mu\lambda_a$$

so that the derivations e_a have the correct dimensions of mass. There are two important length scales, which are related through the integer n. The r sets the global scale of the sphere and k determines the dimensions of the fundamental cells. Since we are considering a noncommutative model of space-time then we are tempted to identify k with the inverse of the square of the Planck mass, $k = \mu_P^{-2}$, and consider space-time as fundamentally noncommutative in the presence of gravity.

A smooth global vector field on S^2 defines and is defined by a derivation of the algebra $\mathcal{C}(S^2)$; the noncommutative analogue of a global vector field on S^2 is a derivation of the algebra M_n. Consider the 3 derivations $e_a = \operatorname{ad} \lambda_a$ of Section 3.1, which we write in terms of the x^a as in Equation (3.1.19); the derivative $e_a x^b$ of x^b is given by (3.1.20). The 1-forms θ^a dual to the derivations e_a are given by Equation (3.1.22). The θ^a satisfy the structure equations (3.1.9) but with $C^c{}_{ab}$ the structure constants of SU_2 as normalized in (3.1.18).

Let $\tilde{f}$ be the limit function of a sequence of matrices f which are polynomials of arbitrary but fixed order. Comparing (2.1.22) with (3.1.20) we can conclude that in general

$$e_a f \to \tilde{e}_a \tilde{f}.$$

We can write then that in the limit when $n \to \infty$,

$$e_a \to \tilde{e}_a, \qquad \theta^a \to \tilde{\theta}^a.$$

The limit $\tilde{\theta}$ of θ is singular. This is necessarily so since otherwise we would have constructed a gauge-invariant connection form on SU_2. It can also be deduced from the fact that the θ^a are dual to the e_a which in the limit satisfy the relation $\tilde{x}^a \tilde{e}_a = 0$.

The limit of $k\theta$ is related to the Dirac-monopole potential A of Example 2.1 by the equation

$$k\theta \to r^2 A.$$

The limit of $k\theta^2$ is given by

$$k\theta^2 = -\frac{1}{k}(x_a \theta^a)^2 = \frac{1}{2i} C_{abc} x^a \theta^b \theta^c \to -r^2 F.$$

Therefore

$$k(d\theta + \theta^2) \to r^2(dA - F).$$

Both sides are in fact identically zero.

We mentioned briefly automorphisms of M_n in Section 4.7. The analogue of a general coordinate transformation in $\mathbb{R}^3$ is a change of generators

$$x^a \mapsto x'^a \tag{7.2.4}$$

of the algebra M_n. This does not necessarily respect the relations of the algebra, for example the Casimir relation, and it does not necessarily possess an extension to an automorphism of M_n. If one imposes the extra conditions

$$g_{ab}x'^a x'^b - g_{ab}x^a x^b = o(\hbar), \qquad [x'^a, x'^b] = o(\hbar)$$

then a sequence of maps (7.2.4) would presumably tend to a general continuous map of the sphere into itself.

A diffeomorphism of S^2 defines and is defined by an automorphism of the algebra of smooth functions on S^2. Let ϕ be a diffeomorphism of S^2. Then ϕ has an extension ϕ^a to $\mathbb{R}^3$ and we can set $\tilde{x}'^a = \phi^a(\tilde{x}^b)$. This defines an automorphism of $C(S^2)$ which is independent of the extension. Conversely such an automorphism ϕ^a restricted to the generators $\tilde{x}^a$ defines a smooth coordinate transformation of $\mathbb{R}^3$ and by restriction, a diffeomorphism of S^2. The noncommutative analogue of a diffeomorphism of S^2 is therefore an automorphism of M_n. Since M_n is a simple algebra all of its automorphisms are of the form $f \mapsto f' = g^{-1}fg$ where g is an arbitrary fixed element of M_n which has an inverse. We have considered complex-valued functions on S^2 and the algebra $C(S^2)$ has a $*$-operation $\tilde{f} \mapsto \tilde{f}^*$ obtained by taking the complex conjugate of $\tilde{f}$. A diffeomorphism of S^2 will define an automorphism of $C(S^2)$ which respects this $*$-operation: $\tilde{f}'^* = \tilde{f}^{*\prime}$. We must therefore require the same condition on the automorphisms of M_n: $(g^{-1}fg)^* = g^{-1}f^*g$. This means that $g^* = g^{-1}$ and therefore that

$$x'^a = g^{-1}x^a g, \qquad g \in SU_n. \tag{7.2.5}$$

A different choice of x'^a not related to x^a by this formula would be equivalent to a different choice of symplectic structure in the commutative limit.

A diffeomorphism of S^2 leaves the set of smooth global vector fields invariant. The change of generators (7.2.5) takes $X = \operatorname{ad} f$ into $X' = \operatorname{ad}(g^{-1}fg)$ and so all automorphisms of M_n are analogues of diffeomorphisms of S^2. If g is near to the identity we can write

$$x'^a \simeq x^a + e^a h, \qquad g \simeq 1 + \frac{h}{i\hbar}. \tag{7.2.6}$$

An important special case is given by $h = h_a x^a$. Then we have

$$x'^a \simeq x^a + C^a{}_{bc}h^b x^c \tag{7.2.7}$$

and therefore in the limit it corresponds to an infinitesimal rotation in $\mathbb{R}^3$ about the axis h_a. The Formula (7.2.7) yields the adjoint action of the Lie algebra of SO_3 on M_n, which contains exactly once the irreducible representation of dimension $2j + 1$ for $0 \leq j \leq n - 1$.

Let $\tilde{x}'^a$ be the limit of the sequence $\{x'^a\}$ defined in (7.2.5). Then the map $\tilde{x}^a \mapsto \tilde{x}'^a$ is a coordinate transformation of $\mathbb{R}^3$ which takes S^2 into itself. Since the map $\tilde{e}_a \mapsto \tilde{e}'_a$ exists the coordinate transformation is smooth. It is a diffeomorphism of S^2. It is not however a general diffeomorphism. The coordinate transformation of $\mathbb{R}^3$ which one obtains as the limit of the automorphisms (7.2.6) is

$$\tilde{x}'^a \simeq \tilde{x}^a + \tilde{e}^a \tilde{h}. \tag{7.2.8}$$

The induced volume 2-form on S^2 is proportional to

$$\tilde{\omega} = C_{abc}\tilde{x}^a \tilde{\theta}^b \wedge \tilde{\theta}^c = \lim_{k \to 0} k\,\omega,$$

where ω is the symplectic form introduced in Example 3.2. Although ω is exact the limit of $k\omega$ is not. Let $\tilde{X} = (\tilde{e}^a \tilde{h})\tilde{e}_a$ be the vector field defined by (7.2.8), considered as a diffeomorphism of S^2. Then $\tilde{\omega}(\tilde{X}) = d\tilde{h}$ and so $L_{\tilde{X}}\tilde{\omega} = 0$. The transformations (7.2.5) define then in the limit symplectomorphisms of S^2 with respect to $\tilde{\omega}$. The vector $\tilde{X}$ is the hamiltonian flow associated to the function $\tilde{h}$. There is no analogue of a general diffeomorphism of the sphere in terms of a transformation of the generators of M_n. This is so because for every finite n the automorphisms of M_n are all inner. In the formal limit M_∞ there would exist other automorphisms which would be more general. The infinitesimal form of an inner automorphism is an inner derivation. On a symplectic manifold this corresponds to a symplectic vector field.

A transformation (7.2.5) defines a new set of derivations

$$e'_a = \frac{1}{ik}\mathrm{ad}\,x'_a$$

which satisfy also the commutation relations (3.1.21). The relation between e_a and e'_a is not as simple however as in the commutative case because the derivations do not form a module over the algebra. The most general derivation then is not a linear combination with matrix coefficients of the derivations e_a. So it is not surprising that in general there is no relation between e'_a and e_a of the form $e'_a = \Lambda'^{-1b}{}_a e_b$. Recall the basis e_r, $1 \le r \le n^2 - 1$, introduced in Section 3.1, of the complete set of derivations of the algebra M_n. We have then

$$e'_a = \Lambda'^{-1r}{}_a e_r,$$

for some complex $3 \times (n^2 - 1)$ matrix $\Lambda'^{-1r}{}_a$. In the particular case (7.2.7) we have $\Lambda'^{-1b}{}_a \simeq \delta^b_a + C^b{}_{ac}h^c$ and only in this case will the $\Lambda'^{-1r}{}_a = 0$ vanish for $r \ge 4$.

Example 7.11 We cannot speak of the position of a particle because of the absence of localization, but the state of a particle on the sphere is described as in quantum mechanics by a state vector ψ. For the matrix algebra M_n then a particle is described by a vector $\psi \in \mathbb{P}^{n-1}(\mathbb{C})$. An observable associated to the particle is an hermitian element of M_n and the value of an observable f is given by the real number $\psi^* f \psi$. The particle is a pure state in the sense of Section 4.3; a mixed state would be given

by Equation (4.3.6) with $\rho \in M_n$. For example, what corresponds to the position of
the particle is given by the 3 numbers $\psi^* x^a \psi$. As acting on the eigenstate of x^3 with
the largest eigenvalue the commutation relations (3.1.18) become, for large n,

$$[x^1, x^2] = i\hbar.$$

We see then that there are elementary Planck cells of area $2\pi\hbar$. The sphere can be
covered by

$$\frac{4\pi r^2}{2\pi\hbar} \simeq n$$

of them. For large n this is also equal to the dimension of the space of pure states,
as in the case of the lattice in Example 2.2. However, whereas the dimension of the
algebra of functions on the lattice was equal to the number of points, the dimension
of the algebra M_n varies as the square of the number of cells. Both cases should in
some sense agree in the singular limit $n \to \infty$. One way to resolve the discrepancy is
to consider the solid-state physics analogy which was touched upon at the beginning
of the chapter. We can consider the n Planck cells as the analog of the n lowest
Landau levels and the $n(n-1)$ remaining as 'excited states'. But since momentum
and position have been exchanged the excitation is in position space and more similar
to screening. The extra degrees of freedom do not acquire a mass, which would lead
to an exponential damping in position space; they become 'non-local' and acquire an
exponential drop-off in momentum space. At the end of Section 2.1 we exhibit some
commutative algebras also with dimension greater than the number of associated
points.   □

Example 7.12 As in quantum mechanics coherent states can be introduced. Let
$|0, 0, \pm r\rangle$ be the two eigenstates of x^3 which minimize the value of $(x^1)^2 + (x^2)^2$.
From the analogy with quantum mechanics the corresponding eigenstates can be
identified with a function on S^2 whose support is concentrated on a surface of area
$2\pi\hbar$, one at the north pole, the other at the south. Consider now an arbitrary point
with coordinates $\tilde{x}^a$ and define the state $|\tilde{x}^a\rangle$ by acting on the state $|0, 0, r\rangle$ by the
representation of the rotation from the north pole to the point $\tilde{x}^a$. Then the set $|\tilde{x}^a\rangle$
is a set of coherent states. It is obvious from the definition that

$$\langle \tilde{x}^a | x^a | \tilde{x}^a \rangle = \tilde{x}^a.$$

This formula is analogous to (4.3.20). The coherent states furnish a map between
the generators x^a of the algebra M_n and the coordinates. The sphere can be covered
by n coherent states as it was with the n cells of the previous example.   □

Example 7.13 One can discuss also quantum mechanics on the fuzzy sphere and
then take separately the two possible limits, the commutative limit $\hbar \to 0$, and the

classical limit $\hbar \to 0$. The Schrödinger equation for a quantum-mechanical particle of mass μ moving in a potential V is given by

$$i\hbar \partial_t \psi - \frac{\hbar^2}{2\mu} \Delta \psi = V\psi. \qquad (7.2.9)$$

If we are considering the fuzzy sphere then ψ and V are elements of M_n for some n and the Laplace operator (3.4.12) is given by $\Delta = -e_a e^a$. If we are considering the commutative limit then ψ and V are functions on the 2-sphere. A solution to equation (7.2.9) on the fuzzy sphere will tend, as $\hbar \to 0$, to a solution to the ordinary Schrödinger equation and we can say that fuzzy quantum mechanics has ordinary quantum mechanics as its limit.

The simplest example is free motion on the sphere, with $V = 0$. In the $\hbar \to 0$ limit the eigenvalues of the equation (7.2.9) are given by the usual angular-momentum contributions to the energy of a particle moving in a spherically symmetric potential. That is,

$$E_j = \frac{\hbar^2}{2\mu} \frac{j(j+1)}{r^2}.$$

The integer j takes all positive integers as values and each energy level E_j has multiplicity $2j + 1$. On the fuzzy sphere with the matrices M_n as algebra of observables, we have exactly the same spectrum but with the added condition that $j \leq n - 1$. There are exactly n^2 eigenvalues corresponding to the n^2 degrees of freedom in the matrix algebra M_n. So in this simple case the spectrum remains unmodified up to the last element and then there is an abrupt cut-off.

In less trivial situations we would not expect this to be so. Let $E_j^{(n)}$ be the jth energy level found using the algebra M_n in a problem with $V \neq 0$. Counting multiplicities there can only be n^2 levels. Let E_j be the levels for the corresponding quantum-mechanical problem on the sphere. There will be an infinite number of these levels. One could expect in general the levels E_j and $E_j^{(n)}$ to coincide only as long as $j \ll n$. That is, the error occasioned by using M_n should be given by

$$|E_j - E_j^{(n)}| = o\left(\frac{\hbar E_j}{\hbar r}\right).$$

Consider now the classical limit. We try the usual Ansatz

$$\psi = \exp(iS/\hbar)$$

and expand in Equation (7.2.9). We find

$$e_a \psi = (i/\hbar)(e_a S)\psi \left[1 + o\left(\frac{\hbar \mu}{\hbar r}\right)\right].$$

The extra terms which we have not written out explicitly have their origin in the fact that if S is a matrix then S and $e_a S$ do not commute in general. For the classical limit to make sense then we must have

$$\tilde{k}\mu \ll \hbar r.$$

□

Example 7.14 Return once more to the graded commutative algebra described in Example 4.6. If, as in Example 4.7, one 'quantizes' the bosonic coordinates then logically one should do the same for the fermionic generators; if the bosonic generators of the algebra no longer commute then the corresponding fermionic ones should no longer anticommute. One could imagine that at the Planck length bosons and fermions become common elements of a noncommutative algebra and are indistinguishable by their commutation relations. A supersymmetric extension of the fuzzy sphere has been explicitly developed which has this property; the fermionic generators anticommute only in the commutative limit. In this model, for all values of the 'cut-off' n the supersymmetric extension of the rotation group is an exact symmetry. In general this need not be the case. The supersymmetric extension of the Poincaré group could be the ordinary one only in the commutative limit. We have already mentioned in Example 4.67 some proposals to modify the Poincaré group itself at short distances.   □

Example 7.15 In the theory of geometric quantization one considers a general compact Kähler manifold V and the Hilbert space of holomorphic sections of a complex line bundle over it. This Hilbert space is of finite dimension and so the algebra of linear operators defined on it is an algebra of matrices. Berezin has given a general prescription of how to associate to each matrix a rational function on V called its covariant symbol. The set $\mathcal{E}$ of these functions can be made into an algebra by defining a new, noncommutative $*$-product as the image of the matrix product. In the particular case of the sphere $S^2 = \mathbb{P}^1(\mathbb{C})$, for each n, $\mathcal{E}$ is a set of rational functions in the complex coordinates z and its conjugate $\bar{z}$, which when expressed in terms of the coordinates $\tilde{x}^a$ coincides with the image of the application ϕ_n.   □

Example 7.16 A noncommutative modification of a compact $2n$-dimensional manifold can be given simply by taking the appropriate tensor product of the algebra associated to the 2-sphere. Higher-dimensional examples can be constructed also by replacing SU_2 by SU_n in the above calculations. However, no satisfactory noncommutative modification of the 4-sphere can be given, at least none which preserves SO_5 invariance. We saw in Section 4.2 that such a modification would induce on S^4 a Poisson structure which in turn would break the invariance. There come to mind two natural manifolds which might be of interest in finding a noncommutative version of S^4. One is the sphere S^7, associated to the octonians in the same way S^3 is associated to the quaternions; it is a principle SU_2 bundle over S^4. The other is the Kähler manifold $\mathbb{P}^3(\mathbb{C})$ which is an S^2 bundle over S^4.   □

7.3   Fuzzy physics

One way to quantize the scalar field ϕ defined in Example 2.14 involves an integration over all field configurations on space-time, that is, over the set of all elements of the algebra $\mathcal{C}(V)$ of classical observables. The integration is ill-defined at best and it is most conveniently described using compactified euclidean space. To define this we shall suppose as in Example 4.31 that space-time is a manifold V which has a metric with Minkowski signature and which can be decomposed into a time direction and a space V_S which is time-independent. If V_S is compact *Wick rotation* of the time coordinate t to $t = i\tau$ we obtain a space which has a metric with euclidean signature. The τ coordinate can be compactified to obtain a compact manifold V without boundary. If V_S is ordinary 3-dimensional euclidean space, the resulting euclidean 4-dimensional space can be compactified to the sphere S^4. The assumption here is that if one wishes to study quantum theory in Minkowski space, one can first Wick-rotate, calculate quantum effects with the euclidean metric and then Wick-rotate back. It is hoped that this process is meaningful at least when there is no gravitational field; it can be shown to be so in the formal perturbation-series approximation.

It is generally believed that all quantum effects can be described by the *partition function $Z[J]$*, which is a functional of the *external source J*. This is defined to be the *path integral*

$$Z[J] = \int e^{-S[\phi, J]} d\phi, \qquad (7.3.1)$$

where the integration is over all field configurations $\phi \in \mathcal{C}(V)$. The euclidean action $S[\phi, J]$ in the presence of the external source J is taken to be of the form (2.2.35). For simplicity we shall suppose that U is a convex function with a minimum at zero. This means that there will be a unique solution to the classical field equation for each value of the external source. The measure $d\phi$ must be defined as a limit of a sequence of measures on finite-dimensional spaces which approximate $\mathcal{C}(V)$. Little is known about the properties of the limit measure or even in general if it can be well defined. However, as a formal power-series expansion in the interaction U it can be shown that the description in terms of $Z[J]$ is equivalent to the description in terms of quantized field variables which was outlined in Example 4.31.

We define the functional $W[J]$ by the equation

$$Z[J] = e^{-W[J]}.$$

The functional derivative of $W[J]$ with respect to J,

$$\phi_c = \frac{\delta W}{\delta J}, \qquad (7.3.2)$$

is called the *classical scalar field*. It is a field configuration which is equal to the solution to the classical field equations to lowest order in $\hbar$. It is not to be confused with the scalar field introduced in Example 2.14. Equation (7.3.2) is the quantum analogue of (2.2.38).

The functional $\Gamma[\phi_c]$ defined as

$$\Gamma[\phi_c] = W[J] - \int J\phi_c dx \qquad (7.3.3)$$

is called the *effective action*. It follows from the definition of ϕ_c that $\Gamma[\phi_c]$ is independent of J if we consider J and ϕ_c as independent variables. It follows also immediately from the definition that

$$\frac{\delta\Gamma}{\delta\phi_c} = -J. \qquad (7.3.4)$$

This equation is the quantum analogue of the classical equations of motion (2.2.36). (It is slightly different because the classical action as we have defined it includes the external source term.) So the classical field satisfies the same equation with respect to the effective action as the solution to the classical field equations satisfies with respect to the classical action. This fact explains the name.

If one were to consider the functional Γ as a classical action for the classical field, the solution would be a perturbation of the solution to the classical field equations which includes quantum corrections. In order to see this we can consider the integral (7.3.1) in the classical limit $\hbar \to 0$. We have set $\hbar = 1$. The correct expression is obtained by the replacement $S \mapsto S/\hbar$. It is clear then that in the classical limit, the integral (7.3.1) will be dominated by those field configurations which minimize the action. These are the classical solutions. In the simple case we are considering there is only one classical solution ϕ_0 for each value of the source J. We find then that in the classical limit

$$W[J] = S[\phi_0, J], \qquad \phi_0 = \phi_0[J]. \qquad (7.3.5)$$

From this it follows that

$$\frac{\delta W[J]}{\delta J(x)} = \int \frac{\delta S[\phi_0, J]}{\delta\phi_0(y)} \frac{\delta\phi_0(y)}{\delta J(x)} dy + \frac{\delta S[\phi_0, J]}{\delta J(x)}.$$

The first term on the right-hand side vanishes by (2.2.36) and by (2.2.38) the second term is equal to ϕ_0. Therefore comparing with (7.3.2) we see that the classical field ϕ_c coincides, to lowest order in $\hbar$, with ϕ_0: $\phi_c = \phi_0 + o(\hbar)$. If we set

$$\Gamma = \Gamma_{(0)} + o(\hbar)$$

then comparing (7.3.3) with (7.3.5) we find that

$$\Gamma_{(0)}[\phi_c] = \frac{1}{2}\int \phi_c(\Delta + \mu^2)\phi_c dx + \int U(\phi_c)dx$$

in the classical limit. The effective action is equal in this limit to the classical action (2.2.35) without the source term. The vacuum-expectation value $\langle \phi \rangle_J$ of the field ϕ in the presence of the source J is defined to be

$$\langle \phi(x) \rangle_J = Z^{-1} \int \phi(x) e^{-S[\phi,J]} d\phi = -Z^{-1} \frac{\delta Z[J]}{\delta J(x)}.$$

It follows then that $Z[J] = \langle 1 \rangle_J$. The classical field can be written therefore

$$\phi_c = \frac{\langle \phi \rangle_J}{\langle 1 \rangle_J}$$

as the vacuum-expectation value of the field ϕ divided by the probability $\langle 1 \rangle_J$ that the vacuum remains the vacuum.

If we take the second functional derivative of $Z[J]$ with respect to the external source we find

$$Z^{-1} \frac{\delta^2 Z[J]}{\delta J(x) \delta J(y)} = \phi_c(x) \phi_c(y) - \frac{\delta^2 W[J]}{\delta J(x) \delta J(y)}.$$

We define

$$\langle \phi(x) \phi(y) \rangle_J = Z^{-1} \int \phi(x) \phi(y) e^{-S[\phi,J]} d\phi = Z^{-1} \frac{\delta^2 Z[J]}{\delta J(x) \delta J(y)}.$$

The *propagator* or *2-point function* $G(x, y)$ is defined to be the second functional derivative of $W[J]$ with respect to to J:

$$G(x, y) = -\frac{\delta^2 W[J]}{\delta J(x) \delta J(y)}. \tag{7.3.6}$$

In the absence of sources the classical field vanishes. So we find in this case that

$$G(x, y) = \langle \phi(x) \phi(y) \rangle_0. \tag{7.3.7}$$

Example 7.17 To make these formal manipulations more transparent we repeat them in the special case where U vanishes. In this case the classical field ϕ_c is equal to the solution to the classical field equations: $\phi_c = \phi_0 = -(\Delta + \mu^2)^{-1} J$. If we define $h = \phi - \phi_c$ we find that the action can be written in the form

$$S = \frac{1}{2} \int \phi_c (\Delta + \mu^2) \phi_c dx + \int J \phi_c dx + \frac{1}{2} \int h (\Delta + \mu^2) h dx. \tag{7.3.8}$$

Therefore we have 'exactly'

$$W[J] = -\frac{1}{2} \int J (\Delta + \mu^2)^{-1} J dx - \log Z_0. \tag{7.3.9}$$

We have defined here $Z_0 = Z[0]$. The second term on the right-hand side is formally of order $\hbar$ but is actually equal to infinity. It is related to the vacuum energy of

Example 4.29 but it is here infinite since there are an infinite number of modes. Neglecting this term we find that $\Gamma = \Gamma_{(0)}$ 'exactly'.

The operator $(\Delta + \mu^2)^{-1}$ can be written using the propagator in the form

$$((\Delta + \mu^2)^{-1}J)(x) = \int G(x,y)J(y)dy.$$

Since

$$\phi_c(x) = -\int G(x,y)J(y)dy \tag{7.3.10}$$

the propagator yields the field at point x due to the source $J(y)$ at point y. If one Wick-rotates back to Minkowski signature the analytic continuation of $G(x,y)$ is the Feynman propagator, which describes the propagation of the field from the space-time point y to the space-time point x. $\qquad\square$

When U does not vanish the terms of first order in $\hbar$ are still infinite in general even when one neglects the $\log Z_0$ infinity. Let ϕ_c be again the classical field and let as before $h = \phi - \phi_c$. Since to lowest order in $\hbar$, ϕ_c satisfies the classical field equation (2.2.37) the action can be written in the form

$$S = \frac{1}{2}\int \phi_c(\Delta + \mu^2)\phi_c dx + \int J\phi_c dx + \frac{1}{2}\int h(\Delta[\phi_c])h dx + o(h^3).$$

This is the same as (7.3.8) except that the operator $\Delta + \mu^2$ has been replaced by

$$\Delta[\phi_c] = \Delta + \mu^2 + \frac{1}{2}U''(\phi_c).$$

It follows directly from the definitions that the effective action $\Gamma[\phi_c]$ can be written to first order in $\hbar$ as

$$\Gamma[\phi_c] = \Gamma_{(0)}[\phi_c] + \hbar\Gamma_{(1)}[\phi_c] + o(\hbar^2),$$

where the first-order correction is given by the formal integral

$$\Gamma_{(1)}[\phi_c] = -\log \int e^{-\frac{1}{2}\int h\Delta[\phi_c]h dx}dh.$$

The operator $\Delta[\phi_c]$ has a discrete positive spectrum ω_j^2 and we can define formally the determinant $\det \Delta[\phi_c]$ as the product of the eigenvalues:

$$\det \Delta[\phi_c] = \prod_j \omega_j^2. \tag{7.3.11}$$

Therefore by formal analogy with the identity

$$\int_{-\infty}^{+\infty} e^{-ax^2}dx = \sqrt{\frac{\pi}{a}}$$

we find

$$\Gamma_{(1)}[\phi_c] = \frac{1}{2} \log \det \left(\frac{\Delta[\phi_c]}{\Delta[0]} \right) - \log Z_0. \qquad (7.3.12)$$

Let ω_{0j}^2 be the spectrum of $\Delta[0]$. As $j \to \infty$ one can show that

$$\frac{\omega_j}{\omega_{0j}} \to 1$$

but the convergence is not fast enough and the first term on the right-hand side of (7.3.12) is infinite as well as the second term. It is a consequence of the general renormalization program of field theory that the infinity can be absorbed in a redefinition of the mass parameter μ and the parameters of U if the theory is renormalizable. If this cannot be done the theory is said to be non-renormalizable. As examples of renormalizable theories we mention the (scalar) theories on 4-dimensional manifolds with interaction terms which are fourth-order polynomials, theories on 3-dimensional manifolds with sixth-order polynomials and theories on 2-dimensional manifolds with arbitrary polynomials. We have presented the calculations as formal manipulations with infinities. They can be made mathematically correct, for example by replacing the differential operator $\Delta[\phi_c]$ with the compact operator $e^{-t\Delta[\phi_c]}$, $t > 0$, and using the 'heat-kernel' methods mentioned in Example 5.12. Extracting meaningful physics from the higher-order corrections in $\hbar$ however requires more work.

The path integral can be made finite if the algebra $\mathcal{C}(V)$ is replaced by an algebra of finite dimension and it can only be defined as the limit of a sequence of finite integrals. We mentioned above that the most widely used prescription is to replace the manifold V by a sequence of lattice approximations. This amounts to replacing $\mathcal{C}(V)$ by a sequence of finite-dimensional commutative algebras. Suppose that we have a sequence of finite-dimensional noncommutative approximations $\mathcal{A}_n$ to the algebra of functions $\mathcal{C}(V)$. We could define then also the partition function given in (7.3.1) as a limit

$$Z = \lim_{n \to \infty} Z_n$$

where each Z_n involves a well defined finite integration over the algebra $\mathcal{A}_n$.

As a model for the compactified version of euclidean space-time we shall choose S^2 and we shall consider the sequence M_n of finite-dimensional matrix approximations to $\mathcal{C}(S^2)$ of the previous section. We rewrite (7.3.1) in the previous notation by replacing ϕ by $\tilde{f}$ and adding a tilde to J to emphasize the fact that it is a function:

$$\tilde{Z}[\tilde{J}] = \int e^{-\tilde{S}[\tilde{f},\tilde{J}]} d\tilde{f}.$$

For simplicity we shall suppose that both $\tilde{f}$ and $\tilde{J}$ are real elements of $\mathcal{C}(S^2)$. We can define $\tilde{Z}$ as the limit

$$\tilde{Z}[\tilde{J}] = \lim_{n \to \infty} Z_n[J]$$

where each $Z_n[J]$ is given using a well defined finite integration over the hermitian elements of the algebra M_n:

$$Z_n[J] = \int e^{-S[f,J]}df.$$

Here df means integration over all of the components of the matrix f.

As geometry the algebra is endowed quite naturally with a metric and a compatible linear connection. We have already mentioned the a complication due to the fact that the 2-sphere is not parallelizable. The differential structure of the sphere has to be studied therefore on a manifold of higher dimension which has a projection onto it. The minimal extension is the 3-sphere and the most natural projection is the Hopf fibration. The resulting matrix geometry is therefore of Kaluza-Klein type with a truncation of the modes in the third direction to the constant zero mode. The Laplace operator on the matrix geometries, although in a certain sense defined on a 'space' of dimension 3, can be considered therefore as a noncommutative approximation to the Laplace operator of the 2-sphere. The 3-sphere is a manifold of constant curvature, given to within numerical factors by the quantity r^{-2}. The most general Laplace operator could contain a term proportional to this curvature, which could be thought of as an effective mass term. We need not consider it since we shall explicitly add a mass term with parameter μ.

The Laplace operator (3.4.12) on the elements of M_n is given by

$$\Delta = -e_a e^a.$$

It is proportional to the Casimir operator of the adjoint representation of SU_2 and its spectrum $\{\omega_j^2\}$ is known. In units of r^{-2} it consists of integers of the form $i(i+1)$ of multiplicity $2i+1$ where $0 \le i \le n-1$. The index j takes the values $0 \le j \le n^2-1$. Consider the *eigenmatrices* f_j of $\Delta + \mu^2$:

$$(\Delta + \mu^2)f_j = (\omega_j^2 + \mu^2)f_j.$$

We shall choose the sequence of matrices f_j orthonormal:

$$\frac{1}{n}\mathrm{Tr}(f_j^* f_l) = \delta_{jl}.$$

The eigenmatrix f_0 is the identity matrix and $\omega_0 = 0$ has multiplicity 1. A general matrix f has then an expansion analogous to that of (4.3.13):

$$f = \sum_{j=0}^{n^2-1}(a_j f_j + a_j^* f_j^*).$$

The a_j are complex numbers.

In the limit $e_a \rightarrow \tilde{e}_a$ considered in Section 7.2, Δ tends to the operator

$$\tilde{\Delta} = -\tilde{e}_a \tilde{e}^a,$$

which coincides with the Laplace operator when acting on functions on the sphere S^2 of radius r. Consider the eigenfunctions $\tilde{f}_j$ of $\tilde{\Delta} + \mu^2$:

$$(\tilde{\Delta} + \mu^2)\tilde{f}_j = (\tilde{\omega}_j^2 + \mu^2)\tilde{f}_j.$$

We shall choose the sequence of functions $\tilde{f}_j(x)$ orthonormal:

$$\frac{1}{4\pi r^2} \int \tilde{f}_j^*(x)\tilde{f}_l(x)dx = \delta_{jl}.$$

From its definition and the equations of motion one sees that when $U = 0$ the propagator $\tilde{G}(x,y)$ is given by

$$\tilde{G}(x,y) = \frac{1}{4\pi r^2} \sum_{j=0}^{\infty} \frac{\tilde{f}_j(x)\tilde{f}_j^*(y)}{(\tilde{\omega}_j^2 + \mu^2)}.$$

After the vacuum energy the most important divergence in field theory arises from the *vacuum fluctuations*, the vacuum-expectation value $\langle \tilde{f}^2 \rangle_0$ of $\tilde{f}^2$ at the point x, in the absence of sources. From (7.3.7) we see that the sum over all vacuum fluctuations is given by the divergent integral

$$\tilde{I} = \int_{S^2} \tilde{G}(x,x)dx.$$

For each n there is matrix version of $\tilde{G}(x,y)$, an element of $M_n \otimes M_n$ defined as

$$G = \frac{1}{n} \sum_{j=0}^{n^2-1} \frac{f_j \otimes f_j^*}{\omega_j^2 + \mu^2}.$$

The limit $y \rightarrow x$ makes now no sense since both x and y are noncommuting operators; it must be defined in terms of eigenvalues. In the commutative case one can introduce a basis $|i\rangle$ of the algebra and define the limit as $y_i \rightarrow x_i$ where y_i and x_i are the respective eigenvalues. Since x and y are in fact two copies of the same operator one can redefine y_i as $x_{i'}$ for some other value of the index and define the limit as $i' \rightarrow i$. This definition makes sense also in the noncommuative case. Consider the generators x^a of M_n introduced in (7.2.1). Each of them has a complete set of eigenvectors in $\mathbb{C}^n$ and corresponding finite set of discrete eigenvalues x_i^a. We denote by $x_{i'}^a$ the eigenvalues of the second copy of x^a. Then the commutative expression of the coincidence limit is again the limit $i' \rightarrow i$. Now however, since the operators x_i^a cannot be simultaneously diagonalized the limit in general cannot be attained in the strict sense. But we do have a notion of distance and we can say that the 'points' coincide when this distance is a minimum. These subtleties are important only when the algebra is of infinite dimension and can be ignored here. There is a map m of $M_n \otimes M_n \to M_n$ given by the product: $f \otimes g \mapsto fg$. The matrix version of the integral $\tilde{I}$ is proportional to the trace of this map:

$$I = \frac{1}{4\pi r^2} \sum_{j=0}^{n^2-1} \frac{1}{\omega_j^2 + \mu^2}, \qquad (7.3.13)$$

For large n the sum is proportional to $\log n$.

For the action $S[f, J]$ we shall choose the matrix functional

$$S[f, J] = \frac{1}{2} k \text{Tr}(f(\Delta + \mu^2)f) + k \text{Tr}(U(f)) + k \text{Tr}(Jf).$$

By (7.2.3) we must include the factor k so that S has the correct dimensions and so that the kinetic term in the commutative limit is correctly normalized. The external source J is here an element of M_n with dimensions of $(\text{mass})^2$. If in the limit $f \to \tilde{f}$ and $J \to \tilde{J}$ then we have also $S \to \tilde{S}$ where

$$\tilde{S}[\tilde{f}, \tilde{J}] = \frac{1}{2} \int \tilde{f}(\tilde{\Delta} + \mu^2)\tilde{f}dx + \int U(\tilde{f})dx + \int \tilde{J}\tilde{f}dx.$$

This is the form of the standard action (2.2.35) for a scalar field in the presence of an external source.

Example 7.18 Consider first the free case $U = 0$ and let $f_c = -(\Delta + \mu^2)^{-1}J$ be the classical field. If we translate the field f to $h = f - f_c$ then the action becomes

$$S = \frac{1}{2} k \text{Tr}(f_c(\Delta + \mu^2)f_c) + k \text{Tr}(Jf_c) + \frac{1}{2} k \text{Tr}(h(\Delta + \mu^2)h).$$

Using the 2-point function G we can write

$$f_c = -\text{Tr}_2(G(1 \otimes J)).$$

This is the matrix analogue of (7.3.10). The subscript on the trace indicates that it is to be taken over the second factor in the tensor product. The partition function Z can be explicitly calculated as usual. We find that $\Gamma[f_c]$ is given by

$$\Gamma[f_c] = \frac{1}{2} k \text{Tr}(f_c(\Delta + \mu^2)f_c) + \frac{1}{2} \log \det (k(\Delta + \mu^2)) - \frac{1}{2} n^2 \log(2\pi). \qquad (7.3.14)$$

□

Example 7.19 In the presence of interactions the effective action is modified exactly as in the commutative case. Consider a potential U of the form

$$U = \frac{\lambda}{4!} f^4, \qquad \lambda > 0.$$

To within terms of order $\hbar$ the classical field f_c is the unique solution to the noncommutative version

$$(\Delta + \mu^2 + \frac{\lambda}{4}f^2)f + J = 0$$

of the nonlinear Klein-Gordon equation (2.2.37). To within terms quadratic in $h = f - f_c$ we find that the action can be written in the form

$$S = \frac{1}{2}k\mathrm{Tr}(f_c(\Delta + \mu^2)f_c) + k\mathrm{Tr}(Jf_c) + \frac{1}{2}k\mathrm{Tr}(h\Delta[f_c]h)$$

where we have defined a positive operator $\Delta[f]$ by the equation

$$\mathrm{Tr}(h\Delta[f]h) = \mathrm{Tr}(h(\Delta + \mu^2)h) + \lambda\mathrm{Tr}(2hffh + hfhf).$$

We find then that $\Gamma[f_c]$ is given by

$$\Gamma[f_c] = \frac{1}{2}k\mathrm{Tr}(f_c(\Delta + \mu^2)f_c) + \frac{1}{2}\log\det\left(k(\Delta[f_c])\right) - \frac{1}{2}n^2\log(2\pi).$$

This is the same as (7.3.14) except for a modification of the operator Δ in the second term.   □

Example 7.20 With no external source the free effective action is given by (7.3.14) with $f_c = 0$:

$$\Gamma[0] = \frac{1}{2}\sum_{j=0}^{n^2-1}\log\left(r^2(\omega_j^2 + \mu^2)\right) - \frac{1}{2}n^2\log(n/2).$$

This can be written as

$$\Gamma[0] = \sum_{s=1/2}^{n-1/2} s\log(s^2 - \frac{1}{4} + r^2\mu^2) - \frac{1}{2}n^2\log(n/2),$$

and for large n it is given by

$$\Gamma[0] \propto n^2\log n.$$

If we consider the mass term as an external source and take the derivative with respect to it we find the relation

$$\mu\frac{\partial}{\partial\mu}\Gamma[0] = 4\pi r^2\mu^2 I.$$

We saw in Example 7.11 that there are n fundamental cells. The action per unit cell is therefore proportional to $n\log n$. We can estimate then that there are this many Bohr cells associated to each fundamental cell in position space.

The probability $\langle 1\rangle_0$ that the vacuum remains the vacuum in the absence of an external source is given by

$$\langle 1\rangle_0 = e^{-\Gamma[0]}.$$

It tends rapidly to zero as $n \to \infty$. This is explained as an effect of 'dressing' of the vacuum and is another manifestation of vacuum fluctuations.   □

Example 7.21 The algebra M_n of $n \times n$ matrices can tend weakly to a large class of algebras as the integer n tends to infinity. If it is considered in its role as irreducible representation of the rotation group $SO(3)$ then the corresponding Casimir operator can be used to introduce a length scale r and the algebra acquires aspects of the (commutative) algebra of smooth functions on the ordinary 2-sphere embedded in $\mathbb{R}^3$. There is now an additional length scale, the ratio r/n, whose limiting value can be used to classify the possible limits of M_n. A measure of the additional length scale is the parameter k of dimensions (length)^2 related to the radius by the relation (3.1.17). We recall that this formula is to be read as stating that the area of the sphere can be divided into n cells of area $2\pi k$. There is already an interesting problem here: the number of cells is not equal to the dimension of the algebra. We touched on this in Example 7.11. We have then three mass scales, which we shall suppose to be ordered as

$$k \ll \mu^{-2} \ll r^2. \tag{7.3.15}$$

In the real world the first length would be of the Planck scale the second of a typical elementary-particle scale and the last of the Hubble scale. With this interpretation the observed ratios of the real world are such that $n \sim 10^{120}$. The number $\mu^2 k$ measures the gravitational force between two particles of mass μ and so to within an order of magnitude or so its inverse can be considered to be the ratio of the gravitational to the electromagnetic coupling and therefore satisfies the relation $\mu^2 k \sim 10^{-40}$.

We have seen that the fuzzy sphere has a laplacian and an euclidean propagator with mass μ whose trace is given by the sum I which was defined above in (7.3.13) The sum can be approximated by an integral:

$$I \simeq \frac{1}{2\pi} \int_0^{n/(r\mu)} \frac{x\,dx}{x^2+1} \simeq \frac{1}{4\pi} \log \frac{n^2}{r^2\mu^2}.$$

There are two limiting cases of interest. The first has r fixed and therefore $k \to 0$ as $n \to \infty$. This is the classical sphere. One sees that

$$I \simeq -\frac{1}{4\pi} \log(k\mu^2).$$

This leading term diverges as it should in the classical limit; the extra terms remain finite.

The second limiting case has $r \to \infty$ and k fixed as $n \to \infty$. This is the noncommutative plane. One sees that

$$I \simeq \frac{1}{4\pi} \log \frac{r}{k\mu} \simeq \frac{1}{4\pi} \log\left(\frac{r\mu}{k\mu^2}\right).$$

It is necessary now to rescale μ so that the product $r\mu$ remains large but finite as $r \to \infty$. If this is not done $k\mu^2$ will remain of order one and the mass μ will be of the order of the Planck mass. After rescaling the situation is the same as the first case only with different units.

A third case has $k \to 0$ and $r \to \infty$ as $n \to \infty$. this is the classical plane. One can write I as in the previous case; one sees that whatever the value of μ the integral diverges as $n \to \infty$.

We can consider the quantity $(r\mu)^2$ as the number of 'de Broglie cells' on the sphere. If the sphere is divided into this number of cells of equal area then each one would be of the dimension of the de Broglie wave-length of the particle of mass μ. The integer n is the number of elementary Bohr cells. What determines the behaviour of I is the number of Bohr cells per de Broglie cell. In the case of a commutative space this number has the cardinality of the continuum; it is the number of points in a de Broglie cell. Even if the 'space' is noncommutative the sum I will diverge in the limit of large n if the density of Bohr cells diverges.   □

Example 7.22 The ideas expressed in Example 7.2 can be made more precise and the integral I of (7.1.7) can be calculated in certain cases. Let $\mathcal{A}$ be an infinite-dimensional algebra as described in Example 4.7. The propagator is an element of the tensor product $\mathcal{H} \otimes \mathcal{H}$ of two copies of the Hilbert space $\mathcal{H} \subset \mathcal{A}$ of modes. We represent $\mathcal{A} \otimes \mathcal{A}$ as an algebra of operators on the tensor product $L^2(V, d\mu) \otimes L^2(V, d\mu)$ of two copies of another Hilbert space $L^2(V, d\mu)$ of functions on a manifold V, square integrable with respect to some measure $d\mu$. We then express $L^2(V, d\mu) \otimes L^2(V, d\mu)$ as the tensor product of a Hilbert space $\mathcal{D} \simeq L^2(V, d\mu)$, which represents the diagonal elements of $\mathcal{A} \otimes \mathcal{A}$, and an extra Hilbert space $\mathcal{F}$, which describes the off-diagonal expansion. This must be done in a way consistent with the commutation relations. Those of $\mathcal{F}$ effectively force the distance from the diagonal in the tensor product to be 'quantized' and exclude the value zero.

Consider the differential d_u of the universal calculus. It is a map of $\mathcal{A}$ into $\mathcal{A} \otimes \mathcal{A}$ given by $d_u f = 1 \otimes f - f \otimes 1$. We define the 'variation' δx^μ of the generator x^μ as

$$\delta x^\mu = \frac{1}{2} d_u x^\mu = \frac{1}{2}(1 \otimes x^\mu - x^\mu \otimes 1). \qquad (7.3.16)$$

We identify $x^\mu = x^\mu \otimes 1$ in the tensor product and we set $x^{\mu\prime} = 1 \otimes x^\mu$. Thus we can write

$$\delta x^\mu = \frac{1}{2}(x^{\mu\prime} - x^\mu).$$

It follows from the commutation rules of the algebra that

$$[\delta x^\mu, \delta x^\nu] = \frac{1}{4} i k (J^{\mu\nu} \otimes 1 + 1 \otimes J^{\mu\nu}).$$

Suppose that a set of elements $\bar{x}^\mu$ of $\mathcal{A} \otimes \mathcal{A}$ can be found such that $\mathcal{A} \otimes \mathcal{A}$ is generated by the set $\{\bar{x}^\mu, \delta x^\mu\}$ and such that

$$[\bar{x}^\mu, \delta x^\nu] = 0. \qquad (7.3.17)$$

Then we can write the tensor product $L^2(V, d\mu) \otimes L^2(V, d\mu)$ in the form

$$L^2(V, d\mu) \otimes L^2(V, d\mu) \simeq \mathcal{D} \otimes \mathcal{F} \tag{7.3.18}$$

where $\bar{x}^\mu$ acts on $\mathcal{D}$ and δx^μ on $\mathcal{F}$. We shall choose accordingly a basis

$$|\bar{i}, k\rangle = |\bar{i}\rangle_D \otimes |k\rangle_F$$

of $L^2(V, d\mu) \otimes L^2(V, d\mu)$. If $J^{\mu\nu}$ lies in the center of the algebra then the elements

$$\bar{x}^\mu = \frac{1}{2}(x^\mu + x^{\mu\prime})$$

are such that Equation (7.3.17) is satisfied. Further one has

$$x^{\prime\prime} = x^\mu - \delta x^\mu, \qquad x^{\mu\prime} = \bar{x}^\mu + \delta x^\mu$$

and with the obvious identifications

$$[\bar{x}^\mu, \bar{x}^\nu] = \frac{1}{2}ik J^{\mu\nu}, \qquad [\delta x^\mu, \delta x^\nu] = \frac{1}{2}ik J^{\mu\nu}. \tag{7.3.19}$$

The tensor product in the definition of G is now to be considered as a tensor product of a 'diagonal' algebra $\bar{\mathcal{A}}$, acting on $\mathcal{D}$ and a 'variation' $\delta\mathcal{A}$, acting on $\mathcal{F}$. That is, we rewrite

$$\mathcal{A} \otimes \mathcal{A} = \bar{\mathcal{A}} \otimes \delta\mathcal{A} \tag{7.3.20}$$

in accordance with (7.3.18). If (7.3.17) is not satisfied the factorization (7.3.18) can still be of interest if δx^μ acts only on $\mathcal{F}$. In general then $\bar{x}^\mu$ will act non-trivially on the complete tensor product $\mathcal{D} \otimes \mathcal{F}$. One would like to suppose that the definition (7.3.16) of δx^μ in terms of the tensor product coincides with the intuitive notion of the 'variation of a coordinate'. One can introduce a new differential calculus $(\bar{\Omega}^*(\mathcal{A}), \bar{d})$ defined by $\bar{d}\bar{x}^\mu = \delta x^\mu$. One would like this new calculus to be isomorphic to the original one if δx^μ and dx^μ are to be thought of as 'infinitesimal variations'.
□

Example 7.23 There is a simple solid-state model which has been used in the study of the fractional quantum Hall effect and which is formally similar in some respects to the preceding example. The coordinates in the planar limit correspond to the cartesian components of the guiding centers of the Landau orbits and the damping factor which is responsible for the finite value of the integral I is replaced by the Debye-Waller factor. □

Quite generally, if one writes

$$Z[J] = 1 + \sum_{n\geq2} \frac{1}{n!} \int G_n(x^1, \cdots, x^n) J(x^1) \cdots J(x^n) dx^1 \cdots dx^n$$

then one can show that the G_n are the analytic continuation into imaginary time of the complete *n-point functions* of the theory, from which the scattering matrix and all physical observables can be constructed. Similarly, if one writes

$$W[J] = \sum_{n\geq 2} \frac{1}{n!} \int G_n^c(x^1, \cdots, x^n) J(x^1) \cdots J(x^n) dx^1 \cdots dx^n$$

one can show that the G_n^c are the *connected n-point functions*, described by connected Feynman graphs. In particular (7.3.6) gives the definition of the connected 2-point function or propagator $G \equiv -G_2^c$. From (7.3.9) we see that when $U = 0$ the connected *n*-point functions vanish identically for $n \geq 3$. If one writes

$$\Gamma[\phi_c] = \sum_{n\geq 2} \frac{1}{n!} \int \Gamma_n(x^1, \cdots, x^n) \phi_c(x^1) \cdots \phi_c(x^n) dx^1 \cdots dx^n$$

then one finds that $\Gamma_2 = -(G_2^c)^{-1}$ and that for $n \geq 3$ the Γ_i are the 1-particle irreducible (1PI) *n-point functions*, described by Feynman graphs which cannot be separated in two disconnected graphs by cutting one line. In (7.3.12) we calculated the Γ_n to first order in $\hbar$.

All of the various *n*-point functions can be given as a power-series expansion in the free propagator, a series which is conveniently represented in terms of Feynman graphs. The classical solution is found if one uses only graphs with no loops. The 1-loop graphs describe the first-order quantum fluctuations, which we calculated in (7.3.12). The power-series can be shown to diverge in most cases of physical interest and each term in the series is infinite. Nevertheless the infinities can be absorbed in redefinitions (renormalization) of the physical parameters and the power series can be used as an asymptotic series to yield meaningful physical predictions.

Example 7.24 Let h_j be the normalized eigenfunctions of $\Delta[\phi_c]$. A general linear field perturbation h about ϕ_c can be written in the form

$$h = \sum_j (\tilde{a}_j h_j + \tilde{a}_j^* h_j^*).$$

Suppose that ϕ_c is time-independent. The operator $\Delta[\phi_c]$ can be written then in the form

$$\Delta[\phi_c] = -\frac{\partial^2}{\partial \tau^2} + \Delta_S[\phi_c],$$

where Δ_S is an operator on V_S with positive discrete eigenvalues $p_j^2 + \mu^2$. The spectrum of $\Delta[\phi_c]$ can be ordered using 2 integers and written in the form

$$\omega_{j,l} = E_l^2 + p_j^2 + \mu^2.$$

The set of eigenvectors is then of the form

$$h_{j,l} = e^{-iE_l\tau} h_{Sj}.$$

If we now Wick-rotate back to real time we find that the spectrum of $\partial^2/\partial t^2$ is continuous:

$$w_{j,E} = -E^2 + p_j^2 + \mu^2.$$

The solutions to the field equations are given by $w_{j,E} = 0$. The linear modes are quantized as in Example 4.31 but this time the quantization is in the presence of ϕ_c considered as a fixed classical external field. □

Example 7.25 Let Δ be a differential operator on V which has a discrete spectrum w_j^2 and suppose that for some real number p we have $w_j^2 \sim j^p$ for large j. For example if Δ is of second order then $p = 2/n$ where n is the dimension of V. The determinant of Δ can be defined using a generalization of the Riemann ζ-function:

$$\zeta_\Delta(s) = \mathrm{Tr}(\Delta^{-s}) = \sum_{j=0}^{\infty} w_j^{-2s}.$$

This defines an analytic function for $\Re(ps) > 1$. Suppose that it has an analytic continuation to $s = 0$. Then by comparison with the formal product (7.3.11) we can define the determinant of Δ by the equation

$$\log \det \Delta = -\frac{d\zeta_\Delta}{ds}(0). \tag{7.3.21}$$

The operator Δ^{-s} can be connected to $e^{-t\Delta}$ by a Mellin transformation and in this way the determinant of Δ can be calculated as a formal power series in its coefficients and their derivatives using the expansion (5.1.23). For this regularization procedure to make sense the physical observables must be independent of a choice of units, which means that they must be invariant under multiplication of the spectrum by a parameter u: $w_j^2 \mapsto u w_j^2$. It is easy to see however that

$$\frac{d\zeta_{u\Delta}}{ds}(0) = \frac{d\zeta_\Delta}{ds}(0) - \zeta_\Delta(0) \log u.$$

There is therefore an ambiguity $\delta\Gamma_{(1)}$ in the expression (7.3.12) for $\Gamma_{(1)}$ given by

$$\delta\Gamma_{(1)} = \frac{1}{2}(\zeta_{\Delta[\phi_c]}(0) - \zeta_{\Delta[0]}(0)) \log u \tag{7.3.22}$$

which corresponds to the ambiguities due to the infinities in the formal calculations and this must be proportional to the original action if the theory is to be renormalizable. So although the definition (7.3.21) is mathematically more correct than the formal definition nothing has been gained physically, except perhaps in cases where the right-hand side of (7.3.22) vanishes. □

Example 7.26 If we neglect the infinite constant on the right-hand side, in (7.3.12) we have expressed the first-order quantum corrections in the form of the trace of a pseudodifferential operator. This is true in general. If Δ is a second-order differential operator which governs the quantum fluctuations around a classical background then the lowest-order quantum correction to the classical action is proportional to $\operatorname{Tr} \log \Delta$. As a simple example consider a quantized Dirac field in the presence of a classical electromagnetic field. In mathematical terms this means that the classical Dirac operator acts on an associated U_1-bundle and the covariant derivative is given by (2.2.31) with $\omega_{\alpha\beta} = 0$. The exact quantum action $\Gamma[A]$ can be written formally as

$$\Gamma[A] \propto \operatorname{Tr} \log(\not{D}^2).$$

The right-hand side diverges as $\log \Lambda$ where Λ is a bound on the spectrum of $\not{D}$ and since the theory is renormalizable the divergent term is proportional to the classical action. We conclude therefore that the classical action S can be written in terms of the *Dixmier trace* introduced in Example 5.16:

$$S[A] \propto \operatorname{Tr}_\omega \log(\not{D}^2).$$

In the preceding example we saw that a regularized value of the complete action can be found using the derivative of a generalized ζ-function at the value $s = 0$. From (7.3.22) we conclude that the classical action can be expressed in terms of the residue of the pole at $s = n/2$ of the same ζ-function.

As a less trivial example consider an arbitrary quantized field in the presence of a classical gravitational field with metric g. If the operator which governs the classical fluctuations is $\Delta[g]$ then the exact quantum action $\Gamma[g]$ can be still written formally as

$$\Gamma[g] \propto \operatorname{Tr} \log \Delta[g] \qquad (7.3.23)$$

but the divergences are now more serious. Again from the expansion (5.1.23) we find a formal series

$$\Gamma[g] = c_0 \Lambda^4 \operatorname{Vol}(V)[g] + \Lambda^2 S_1[g] + (\log \Lambda) S_2[g] + \sum_{i \geq 3} S_i[g].$$

The large-Λ behaviour of the first three terms corresponds to the small-t behaviour of the corresponding terms in the expansion (5.1.23). The S_i are integrals over V of complicated expressions involving the Riemann tensor. The S_1 is proportional to the Einstein-Hilbert action mentioned in Example 2.18. The first term defines the cosmological constant. The coefficient c_0 depends on the operator Δ as do the details of the terms S_i. If Δ is an operator which governs a mixture of an equal number of fermionic and bosonic modes then c_0 vanishes. This would be the case for example in a supersymmetric theory. Due to the presence of the term which diverges as $\log \Lambda$ the divergent terms are not proportional to the original classical action. The theory is not renormalizable.   □

Example 7.27 The vague idea expressed in Example 7.6 can be given more mathematical substance. Let V be compactified euclidean space-time and let Δ be some differential operator which depends on a metric g. Then the regularized version of (7.3.23) can be considered as the action of the gravitational field. From this point of view the gravitational field is not a fundamental field but is induced by the quantum fluctuations of all other fields. The bound Λ must be chosen to be of the same order as the Planck mass (2.2.40) and a reason must be found to suppress the cosmological-constant term. Let $\mathcal{A}$ be an associative algebra which in some sense is a noncommutative version of the algebra of functions on space-time. Let Δ be an operator which acts on an $\mathcal{A}$-module which in some sense contains the tangent bundle as factor. Then (7.3.23) would be a possible generalization of the gravitational action to $\mathcal{A}$ provided the operator $\log \Delta$ is of trace class.   □

Example 7.28 A noncommutative version of the idea of Sakharov has been also proposed. One possible expression of the *action principle* is that the spectral lines of some operator be as 'normally' spaced as possible at equilibrium. For example one might suppose that the generators of an algebra which describes Minkowski space be distributed with uniform density and that the gravitational field manifest itself by a departure from this configuration. The action would then increase as the density becomes non-uniform. This is very much like the idea expressed in Example 7.6. The problem is that it is difficult to find an appropriate 'action'. One elegant possibility which might be relevant on compact 'spaces' with metric of euclidean signature is to choose an action based on the Dirac operator. The vacuum action principle would then require that the (discrete) spectrum of the Dirac operator be as close as possible to that defined on a space of uniform curvature.   □

We are now in a position to examine in more detail the remark made after Example 3.4 concerning the definition of topology at the Planck length. Since the early efforts of Wheeler in this direction it has often been speculated that the topology of space might not be a well defined dynamical invariant. There can of course be no smooth time evolution of a space of one topology into that of another; a classical space cannot change topology without the formation of a singularity. However, the 'true' description of space and of space-time should reasonably include quantum fluctuations and it is possible that a quantum space-time exists which seems in a quasi-classical approximation to evolve from a space of one topology to that of another, the 'exact' quantum space-time being a sum over many topologies. Noncommutative geometry furnishes a possible alternative mathematical language in which one can also discuss this question. A change in topology is possible between space-times of different topology simply because an individual space-time is never completely in a 'pure' topological state.

Example 7.29 Referring back to Example 2.1 and Example 2.2 we notice that
the difference in topology between the two surfaces is expressed in a discontinuity
in the functions $\tilde{x}^a = \tilde{x}^a(\tilde{u}^\alpha)$. But as long as $\tilde{k}$ is not equal to zero the correct
description of every surface is given in terms of a filtration of the matrix algebra M_n
for some (very large) integer n. A transition occurs when one filtration becomes more
appropriate than another. In Section 3.1 we introduced two differential calculi on
M_n. In Section 7.2 we showed that the first made the algebra look like a sphere. A
similar argument can be given to show that the second makes the algebra look like a
torus. We refer to the filtration of the algebra of functions on the torus by a sequence
of matrix algebras as a *fuzzy torus*. In both cases we are in a position to speak of
the noncommutative analogue of a smooth scalar field. A generic such field $\tilde{f}$ on a
surface Σ_h of genus h must have finite action $\tilde{S}_h(\tilde{f})$ and every other action $\tilde{S}_{h'}(\tilde{f})$
must be 'almost always' infinite. If during the time evolution the action changes so
that $\tilde{f}$ has finite action for the genus $h' \neq h$ then this means that the surface has
evolved towards a different topology.

For each genus h we conjecture that one can define a differential structure over
M_n and that one can speak of the noncommutative analogue of a smooth scalar field.
We can then define an action $S_{h,n}$ which tends to the action $\tilde{S}_h$ of a complex classical
field on Σ_h. Let f be an element f of M_n which tends to a function $\tilde{f}$ on Σ_h. Then
we have

$$\lim_{n\to\infty} S_{h,n}(f) = \tilde{S}_h(\tilde{f}).$$

We shall use the action to define a Sobolev-like norm on the matrices and a Sobolev
norm on the limit functions.

To describe a topological transition from the sphere to the torus one introduces
a 'temperature' β and an action $\tilde{S}_{h(\beta)}$ such that $h(\beta) = 0$ for $\beta < \beta_c$ and $h(\beta) = 1$
for $\beta > \beta_c$. The transition will be of first order. It can be made to be of infinite
order by choosing $h(\beta) = 0$ for $\beta < \beta_c - \epsilon$ and $h(\beta) = 1$ for $\beta > \beta_c + \epsilon$ and choosing
as action a smooth functional

$$\tilde{S}_{h(\beta)} = (1 - p(\beta))\tilde{S}_0 + p(\beta)\tilde{S}_1$$

in the region $\beta_c - \epsilon \leq \beta \leq \beta_c + \epsilon$. One task of a noncommutative version of gravity
would be to motivate this *ad hoc* change of action functional, to calculate, that is,
the function $p(\beta)$.

The partition function for a complex scalar field over a surface of genus $h = h(\beta)$
is given by

$$\tilde{Z}_{h(\beta)} = \int e^{-\tilde{S}_{h(\beta)}[\tilde{f}]} d\tilde{f}. \tag{7.3.24}$$

The matrix approximation is given by

$$Z_{h(\beta),n} = \int e^{-S_{h(\beta),n}[f]} df \tag{7.3.25}$$

where the path integral is now a well defined integration over matrices. We suppose that the 'real' value of n is 'large' but not infinite, given by (3.1.17) or (3.1.27). We can then claim that the expression (7.3.25) is the 'correct' one and (7.3.24) is the approximation. For $\beta < \beta_c$ the contributions from almost all those matrices f which approximate functions on the torus (and other genera) are suppressed since $S_{1,n}[f] = 0(n)$. On the other hand for $\beta > \beta_c$ the contributions from almost all those matrices f which approximate functions on the sphere (and other genera) are suppressed since $S_{0,n}[f] = 0(n)$. $\Box$

It has been argued, for conceptual as well as practical, numerical reasons, that a lattice version of space-time or of space is quite satisfactory if one uses a random lattice structure or graph. The most widely used and successful modification of space-time is in fact what is called the lattice approximation. On the 2-torus this amounts to replacing the infinite algebra $\mathcal{P}$ of Example 2.2 by the finite-dimensional $\mathcal{P}_n$. From this point of view the Lorentz group is a classical invariance group and is not valid at the microscopic level. Historically the first attempt to make a finite approximation to a curved manifold was due to Regge and this developed into what is now known as the Regge calculus. The idea is based on the fact that the Euler number of a surface can be expressed as an integral of the gaussian curvature. If one applies this to a flat cone with a smooth vertex then one finds a relation between the defect angle and the mean curvature of the vertex. The former is encoded in the latter. In recent years there has been a burst of activity in this direction, inspired by numerical and theoretical calculations of critical exponents of phase transitions on random surfaces. One chooses a random triangulation of a surface with triangles of constant fixed length, the lattice parameter. If a given point is the vertex of exactly six triangles then the curvature at the point is flat; if there are less than six the curvature is positive; it there are more than six the curvature is negative. Non-integer values of curvature would appear through statistical fluctuation. Attempts have been made to generalize this idea to three dimensions using tetrahedra instead of triangles and indeed also to four dimensions, with euclidean signature. The main problem, apart from considerations of the physical relevance of a theory of euclidean gravity, is that of a proper identification of the curvature invariants as a combination of defect angles. On the other hand, inspired by the 2-point model which we briefly mentioned in Example 3.9 some authors have investigated random lattices from the point of view of noncommutative geometry. There has not yet been an attempt to conciliate the two points of view.

Example 7.30 Let V be a set of N points and let f_i be the function which is equal to one on the point i and zero elsewhere. Then $f_i f_j = \delta_{ij} f_j$. Let $\mathcal{A}$ be the (commutative) algebra generated by the f_i. It is the cartesian product of N copies of $\mathbb{C}$. Although, as we saw in Example 2.2, $\mathcal{A}$ has no derivations one can always construct over it the universal calculus. Define

$$f_{ij} = f_i d_u f_j = f_i(1 \otimes f_j - f_j \otimes 1) = f_i \otimes f_j - \delta_{ij} f_j \otimes 1.$$

Since $\sum_i f_i = 1$ we have $\sum_i d_u f_i = 0$ and so the f_{ij} are not all independent:

$$\sum_j f_{ij} = 0.$$

They generate however $\Omega_u^1(\mathcal{A})$ as a vector space over the complex numbers. In particular

$$d_u f_j = \sum_i f_{ij}.$$

Introduce the 1-form

$$\theta = -\sum_{i \neq j} f_{ij} = -\sum_{i \neq j} f_i \otimes f_j = -1 \otimes 1 + \sum_i f_i \otimes f_i.$$

Then for any function f we have $d_u f = -[\theta, f]$. The 2-form $\Omega = d_u \theta + \theta^2 = -\sum_i f_i \theta^2 f_i$ commutes with the elements of $\mathcal{A}$.

To make the set V look like a periodic lattice of, for example, dimension 2 we choose $N = n^2$ and write each index i, j etc. as the union of two integers taken modulo n. We impose then the relations $f_{ij} = 0$ whenever $j \neq i$ and j is not obtained from i by adding $+1$ modulo n to exactly one coordinate, which corresponds to oriented nearest-neighbour interactions. In this case $\Omega = 0$. The resulting differential algebra is a quotient of $\Omega_u^*(\mathcal{A})$ by the differential ideal generated by the relations. $\square$

Since we are not in a position to study physics on a realistic noncommutative version of Minkowski space it is of interest to examine some simple physical problems in model space-times. We end this section with a few examples of this. Except for the last, they are all based on the algebra $\mathbb{R}_q^1$ introduced in Example 4.18.

Example 7.31 Consider the algebra $\mathcal{A}_q$ obtained by adding a time parameter $t \in \mathbb{R}$ to the algebra $\mathbb{R}_q^1$ of Example 4.18: $\mathcal{A}_q = C(\mathbb{R}) \otimes \mathbb{R}_q^1$. Choose $\mathcal{H}$ as an $\mathcal{A}_q$-bimodule which is free of rank one as a left or right module. Introduce a differential calculus over $\mathcal{A}_q$ by choosing the ordinary de Rham differential calculus over the time parameter and the calculus $\Omega_r^*(\mathbb{R}_q^1)$ of Example 4.18 over the factor $\mathbb{R}_q^1$. According to the general theory of Section 3.5 one defines a covariant derivative of $\psi \in \mathcal{H}$ as a map (3.5.1) which satisfies the left Leibniz rule (3.5.2). We shall henceforth drop the tensor product symbol and write

$$D\psi = dt D_t \psi + D_r \psi$$

where D_r is defined in Example 4.20. Recall from Section 7.2 that on a curved manifold with metric $g_{\mu\nu}$ the Laplace operator acting on scalars becomes the hermitian operator

$$\Delta = -g^{\mu\nu}D_\mu D_\nu = -\frac{1}{\sqrt{g}}\partial_\mu(\sqrt{g}g^{\mu\nu}\partial_\nu).$$

Because of the identity (4.1.58) on the geometry defined by $\Omega_r^*(\mathbb{R}_{qr}^1)$, with metric (4.1.54), this becomes

$$\Delta_r = -D_{r1}^2 \tag{7.3.26}$$

where we have written $D_r = \theta_r^1 D_{r1}$.

We shall suppose that the gauge-covariant Schrödinger equation has the usual form

$$iD_t\psi = -\frac{1}{2m}\Delta_r\psi. \tag{7.3.27}$$

There is a conserved current which we write in the form

$$\partial_t\rho = D_{r1}J^{r1} \tag{7.3.28}$$

with as usual

$$\rho = \psi^*\psi, \qquad J_{r1} = \frac{i}{2m}(\psi^* D_{r1}\psi - D_{r1}\psi^*\psi).$$

The conservation law follows directly from the field equations.

Consider the relativistic case and assume the usual form

$$-\partial_t^2\psi = \Delta_r\psi + m^2\psi \tag{7.3.29}$$

for the Klein-Gordon equation. Suppose that $A_r = 0$. The Laplace operator has then a set of 'almost' eigenvectors. From the commutation relations

$$e^{iky}\Lambda = e^{ik}\Lambda e^{iky}$$

one finds that

$$e_1 e^{iky} = z^{-1}(e^{ik} - 1)\Lambda e^{iky}, \qquad \bar{e}_1 e^{iky} = z^{-1}(1 - e^{-ik})\Lambda^{-1}e^{iky}$$

from which it follows that

$$e_{r1}e^{iky} = ikLe^{iky}$$

where

$$L = \frac{1}{2ikz}\left((e^{ik} - 1)\Lambda, \ (1 - e^{-ik})\Lambda^{-1}\right).$$

From the expression (7.3.26) one concludes then that

$$\Delta_r e^{iky} = k^2 L^2 e^{iky}.$$

We could renormalize the space unit as in Equation (4.1.40) to laboratory units. If we keep the Planck units we must renormalize the time unit so it will be also in Planck units. We do this by the transformation

$$z^{-1}t \mapsto t.$$

We find then that

$$\psi = e^{-i(\omega Lt - ky)}$$

is a solution to (7.3.29) provided the dispersion relation

$$(\omega^2 - k^2)z^2L^2 = m^2 \tag{7.3.30}$$

is satisfied. However, the above dispersion relations are misleading since ω cannot be identified with the energy; the coefficient of the time coordinate is in fact the product ωL and we must set therefore

$$E = \omega L.$$

We argued in Example 4.47 that $\Lambda \to 1$ in the limit $q \to 1$ but at the same time $z \to 0$ so the following argument is subject to caution. We supposed that as $q \to 1$ we have $\bar{e}_1 \to e_1$. In this rather singular limit we can identify then

$$e_{r1} = \frac{1}{2}(e_1 + \bar{e}_1)(1, 1) + o(z)$$

and in this limit

$$L = z^{-1}\frac{\sin k}{k}(1, 1) + o(1).$$

This equality seems to follow from the preceding one but it is rather difficult to justify. If we accept it however then with the new time unit we find that

$$E^2 = \omega^2\frac{\sin^2 k}{k^2}$$

and the dispersion relation (7.3.30) in the relativistic case becomes

$$E^2 = m^2 + \sin^2 k. \tag{7.3.31}$$

If $k = \pi n$, with $n \in \mathbb{Z}$ then $E = 0$ and one has

$$e_{r1}e^{-iky} = 0.$$

In the massless case this yields a set of 'stationary-wave' solutions to the field equations.

When $k \ll \pi/2$ in Planck units one obtains the usual dispersion relation $E^2 = m^2 + k^2$. In the case $m \ll 1$ as $k \to \pi/2$ then E tends to a maximum value equal to 1, again in Planck units. Values of k greater than $\pi/2$ would be difficult to interpret physically. For comparison we recall that, neglecting the gap corrections, the dispersion relation for acoustical phonons on a lattice is of the form

$$E^2 = \sin^2 \frac{k}{2}.$$

Here E is the phonon energy and k is the wave number. This has the same form as (7.3.31) when $m = 0$. The factor $1/2$ is a convention. The first Brillouin zone is the range $-\pi \leq k \leq \pi$. On the lattice there are also optical phonons which are similar to the case $m > 0$ but they have a different dispersion relation. The 'space' $\mathbb{R}_q^1$ is not an ordinary crystal. One can compare the dispersion relation found here with that of Example 7.13. □

Example 7.32 One can construct a 'phase space' associated to Example 4.18. If we wish to construct a real phase 'space' associated to the position 'space' we must define two hermitian operators which can play the roles of 'position' and 'momentum'. We have already remarked in Example 4.49 that the distance operator s can be identified with the element y introduced in (4.3.31). As 'position' operator we choose then the renormalized y given by (4.1.40). A short calculation shows that

$$e_{r1} y = z K^{-1}$$

where the matrix K was introduced in (4.1.51). If consider then e_{r1} as an operator we find the commutation relation

$$[e_{r1}, y] = (\Lambda, \Lambda^{-1}).$$

By equation 4.3.35 we see that e_{r1} is anti-hermitian. We define then the momentum associated to y to be

$$p_y = -i e_{r1} = i z^{-1} (\Lambda, \Lambda^{-1}).$$

We have not written the extra constant term c_r of Equation (4.3.35) since it does not contribute to the commutation relation:

$$[p_y, y] = -ih. \tag{7.3.32}$$

We have here introduced the hermitian element

$$h = (\Lambda, \Lambda^{-1})$$

of $\mathbb{R}_q^1 \times \mathbb{R}_q^1$. Since we suppose (See Example 4.47) that $\Lambda \to 1$ as $q \to 1$ we see that the commutation relation (7.3.32) becomes the ordinary one in this limit.

We introduce the 'annihilation operator'

$$a = \frac{1}{\sqrt{2}}(y + i p_y). \tag{7.3.33}$$

Then from (7.3.32) follows the commutation relation

$$[a, a^*] = h. \tag{7.3.34}$$

It is not possible to express h in terms of a and a^*. The operator e_{r1} was taken as the anti-hermitian part of e_1; the operator h depends also on the hermitian part. From (4.1.39) we find however that

$$[a, h] = \frac{1}{2}z^2(a^* - a). \tag{7.3.35}$$

To define a vacuum and a number operator we must 'dress' the operator a, introduce an operator b so that the standard relations $[b, b^*] = 1$ hold. It does not seem to be possible to do this exactly but it can be done as a perturbation series in z. One finds from (7.3.34) and (7.3.35) that

$$b = h^{-1/2}a + \frac{1}{4}z^2a + \frac{1}{6}z^2(a - a^*)^3 + o(z^4).$$

The vacuum is chosen then as usual by the condition $b|0\rangle = 0$, the number operator is given by $N = b^*b$ and the number representation $|n\rangle$ for $n \in \mathbb{N}$ by

$$|n\rangle = \frac{1}{\sqrt{n!}}(b^*)^n|0\rangle.$$

From (7.3.35) we find that

$$[N, h] = \frac{1}{2}z^2((b^*)^2 - b^2) + o(z^4).$$

As hamiltonian for the harmonic oscillator we choose

$$H = \frac{1}{2}(\Delta_r + y^2)$$

in Planck units. This can be written also as $H = a^*a + \frac{1}{2}h$ and in terms of b it is given by

$$H = b^*hb + \frac{1}{2}h - \frac{1}{2}z^2b^*b + \frac{1}{6}z^2((b - b^*)^3b - b^*(b - b^*)^3) + o(z^4).$$

We see then that in terms of the 'dressed' annihilation and creation operators the 'bare' hamiltonian is rather complicated. In particular the 'physical vacuum' is no longer an eigenvector of the 'bare' hamiltonian:

$$H|0\rangle = \frac{1}{2}|0\rangle + \frac{1}{6}z^2|1\rangle + \frac{1}{\sqrt{2}}z^2|2\rangle - \frac{1}{\sqrt{6}}z^2|3\rangle + o(z^4).$$

$\square$

Example 7.33 One can also write the Laplace operator Δ using the non-local metric introduced in Example 4.19. There are two possible forms. In the absence of a gauge potential one can choose either

$$\Delta = -g^{11}D_1D_1 = -q\Lambda^{-2}x^{-2}e_1^2 + q\Lambda^{-1}x^{-2}e_1$$

or

$$\Delta = -\frac{1}{\sqrt{g}}(e_1\sqrt{g}g^{11}e_1) = -q\Lambda^{-2}x^{-2}e_1^2 + \Lambda^{-1}x^{-2}e_1.$$

The two coincide when $q = 1$. We shall choose the latter. If we introduce then the current 'density'

$$\sqrt{g}J^1 = \frac{i}{2m}\Lambda^{-1}(\psi^*\Lambda\partial_1\psi - \Lambda\partial_1\psi^*\psi)$$

the right-hand side of (7.3.28) becomes

$$\frac{1}{\sqrt{g}}e_1(\sqrt{g}J^1) = \frac{i}{2m}q\Lambda^{-1}\partial_1(\psi^*\Lambda\partial_1\psi - \Lambda\partial_1\psi^*\psi).$$

We have here used the relations (4.4.29). The conservation law becomes then

$$\partial_t\rho = \frac{1}{\sqrt{g}}e_1(\sqrt{g}J^1).$$

This is the equivalent of (7.3.28) in the new metric. $\square$

Example 7.34 With the non-local metric introduced in Example 4.19 the space $\mathbb{R}_q^1$ looks somewhat like a lattice with an impurity at the origin. If one considers the 'free' hamiltonian then numerical integration seems to show that the wave function is confined to the origin. In Example 4.34 we noticed that on a real crystal electrons tended to be trapped by impurities. $\square$

Example 7.35 In the spirit of noncommutative geometry the 'state vectors' play the role of the set of points. The eigenvalues of an observable of the algebra, in a given representation, are the noncommutative equivalents of the values which its classical counterpart can take. An eigenvector associated to a given eigenvalue describes a set of states in which the given observable can take the prescribed value. This is exactly like quantum mechanics but in position space. Consider now a field configuration, for example an element of the initial algebra $\mathcal{A}$ if it is a scalar field or an element of an algebra of forms over $\mathcal{A}$ if it is a Yang-Mills field. Suppose that both of these algebras have a representation on some Hilbert space and suppose that there exists a well defined energy functional which is also represented as an operator on the Hilbert space. A vacuum configuration would be then an element of the algebra which is such that the expectation value of the corresponding value of the energy functional in any state vanishes. This is the same as saying that a field is equal to zero if the value of its energy is equal to zero at every point of space.

In the 'classical' noncommutative case a derivation, if it exists at all, is a map of the algebra into itself; it is not an element of the algebra. In Example 7.31 this is not the case. The algebra $\mathbb{R}_q^1$ is a position space described by the subalgebra generated by x extended by Λ which is an element of the associated phase space. The differential calculus however is somehow restricted to the position space by the condition $d\Lambda = 0$. Both the initial algebra and the algebra of forms contain then operators which correspond to derivations. We have in fact given the representation of these elements on $\mathcal{R}_q$, the same Hilbert space on which the 'position' variables and the forms are represented. A vacuum configuration is then something different from what it is in the 'classical' case.

Consider, for example, a scalar field $\psi(x) \in \mathcal{H}$ and suppose that the energy functional is of the simple form $\mathcal{E} = (e_{r1}\psi)^*(e_{r1}\psi)$. If e_{r1} is considered as partial derivative then $\mathcal{E} = \mathcal{E}(x)$ depends only on the position variable x and a vacuum configuration would be one in which the expectation value of $\mathcal{E}$ vanishes for all state vectors. This would normally be one with $\psi = \psi_0$ for some ψ_0 with $e_{r1}\psi_0 = 0$. However, e_{r1} as operator belongs also to $\mathbb{R}_{qr}^1$ and the expression for the energy functional could be interpreted as one quadratic in this element. In this case the only possible vacuum configuration would be $\psi = 0$. There exist particular state vectors for which the energy functional of more complicated configurations vanish. As an example of this we return to the Yang-Mills case. One would like a vacuum to be given as usual by $\psi = 1$ (the unit cyclic vector of $\mathcal{H}$) and $A_t = A_r = 0$. One finds then as condition that

$$D_{r1}\psi(|k\rangle + \overline{|k\rangle}) = e_{r1}(|k\rangle + \overline{|k\rangle}) = 0.$$

To be concrete we shall suppose that c_r is given by (4.3.36). From (4.1.45) one sees that the vacuum equation leads to the conditions

$$(\Lambda - 1)\sum_k a_k|k\rangle = 0, \qquad (\Lambda^{-1} - 1)\sum_k \bar{a}_k|k\rangle = 0$$

on the two copies of $\mathcal{R}_q$. The vacuum state vectors form then a subspace of $\mathcal{R}_q$ of dimension 2 spanned by the vectors given by $a_k = 1$, $\bar{a}_k = 1$. These values depend of course on our choice of c_r. All vacuum state vectors have infinite norm. The vacuum state vectors would be the analogue of the vacuum of quantum field theory which is defined as the vector in Fock space which is annihilated by the energy-momentum vector of Minkowski space. The Fock-space vector is taken to be of unit norm. $\quad\Box$

Example 7.36 The action of the classical Schwinger model is

$$\tilde{S} = \int \tilde{\psi}^* \slashed{D} \tilde{\psi} dx - \frac{1}{4e^2} \int \tilde{F}_{ab} \tilde{F}^{ab} dx.$$

To define the corresponding action on the fuzzy sphere we would have to use the definitions on it of the equivalents of spinors, the electromagnetic 293 potential and the electromagnetic field strength. A general hermitian 1-form ω has an expansion $\omega = \omega_a \theta^a$. In the commutative limit this tends to a 1-form $\tilde{\omega} = \tilde{\omega}_a \tilde{\theta}^a$ where the $\tilde{\omega}_a$ are functions on S^2. From them one can extract an electromagnetic potential and a scalar field. The scalar field appears for the same reason that scalar fields appear in the theories which we shall examine in Section 8.3. □

Example 7.37 An important tool in the study of the topological properties of gauge fields is the set of maps from space (or space-time) into the corresponding gauge group. In the language of local coordinate charts this can be expressed as a mixing of coordinate indices with Lie-algebra indices; the gauge potential has one of each. Since the Lie-algebra is noncommutative it is natural that this mixing induce a noncommutative structure on space. In the closely related calculations of the Example 7.39 below we see that the catalyst of the conversion is a boundary condition. This can be also seen in an simpler calculation in which the fuzzy torus is 'constructed' from a lattice approximation to the ordinary torus using a SU_n gauge potential. To see this one starts with the approximation

$$S_{EK} = \sum_{i,j} \left(1 - n^{-1} \Re \operatorname{Tr}(U_i U_j U_i^* U_j^*) \right)$$

to the Yang-Mills action which is used on a 'periodic 1-point' lattice. The U_i is the group element which parallely transports the matter field (which we neglect here) from the site back to itself using as path the ith potential. The relation between a group element obtained by parallel transport around a 'small' loop and the Yang-Mills field strength on the *plaquette* whose boundary is the loop is related to holonomy, a word which was mentioned in Section 2.2.

At the beginning of the chapter we noticed the analogy between the fuzzy sphere and a quantized spin at a point. Considering a sphere in a matrix approximation is like considering a quantum spin as an approximation to the classical limit; the approximation is valid only for large values of the angular momentum. For the torus it is the Poynting vector which plays the role of the spin. The action S_{EK} describes a quantized electric field over the torus; we can form a Poynting vector if we add a magnetic field. This can be done by twisting the loops, by including at each site an extra factor

$$q_{ij} = \exp ih_{ij}, \qquad h_{ij} = 2\pi n^{-1} \epsilon_{ij}, \qquad q_{ij}^n = 1.$$

A particular vacuum solution is given by $U_i = u_i$ with $u_i u_j = q_{ij}^{-1} u_j u_i$. What especially interests us is the fact that $u = u_1$ and $v = u_2$ satisfy the Weyl commutation relations (3.1.25) with $q = q_{12}^{-1}$. Since we can suppose also that $u^n = 1$ and $v^n = 1$ we conclude that the algebra generated by u and v is the matrix algebra M_n. We shall see in Section 8.2 that the 'local' gauge group of electromagnetism on the noncommutative geometry described by this algebra can be identified with SU_n.

In the perturbation expansion of S_{EK} it is found that the Feynman diagrams can be ordered by their topological complexity with the *planar graphs*, the ones which have no lines which cross, dominating. The remaining graphs can be ordered in a *topological expansion*, which coincides with an expansion in powers of n^{-1}. This expansion is formally like a *loop expansion* of quantum field theory with n^{-1} replacing $\hbar$ in the limit $n \to \infty$. In Section 7.3 we considered briefly the first two terms in the loop expansion. In the topological expansion the planar graphs are of highest order, as are the classical solutions in the loop expansion. This analogy must be used with care however since it can be shown that the approximation obtained by considering only the dominant powers in the n^{-1} expansion can have no relation to a classical solution. It can also be shown that the planar graphs remain unmodified by the noncommutativity. In Example 7.21 we considered a similar situation with a massive field theory. We are here considering a massless theory so the only unit of mass is the lattice spacing; the continuum limit is more difficult to control. □

Example 7.38 The world sheet of an ordinary string moving freely in space-time is a smooth time-like surface. We define accordingly a (bosonic) *string* as a map of a 2-surface Σ into a *target space* V, which is usually either $\mathbb{R}^d, d = 10, 11, 12, 26$ or a product space of the form $\mathbb{R}^d \times K, d = 4$ and with K compact. Let $\tilde{x}^i$ be local coordinates of V and let $\tilde{\theta}^a$ be a moving frame; let $\tilde{x}^\mu$ be coordinates of Σ and let $\tilde{\theta}^\alpha$ be a moving frame. Let $\tilde{g}^{ab} = \eta^{ab}$ be the components of the standard euclidean or Minkowski metric in the dimension of V and let $\tilde{g}^{\alpha\beta} = \eta^{\alpha\beta}$ be the same in the dimension of Σ. For simplicity we shall consider only the case with a flat target space and with $\tilde{\theta}^a = d\tilde{x}^a$. The string action can be given in the form

$$S = \int_\Sigma \tilde{g}^{\alpha\beta}\tilde{g}_{ab}\tilde{e}_\alpha\tilde{x}^a\tilde{e}_\beta\tilde{x}^b\tilde{\eta}_\Sigma. \tag{7.3.36}$$

The $\tilde{\eta}_\Sigma$ is the invariant volume element on Σ. We have written this in a form which is also valid in the noncommutative case, which we distinguish by dropping the tilde. Compare with (3.4.10). The expression

$$\tilde{g}_{V\alpha\beta} = \tilde{g}_{ab}\tilde{e}_\alpha\tilde{x}^a\tilde{e}_\beta\tilde{x}^b$$

is the matrix of components of the induced metric on Σ defined by its embedding in V. Since any two metrics on Σ are conformally equivalent, at least locally, we can write

$$\tilde{g} = e^{2\phi}\tilde{g}_\Sigma$$

for some scalar field ϕ on V. If $\phi = 0$ then the action is to within a constant the volume of Σ; it is called the *Nambu action*. If $\phi \neq 0$ then it is referred to as the *Polyakov action*.

To both Σ and V one can add anti-commuting coordinates and extend them to *supermanifolds*. In the former case one speaks of world-sheet supersymmetry, in the later of target-space supersymmetry. We would like as far as possible to consider the more general case where $\mathcal{C}(\Sigma)$ and $\mathcal{C}(V)$ are replaced by noncommutative algebras $\mathcal{A}_\Sigma$ and $\mathcal{A}_V$ and the embedding is described by an algebra projection of $\mathcal{A}_V$ onto $\mathcal{A}_\Sigma$.

The metrics g_Σ and g_V should be chosen symmetric of course, but σ-symmetric, such that the condition (3.4.2) is satisfied. It implies that there is in general a decomposition

$$g_{ab} = \eta_{ab} + B_{ab}, \qquad g_{\alpha\beta} = \eta_{\alpha\beta} + \epsilon_{\alpha\beta}$$

of $g_V = (g_{ab})$ and $g_\Sigma = g_{\alpha\beta})$ into a symmetric (in the usual sense of the word) part η and an antisymmetric (in the usual sense of the word) part which we have written respectively as $\epsilon_{\alpha\beta}$ and B_{ab}. The components are here with respect to a frame. The symmetric part is chosen equal to the set of components of the standard euclidean (Minkowski) metric; the antisymmetric part ϵ is chosen equal to the components of the volume element on Σ and the antisymmetric matrix B are the components of a new field, the *B-field*. The η and ϵ will be chosen in their standard form with constant coefficients. Accordingly, we shall suppose that the B-field is such that the components B_{ab} are constants. This can be considered as a restriction on B or on the frame. We introduce the 2-form

$$B = \frac{1}{2} B_{ab} \theta^a \theta^b$$

In the commutative case one can always choose a moving frame so that the B_{ab} are constants provided that $\det B$ is constant. It is not clear to what extent this is true in the noncommutative case. $\quad\square$

Example 7.39 A *D-brane* or *Dp-brane* is a p-dimensional submanifold of V on which an open string can end. In the noncommutative case one must replace the submanifold by a factor algebra $\mathcal{A}_D$ of $\mathcal{A}_V$. The following simple formulae are formally valid in both cases. The action (7.3.36) of a string must be supplemented by boundary conditions. Introduce the 1-form $J = g_{ab} x^a \theta^b$ and let J_Σ be its image in the differential calculus over $\mathcal{A}_\Sigma$. The equations of motion of a string can be written as

$$\delta J_\Sigma = 0$$

where δ is the dual to the differential over $\mathcal{A}_\Sigma$. In the particular simple situation when the frame is exact one calculates, using (3.4.7) that the commutators of the target-space coordinates do not vanish on the brane:

$$[x^a(0), x^b(0)] = 2 g_A^{ab}.$$

The external field B can be in this way identified as the source of the commutation relations. It follows that as $B \to \infty$ the 1-form J tends to the Dirac operator for the calculus $\Omega^*(\mathcal{A}_V)$. $\quad\square$

Example 7.40 The geometric ideas behind the construction of *solitons* and *instantons* can be to a certain extent extended to noncommutative geometries, at least to some of them. Within the present context the word soliton refers to a finite-energy smooth solution with no reference to conserved quantities; an instanton is a smooth finite-action solution with euclidean signature, generally involving Yang-Mills fields whose field equations reduce to the self-duality conditions we mentioned in Example 5.11. In ordinary geometry they are both stable because of a topological obstruction to their decay. We saw in Example 7.29 that topology now becomes ill-defined and in Examples 7.2 and 7.22 that in some geometries propagators are regular. Soliton solutions thus loose somewhat their specificity. It is nevertheless interesting to study their structure on a fuzzy space. As an example we consider the 't Hooft-Polyakov monopole which, we recall, is a solotonic version of the Dirac monopole. It is stable because of a triplet of scalar fields with values in the Lie algebra of SU_2 and whose asymptotic values define a non-contractible map of the 2-sphere at infinity onto the quotient SU_2/U_1. Also present is a triplet SU_2 Yang-Mills field two of whose components describe massive particles. We first consider a 'semi-commutative' problem with the structure algebra

$$\mathcal{A} = \mathcal{C}(\mathbb{R}^3) \otimes M_2,$$

which we have encountered in Equation 3.3.4 and to which we shall return in the next chapter. If we consider Maxwell theory on this algebra then the extra matrix factor transforms the Maxwell potential into a U_2-gauge multiplet and a triplet of Higgs scalars. There are more elegant ways of doing this but we shall simply force a reduction of the gauge group to SU_2. The 't Hooft-Polyakov solution can be considered then as a 'Maxwell' solution on the extended algebra $\mathcal{A}$ and in this sense one can say that with the extra noncommutative factor the core region of the Dirac monopole has been regularized. This is not however the sense in which we used the word in Example 7.2.

As a noncommutative version of the space $\mathbb{R}^3$ in which the monopole lives we shall choose the simplest. We shall make the replacement of Section 7.2 but with no restriction on the representation. We obtain the direct sum of all matrix algebras. Space acquires an onion-like structure with an infinite sequence of concentric fuzzy spheres at the radii given by the Casimir relation $4r^4 = (n^2 - 1)k^2$. The algebra becomes

$$\mathcal{A} = \bigoplus_1^\infty (M_n \otimes M_2)$$

where to make matters simple we shall assume that each element in $\mathcal{A}$ has only a finite number of non-vanishing components. The differential calculus is the product calculus described in Section 3.3; on the first factor it is that of the fuzzy sphere plus the universal one along the generator r. We restrict our attention to the Higgs field.

Its vacuum value is the 'Dirac operator' θ and as we noticed in Section 7.2 it tends at infinity to the Dirac-monopole solution.   □

Notes

The idea that noncommutative (differential) geometry might be of interest in classical and quantum field theory was first explicitly expressed by Connes (1988) although the idea was implicit in the papers of Markov (1940) and Snyder (1947b, 1947a). These in turn were inspired by closely related ideas concerning the existence of an elementary length scale in particle physics (Heisenberg 1938) and the possibility that fundamental interactions are intrinsically non-local at a sufficiently high energy (Yukawa 1949; Pais & Uhlenbeck 1950). See also Hellund & Tanaka (1954), Jordan (1968), Kadysheveky (1978), Szabó (1989) and Coquereaux (1989). We refer for example to Blokhintsev (1970) or to Prugovečki (1995) for a brief history of previous ideas on the micro-structure of space-time. The possibility of applications to quantum field theory was one of the motivations for the introduction of quantum groups (Woronowicz 1980; Woronowicz 1987b; Majid 1988; Majid 1997). On the use of the word 'fuzzy' we refer to Madore (1992). The word 'quantum' has been used (Snyder 1947b; Snyder 1947a; Doplicher et al. 1995; Madore & Mourad 1998) as well as the word 'lattice' ('t Hooft 1996) and 'noncommutative' (Carow-Watamura & Watamura 1997) to designate what we here qualify as 'fuzzy'. A 'quantum sphere' has been introduced by Podleś (1987) which is not invariant under the action of SO_3. In Example 7.3 we argue that, in general, commutation relations will have to be modified if space-time becomes fuzzy. For a different approach to this idea we refer to Kempf *et al.* (1995) and Kempf & Mangano (1997). Pusz & Woronowicz (1989) were the first to consider the particularities which arise when one q-deforms the canonical commutation relations. By choosing appropriate values of the parameter q, either a real number or a root of unity, it is possible (Alekseev et al. 1999; Grosse et al. 2000) to introduce finite models with two parameters, the integer n as well as q.

A discussion of the role of gravity as a universal regulator is to be found for example in the review article by Garay (1995). More details of Example 7.6 can be found in Wheeler (1957) and in Misner *et al.* (1973). For a recent overview of the problems encountered in attempts to define a quantum theory of gravity we refer to Isham (1997). Quantum-mechanical fluctuations of the gravitational field were proposed some time ago by Bohr & Rosenfeld (1933) as a source of uncertainties in the measurement of the space-time coordinates and more recently (Madore & Mourad 1995; Ashtekar et al. 1998) as a source of a noncommutative space-time structure. Example 7.8 can be found in the review article by Várilly (2000) or the book by Figueroa *et al.* (2000). There have been several attempts to modify Connes' axioms so as to include the fuzzy sphere (Carow-Watamura & Watamura 2000). Details of the finite models can be found in the articles by Paschke & Sitarz (1998) and

Krajewski (1998). The Example 7.9 is taken from Madore & Mourad (1998) and the
Example 7.10 is taken from Cerchiai & Wess (1998). See also Kehagias *et al.* (1995),
Podleś (1996), Podleś & Woronowicz (1996), Aschieri, Castellani (1996), Lukierski
et al. (1996), Lorek *et al.* (1997) and Aschieri, Castellani & Scarfone (1999). The
differential calculus of Example 7.10 is due to Carow-Watamura *et al.* (1991) and
to Ogievetsky *et al.* (1992). Explicit representations were first found by Lorek *et
al.* (1997) and by Cerchiai & Wess (1998). Gravity was studied in this context for
example by Castellani (1995).

   Although we are interested in the matrix version of surfaces primarily as a model
of an eventual noncommutative theory of gravity they have a certain interest in
other, closely related, domains of physics. Without the differential calculus the
fuzzy sphere is essentially an approximation to a classical spin r by a quantum spin
r with $\hbar$ in lieu of $\check{k}$. It has been extended in various directions under various names
and for various reasons (Berezin 1975; Collins & Tucker 1976; de Wit et al. 1988;
Hoppe 1987; Hoppe 1989; Cahen et al. 1990; Bordemann et al. 1991). In order
to explain the finite entropy of a black hole it has been conjectured, for example
by 't Hooft (1996), that the horizon has a structure of a fuzzy 2-sphere since the
latter has a finite number of 'points' and yet has an SO_3-invariant geometry. The
horizon of a black hole might be a unique situation in which one can actually 'see'
the cellular structure of space (Maggiore 1994). The relation between the limit of
SU_n when $n \rightarrow \infty$ and symplectomorphisms first appears in the thesis (1982) of
Hoppe. It was published later by him (Hoppe 1987; Hoppe 1989) and by de Wit *et
al.* (1988) by Floratos *et al.* (1989) and Bordemann *et al.* (1991). A different, and
simplar, analysis was made for the torus by Fairlie *et al.* (1989) and by Fairlie and
Zachos (1989). For a recent review we refer to Zachos *et al.* (1997). The fuzzy sphere
was first studied using coherent states by Grosse & Prešnajder (1993). Versions with
arbitrary topological twist were developed by Grosse *et al.* (1997c). Example 7.13
is taken from Madore (1991b). Details of Example 7.14 can be found in Grosse *et al.*
(1997a). More details of Example 7.15 are to be found in Berezin (1975) and Cahen
et al. (1990). See also Bayen *et al.* (1978). For a further discussion of Example 7.16
we refer to Madore (1996) and Grosse *et al.* (1997b).

   Examples 7.21 and 7.22 touch on the problem of the reularization of a quantum
field theory defined on a geometry described by an infinite algebra. This subject
has been studied from several points of view in the past few years (Doplicher et al.
1995; Kehagias et al. 1995; Filk 1996; de Azcárraga et al. 1997; Chaichian et al.
2000; Várilly & Gracia-Bondía 1999; Oeckl 1999; Kosinski et al. 2000; Krajewski &
Wulkenhaar 2000; Martín & Sánchez-Ruiz 1999; Cho et al. 2000). The analysis has
also been partially extended to the Lobachevsky plane (Cho et al. 2000; Madore &
Steinacker 2000). Models in '1-dimension' (Fichtmüller et al. 1996; Kempf et al.
1995; Kempf & Mangano 1997; Cerchiai et al. 1999b) have also added to our
understanding of the 'lattice' structure. In some approaches (Cho et al. 2000; Madore

& Steinacker 2000) the effective propagator appears to be associated to a non-local differential operator (Yukawa 1949; Pais & Uhlenbeck 1950). A majority of authors find that noncommutative geometry has no virtues as a universal regulator. This point of view had been stressed in the review articles by Seiberg & Witten (1999) and by Minwalla *et al.* (1999) as well as in that of Connes (2000). We refer to standard text books (Misner, Thorne, & Wheeler 1973) for a discussion of Dirac's speculations concerning the large number 10^{40}. More details on Example 7.23 are to be found in Meissner (1993).

For more details on the generating functionals of the n-point functions and on quantum field theory in general we refer to Ryder (1985), Ramond (1989), Zinn-Justin (1996) or Peskin & Schroeder (1995). It has been suggested (Mangano 1998) that 'n-point' functions of a noncommutative version of space-time could also be expressed as a path integral.

Example 7.27 is an idea due to O. Klein which was developed by Sakharov (1967, 1975). The noncommutative version of Sakharov's idea mentioned in Example 7.28 has been developed by Kalau & Walze (1995), Kastler (1995), Ackermann & Tolksdorf (1996), Connes (1996), Chamseddine & Connes (1996) and Carminati *et al.* (1997). More details of Example 7.29 can be found in Madore & Saeger (1998). For a recent discussion of the possibility of a change in topology within the framework of quantum gravity we refer, for example, to Balachandran (1996). For an introduction to the lattice theory of gravity from the two completely different points of view we briefly outlined at the end of Section 7.3 we refer to the books by Ambjørn & Jonsson (1997) and by Landi (1997). Example 7.30 is taken from Dimakis & Müller-Hoissen (1994). See also, for example, Bimonte *et al.* (1994, 1996) and Balachandran *et al.* (1998) and compare the loop-space approach to quantum gravity, for example in the monographs by Baez & Muniain (1994) and by Gambini *et al.* (1996). The q-deformed harmonic oscillator was first studied by Macfarlane (1989) and Biedenharn (1989). See also Kulish & Damaskinsky (1990), Fiore (1994), Fichtmüller *et al.* (1996) and Lorek *et al.* (1997). Examples 7.31 and 7.32 are taken from Cerchiai *et al.* (1999b). A standard introduction to dispersion relations in solid-state physics is to be found in Kittel (1996). Example 7.33 is taken from Cerchiai *et al.* (1999b, 1999a). The details of Example 7.34 can be found in Schwenk & Wess (1992) and Fichtmüller *et al.* (1996).

More details on Example 7.36 are to be found in Grosse & Madore (1992). Example 7.37 is based on the 'large-N expansion' as introduced by 't Hooft (1974) and developed by Eguchi & Kawai (1982) and others. The key idea of introducing the twist in the action over a point is due to Gonzalez-Arroyo & Okawa (1983). We refer to Gonzalez-Arroyo & Korthals Altes (1983, 1988) for further details. Although the 'philosophy' is somewhat different this example is the first explicit calculation to use a noncommutative structure of space-time to discuss regularization in which the differential structure plays an important role. As we showed in Section 3.1 the momentum operators could be replaced by a differential calculus. It anticipates in many respects the calculations to which we refer in the notes to Examples 7.21 and 7.22. For further details of Examples 7.38 and 7.39 we refer to the book by Polchinski (1998) and the review article by Seiberg & Witten (1999). For more details on Example 7.40 we refer to the literature (Cho et al. 2000; Gopakumar et al. 2000; Harvey et al. 2000; Gross & Nekrasov 2000). We have outlined only the construction of monopole charge ± 1; for greater generality we refer to Grosse *et al.* (1997c). Some rather exotic solutions have been found to which their authors (Dubois-Violette et al. 1989a; Dabrowski et al. 2000) have refered as instantons.

8   Extensions of Space-Time

You never know what is enough
unless you know what is more than enough. *

In this chapter we shall consider the idea of Kaluza and Klein that space-time has more than 4 dimensions. We shall slightly reformulate this and suggest that what appears to be a point will, at a sufficiently small length scale, be seen to possess an algebraic structure which can be described by a noncommutative geometry. It would seem that the particles which appear in nature do so in finite multiplets and there is no indication that this is due to a lack of the accelerator energy which would be necessary to excite an as yet unseen infinite multiplet structure. We shall take this fact as motivation for considering as extra structure the finite noncommutative geometries based on matrix algebras which were studied in Chapter 3. From a practical point of view in any case a finite structure will always be a sufficient if not the most esthetic description of the known particle spectrum.

In the first section we shall show that to a certain extent spin can be interpreted as an additional noncommutative structure to space-time. Since this will involve choosing a unitary representation of the Lorentz group, the structure will necessarily be an infinite one. In the next section we shall examine the physical theories which would appear if one were to study ordinary electrodynamics in a space-time with an additional algebraic structure. We shall see that these contain some but not all of the properties one might require of an extension of the standard model of the electroweak interactions. In the last section we shall examine the theory of gravity in the same extended geometry. This yields for the appropriate algebraic structure one of the truncated versions of ordinary Kaluza-Klein theory.

8.1   The spinning particle

As examples of noncommutative extensions of space-time we shall choose algebras which are products of the algebra of smooth functions on Minkowski space by an additional noncommutative factor. Let $\tilde{x}^\mu$ be cartesian coordinates. As in Example 4.7 we wish to replace $\tilde{x}^\mu$ by four hermitian generators x^μ, elements of an abstract *-algebra $\mathcal{A}$ which do not commute:

$$[x^\mu, x^\nu] \neq 0, \qquad x^{\mu*} = x^\mu.$$

Consider first of all the non-relativistic case, with the three space coordinates $\tilde{x}^a$. If we wish the absence of commutativity to describe an intrinsic fuzz around each

---
*William Blake

point in space, it seems natural to require that this fuzz have a certain characteristic dimension and that its structure be invariant under rotations. If we require further invariance under translations we are led directly to a necessary simple form for the x^a:

$$x^a = \tilde{x}^a + \kappa\lambda^a \tag{8.1.1}$$

where κ is a length parameter and the λ^a are three non-commuting quantities which do not depend on $\tilde{x}^a$. The algebra generated by the x^a is a product of $\mathcal{C}(\mathbb{R}^3)$, the algebra of polynomials on 3-dimensional space, and a matrix algebra generated by a representation of the rotation group SO_3.

The relativistic generalization leads to something which is less trivial. Consider the special case where the λ^a are equal to the three Pauli matrices. A natural relativistic generalization of (8.1.1) which respects all reflection symmetries would be given by

$$x^\mu = \tilde{x}^\mu + \kappa\gamma^\mu.$$

The Dirac matrices γ^μ are however not self-adjoint and the above defined x^μ cannot all have real eigenvalues. We are naturally led to introduce an operator-valued Dirac spinor z and to consider the algebra $\mathcal{A}$ generated by

$$x^\mu = \tilde{x}^\mu + \kappa J^\mu, \qquad J^\mu = \bar{z}\gamma^\mu z. \tag{8.1.2}$$

We shall impose on z the following commutation relations:

$$[z, z] = 0, \quad [z, \bar{z}] = 1, \quad [\bar{z}, \bar{z}] = 0. \tag{8.1.3}$$

The unit on the right-hand side of these equations is the tensor product of the unit in the Clifford algebra and the unit in the operator algebra. Written out in terms of components of the Dirac spinors Equations (8.1.3) become

$$[z^\alpha, z^\beta] = 0, \quad [z^\alpha, \bar{z}_\beta] = \delta^\alpha_\beta, \quad [\bar{z}_\alpha, \bar{z}_\beta] = 0.$$

From the commutation relations (8.1.3) follow the commutation relations

$$[J^\mu, \bar{z}] = \bar{z}\gamma^\mu, \qquad [J^\mu, z] = -\gamma^\mu z$$

and from these we can easily find the commutation relations for the generators:

$$[x^\mu, x^\nu] = -i\hbar S^{\mu\nu}, \quad \hbar = 2\kappa^2$$

where we have set

$$S^{\mu\nu} = \bar{z}\sigma^{\mu\nu}z, \qquad \sigma^{\mu\nu} = \frac{i}{2}[\gamma^\mu, \gamma^\nu].$$

This is of the general form described in Example 4.7. The elements $(\gamma^\lambda, \sigma^{\mu\nu})$ of the Clifford algebra generate the Lie algebra of the de Sitter group $SO(3,2)$. So the de Sitter group acts on $\mathcal{A}$. It would act also on the extension $\mathcal{A}^c$ of $\mathcal{A}$ obtained by adding the generator $\bar{z}\gamma^5 z$. If we define

$$C_\lambda = \frac{1}{2}\mathrm{ad}\, J_\lambda, \qquad C_{\mu\nu} = \frac{1}{2}\mathrm{ad}\, S_{\mu\nu}, \tag{8.1.4}$$

then the derivations $\mathrm{Der}(\mathcal{A})$ are generated by the elements $(\tilde{\partial}_\lambda, C_\mu, C_{\nu\rho})$. It follows also that both $\tilde{x}^\lambda$ and J^λ belong to $\mathcal{A}$ and therefore, as in the non-relativistic case, each x^λ is the sum of two elements, the first of which belongs to the center and the second is invariant under the action of the translations. The $(J^\lambda, S^{\mu\nu})$ satisfy the commutation relations of the Lie algebra of the de Sitter group and they generate an infinite-dimensional algebra, as do the $\tilde{x}^\lambda$.

We can consider the Dirac spinor as an element of the quantized version of an algebra of functions over the classical phase space $(z, \bar{z})$ with Poisson bracket $\{z, \bar{z}\} = i$. There are therefore two distinct quantization procedures, the ordinary one involving $\hbar$ and this new one. As a mathematical simplification we shall 'dequantize' z and consider the classical phase space $(z, \bar{z})$. The C^λ defined by (8.1.4) becomes the derivation

$$C^\lambda = \frac{i}{2}((\bar{z}\gamma^\lambda)_\alpha\bar{\partial}^\alpha - (\gamma^\lambda z)^\alpha\partial_\alpha), \quad \partial_\alpha = \partial/\partial z^\alpha.$$

Consider the condition

$$\tilde{\partial}_\lambda C^\lambda f = 0. \tag{8.1.5}$$

This is of second order in all the derivatives but of first order in $\tilde{\partial}_\lambda$. So it resembles a constraint. If f depends only on the quantity x^λ defined in (8.1.2) then (8.1.5) is identically satisfied. However, the converse is not true. For a general $f \in \mathcal{A}$ we have

$$\tilde{\partial}_\lambda C^\lambda f = \frac{1}{2}(\phi^\alpha\partial_\alpha - \bar{\phi}_\alpha\bar{\partial}^\alpha)f,$$

where we have introduced the expression

$$\phi = (\mu - p_\lambda\gamma^\lambda)z, \qquad p_\lambda \equiv i\tilde{\partial}_\lambda.$$

So the condition that (8.1.5) be satisfied for all $f \in \mathcal{A}$ is equivalent to the constraint

$$\phi = 0 \tag{8.1.6}$$

on the internal phase space.

To the $(\tilde{x}^\mu, z, \bar{z})$ we add p_λ to form a phase space. We extend the bracket by requiring that $(p_\lambda, \tilde{x}^\mu)$ Poisson-bracket commute with $(z, \bar{z})$. It is not this full phase space which interests us but rather the reduced phase space given by the $(p_\lambda, \tilde{x}^\mu, z, \bar{z})$ which satisfy the constraints (8.1.6). This reduced phase space describes the motion of a spinning particle. Designate as γ^5 the ϵ of (2.2.16). Define S^λ by

$$S^\lambda = \bar{z}\gamma^\lambda\gamma^5 z.$$

Then the constraints (8.1.6) are equivalent to the conditions

$$p^2 - \mu^2 = 0, \quad p_\mu S^\mu = 0, \qquad \bar{z}\gamma^5 z = 0,$$
$$\mu J^\lambda = \bar{z} z p^\lambda, \quad \mu S^{\mu\nu} = \epsilon^{\mu\nu\rho\sigma} p_\rho S_\sigma.$$

The parameter μ is a mass parameter. It is possible to set $\mu = 0$ and study the massless case.

We have used the Poisson bracket $\{z, \bar{z}\} = i$. There is another obvious possibility: $\{z, \bar{z}\} = \gamma^5$. Using this one finds that the constraints are all of first class, the reduced phase space is of dimension 6 and the constraint $\phi = 0$ means that the spin vanishes. In fact with the second possibility it is not $S^{\mu\nu}$ but rather its dual which is the correct spin tensor, that is, which satisfies the correct commutation relations.

The quantities $(i\bar{z}\gamma^5 z, S^\lambda, J^\lambda, S^{\mu\nu})$ generate the Lie algebra of the conformal group $SO(4,2)$. The constraint (8.1.6) reduces this to the Poincaré Lie algebra. The constraint $\phi' = \bar{z}z - 1$ commutes with ϕ. If we impose it also we obtain a reduced phase space of dimension 8. On this reduced phase space we have $p^\lambda = \mu J^\lambda$. The generators of the translations in the conformal algebra are $P^\lambda = \mu(J^\lambda + S^\lambda)$. Therefore on the 8-dimensional reduced phase space we have $P^\lambda = p^\lambda + \mu S^\lambda$.

The natural choice of hamiltonian is

$$H = \bar{z}\phi.$$

The interaction with an external electromagnetic potential is given by the minimal-coupling prescription, $p_\lambda \to \pi_\lambda = p_\lambda + eA_\lambda$, so that H becomes

$$H = \bar{z}(\mu - \pi_\lambda\gamma^\lambda)z.$$

Therefore the equations of motion are

$$\frac{d\pi^\lambda}{dt} = -eF^\lambda{}_\mu J^\mu.$$

The time evolution of the algebra $\mathcal{A}$ is given by the automorphism whose infinitesimal generator is H.

Example 8.1 One of the best attempts to give a satisfactory noncommutative generalization of space-time was the first. The de Sitter space has a 10-dimensional invariance group, the de Sitter group, which is a deformation of the Poincaré group. One can identify the 4 generators which tend to the translations as the 4 generators x^μ of a noncommutative algebra. The 6 commutators of the x^μ can be identified with the generators of the Lorentz group. The resulting algebra is Poincaré invariant when the radius of the de Sitter space tends to infinity.   □

Example 8.2 Let $J^{\mu\nu}$ be a (real) antisymmetric Lorentz tensor and define

$$x^\mu = \tilde{x}^\mu + \frac{1}{2}J^{\mu\nu}p_\nu, \qquad p_\nu \equiv i\tilde{\partial}_\nu. \qquad (8.1.7)$$

Then one sees that

$$[x^\mu, x^\nu] = iJ^{\mu\nu}.$$

The tensor $J^{\mu\nu}$ contains 6 variables. One can reduce this to 4 by requiring, for example, that $J_{\mu\nu}J^{\mu\nu}$ and $\epsilon_{\lambda\mu\nu\sigma}J^{\lambda\mu}J^{\nu\sigma}$ take specific values. These are Lorentz-invariant conditions. It is obvious that x^λ and $J^{\mu\nu}$ commute:

$$[x^\lambda, J^{\mu\nu}] = 0. \qquad (8.1.8)$$

The algebra generated by the x^μ is a noncommutative version of Minkowski space-time. Although (8.1.7) is formally similar to (8.1.2) the two formulae differ in the way in which they give rise to noncommutativity. Whereas here it is the two terms which do not commute, there it is the components of the second term. One can ignore the realization (8.1.7) and consider an abstract algebra generated by the x^μ subject to the condition (8.1.8) that the $J^{\mu\nu}$ belong to the center. Equation (8.1.8) is a particular case of the models of Example 4.7 and has already been studied in Example 7.9. $\qquad \square$

8.2   Noncommutative electrodynamics

A restricted version of Kaluza-Klein theory consists of writing the Maxwell-Dirac action in the geometry defined by the algebra $\mathcal{A} = \mathcal{C}(V) \otimes M_n$ of Section 3.3. In this and the following section V will be the 4-dimensional space-time of general relativity with Minkowski signature but it could be any smooth manifold with a metric of any signature. On ordinary space-time the *Maxwell action* is given by

$$S_B = \frac{1}{4}\int F_{\alpha\beta}F^{\alpha\beta}dx$$

and the Dirac action without an explicit classical mass term is given by

$$S_F = \mathrm{Tr}\int \bar{\psi}\slashed{D}\psi dx.$$

The integration here is with respect to the invariant volume element. Using the 1-forms θ^i introduced in Section 3.3 we can consider the algebra $\mathcal{A}$ in certain aspects as the algebra of functions over a formal manifold of dimension $4 + m^2 - 1$, a product of an ordinary manifold of dimension 4 and an algebraic structure of dimension $m^2 - 1$. We can therefore write the Maxwell-Dirac action in the extended geometry defined by $\mathcal{A}$. We shall use the notation of Section 3.3 except that as in Section 7.2 we shall include a parameter μ with the dimensions of mass in the definition of the matrices λ_a. The matrices λ_a as well as the derivations e_a have then the dimensions of mass and the forms θ^a have the dimensions of length.

A connection is an anti-hermitian element ω of $\Omega^1(\mathcal{A})$. By (3.3.5) it can be split as the sum of a horizontal part and a vertical part. We write then

$$\omega = A + \omega_v, \tag{8.2.1}$$

where A is an element of Ω^1_h and ω_v is an element of Ω^1_v. Just as we decomposed the ω of Section 3.5 we write

$$\omega_v = \theta + \phi. \tag{8.2.2}$$

For reasons which will be explained below the field ϕ is called the *Higgs field*.

We noted previously that θ resembles a Maurer-Cartan form. Formula (8.2.1) with $\phi = 0$ is therefore formally similar to the connection form on a trivial principal U_1-bundle. We have in fact a bundle over a space which itself resembles a bundle. The $\mathcal{A}$-modules which we shall consider are the natural generalizations of the space of sections of a trivial U_1-bundle since M_n has replaced $\mathbb{C}$ in our models. So the U_n gauge symmetry we shall find comes not from the number of generators of the module, which we shall always choose to be equal to 1, but rather from the factor M_n in our algebra $\mathcal{A}$.

Let U_n be the unitary elements of the matrix algebra M_n and let $\mathcal{U}_n$ be the group of unitary elements of $\mathcal{A}$, considered as the algebra of functions on space-time with values in M_n. We shall choose $\mathcal{U}_n$ to be the group of local gauge transformations. A gauge transformation defines a map of $\Omega^1(\mathcal{A})$ into itself of the form

$$\omega' = g^{-1}\omega g + g^{-1}dg.$$

We define

$$\theta' = g^{-1}\theta g + g^{-1}d_v g,$$
$$A' = g^{-1}Ag + g^{-1}d_h g,$$

and so ϕ transforms under the adjoint action of $\mathcal{U}_n$:

$$\phi' = g^{-1}\phi g.$$

Because of (3.5.6) the transformed potential ω'_v is again of the form (8.2.2).

We define the curvature 2-form Ω and the field strength F as usual:

$$\Omega = d\omega + \omega^2, \quad F = d_h A + A^2.$$

In terms of components, with $\phi = \phi_a \theta^a$ and $A = A_\alpha \theta^\alpha$ and with

$$\Omega = \frac{1}{2}\Omega_{ij}\theta^i\theta^j, \quad F = \frac{1}{2}F_{\alpha\beta}\theta^\alpha\theta^\beta,$$

we find

$$\Omega_{\alpha\beta} = F_{\alpha\beta}, \quad \Omega_{\alpha a} = D_\alpha \phi_a, \quad \Omega_{ab} = [\phi_a, \phi_b] - C^c{}_{ab}\,\phi_c.$$

The last of these equations is identical to (3.5.9) except that now the ϕ_a are matrix functions.

The analogue of the Maxwell action is given by

$$S_B = \int \mathcal{L}_B dx \tag{8.2.3}$$

where

$$\mathcal{L}_B = \frac{1}{4}\mathrm{Tr}(\Omega_{ij}\Omega^{ij}) = \frac{1}{4}\mathrm{Tr}(F_{\alpha\beta}F^{\alpha\beta}) + \frac{1}{2}\mathrm{Tr}(D_\alpha\phi_a D^\alpha\phi^a) - V(\phi). \tag{8.2.4}$$

The first term on the right-hand side is the kinetic term for the gauge potential. It is quadratic in the derivatives, as is the corresponding expression for the scalar field. It contains however also cubic and quartic interaction terms.

Suppose that in a vacuum configuration ϕ_a assumes a non-vanishing value $\phi_a = \mu_a$. Then the second term on the right-hand side of (8.2.4) becomes

$$\mathcal{L}_M = \frac{1}{2}\mathrm{Tr}([A_\alpha, \mu_a][A^\alpha, \mu^a]).$$

This expression is quadratic in the potential and for the same reasons as (2.2.34) it gives a mass to the Higgs fields. The mechanism by which a non-vanishing vacuum-expectation value of one field produces a mass term for another is known as the *Higgs mechanism*. The idea comes from a similar situation in solid-state physics where a magnetic field (the gauge field) is suppressed (acquires a mass) in a super-conducting fluid of electrons (the Higgs field).

The *Higgs potential* $V(\phi)$ is given by

$$V(\phi) = -\frac{1}{4}\mathrm{Tr}(\Omega_{ab}\Omega^{ab}). \tag{8.2.5}$$

It is a quartic polynomial in ϕ which is fixed and has no free parameters apart from the mass scale μ. The trace is the integration on the matrix factor in the algebra. We have put the gauge coupling constant equal to 1. We see then that the analogue of the Maxwell action describes the dynamics of a U_n-gauge field unified with a set of Higgs fields which take their values in the adjoint representation of the gauge group. The Ω_{ab} are the components of the curvature Ω_v of the connection (8.2.2):

$$\Omega_v = d\omega_v + \omega_v^2 = \frac{1}{2}\Omega_{ab}\theta^a\theta^b.$$

We have already encountered the potential $V(\phi)$ in Example 3.31.

Choose $m = 2$ and recall the algebra of derivations $\mathrm{Der}_2(M_n)$ introduced in Section 3.1 with the λ_a generators of an n-dimensional representation of $\underline{su}_2$. With it the geometry of the algebra $\mathcal{A}$ resembles in some aspects ordinary commutative geometry in dimension 7. As $n \to \infty$ it more and more resembles ordinary commutative geometry in dimension 6 and the frame θ^i becomes a redundant one in the limit. Let g_{kl} be the Minkowski metric in dimension 7 and γ^k the associated Dirac matrices which we shall take to be given by

$$\gamma^k = (1 \otimes \gamma^\alpha, \sigma^a \otimes \gamma^5).$$

The space of spinors must be a left module with respect to the Clifford algebra. It is therefore a space of functions with values in a vector space $\mathcal{H}'$ of the form

$$\mathcal{H}' = \mathcal{H} \otimes \mathbb{C}^2 \otimes \mathbb{C}^4$$

where $\mathcal{H}$ is an M_n-module.

The Dirac operator acting on a bundle over a compact riemannian manifold is introduced in Section 2.2. We are now considering an extension of space-time and the signature is not the same. With the conventions we are using the Dirac operator is now written as

$$\displaystyle{\not{D}} = \gamma^k D_k,$$

where D_k is the appropriate covariant derivative, which we must now define. The space-time components are the usual ones which we wrote in Equation (2.2.31). We rewrite this equation in terms of components as

$$D_\alpha \psi = e_\alpha \psi + A_\alpha \psi + \frac{1}{4} \omega_\alpha{}^\beta{}_\delta \gamma_\beta \gamma^\delta \psi, \qquad \omega^\alpha{}_\beta = \omega_\gamma{}^\alpha{}_\beta \theta^\gamma. \tag{8.2.6}$$

By analogy we have to add to the covariant derivative given in Section 3.5 a term which reflects the fact that the algebraic structure resembles a curved space with a linear connection given by (3.6.31). We make then the replacement

$$D_\alpha \psi \to D_\alpha \psi - \frac{1}{8} C^b{}_{ca} \gamma_b \gamma^c \psi. \tag{8.2.7}$$

From what we have seen in Section 3.5 it is natural to choose $\mathcal{H}$ an M_n-bimodule with the group action for example on the left and the algebra action on the right as in (3.5.14). With this convention

$$D\psi = D_i \psi \theta^i$$

is a covariant derivative as we defined it in Section 3.5.

The analogue of the Dirac action is given by

$$S_F = \int \mathcal{L}_F dx \tag{8.2.8}$$

where

$$\mathcal{L}_F = \mathrm{Tr}(\bar{\psi} \displaystyle{\not{D}} \psi).$$

We have therefore defined a set of theories which are generalizations of electrodynamics to the algebra $\mathcal{A}$. In order to restrict the generality we shall make three assumptions. First, we shall suppose there is no explicit mass term in the classical Dirac action. We have already supposed that the derivations to be used are the algebra $\mathrm{Der}_2(M_n)$. Last, we shall suppose also that $\mathcal{H}$ has the module structure which leads to the covariant derivative defined by (3.5.14) to which we add the curvature term as in (8.2.7). The last two assumptions can be motivated by showing that in the limit for large n the covariant derivative tends in a sense which can be made explicit to that used in the Schwinger model.

With the restrictions we have a set of classical models which for each integer n depend only on the Yang-Mills coupling constant (which we have set equal to 1) and the mass scale μ. They are given by the classical action

$$S = S_B + S_F$$

where S_B is defined by (8.2.3) and S_F is defined by (8.2.8). As was mentioned in the Introduction, one of the more important problems with this model is the lack of any generalized quantization procedure which would insure that the constraints on the models which come from the noncommutative structure remain valid after quantum corrections. For this reason they can at best be considered as phenomenological models. The complete local invariance group is formally $SO(6,1)$ but we have broken it to $SO(3,1) \times SO_3$. By introducing a finite-dimensional algebra to describe the internal structure we have kept the dimension of the manifold at 4 instead of 7. But by so doing we have reduced the original symmetry group and therefore lost the additional constraints on the quantum corrections which it would have implied.

Example 8.3 The scalar field serves as a good first example of what happens when one replaces $\mathcal{C}(V)$ by the algebra $\mathcal{A}$ defined in (3.3.4). In the ordinary classical field theory of Example 2.14 we defined a scalar field as an element of $\mathcal{C}(V)$. Therefore we define a scalar field f to be an element of $\mathcal{A}$. Because f can be written as

$$f = \lambda_a f^a + \frac{i}{\sqrt{n}} f^0,$$

where f^a and f^0 are real-valued functions, f is in fact a multiplet of n^2 ordinary scalar fields. The generalization of the action of a free massless scalar field is given by the expression

$$S_S = \frac{1}{2}\mathrm{Tr}\int e_i f e^i f dx. \qquad (8.2.9)$$

The integral is over the factor $\mathcal{C}(V)$ in $\mathcal{A}$ and the trace is over the factor M_n. The derivatives e_i were defined in Section 3.3. In this example we shall choose $m = n$ arbitrary. There are therefore two phases. It is straightforward to see that on elements of $\mathcal{A}$ the Laplace operator is the operator

$$\Delta = -e_i e^i.$$

Using it the equations of motion which follow from the action (8.2.9) can be written as

$$\Delta f = 0.$$

We split Δ as the sum of a term Δ_h containing only the space-time partial derivatives e_α and an algebraic term Δ_v containing the derivations e_a:

$$\Delta = \Delta_h + \Delta_v.$$

The complete spectrum of Δ_v can be readily calculated. It is given by

$$\Delta_v 1 = 0, \qquad \Delta_v \lambda_a = 2n\mu^2 \lambda_a.$$

This is the broken phase. The scalar mode f^0 is massless and the other modes f^a have mass given by the formula

$$m_f^2 = 2n\mu^2. \tag{8.2.10}$$

The derivative we have used in the expression (8.2.9) for the action is the ordinary derivative. To make the action invariant under a local gauge transformation we introduce a coupling to a gauge potential through the normal minimal-coupling procedure and add (8.2.3) to the action. Since we wish the gauge transformation to be compatible with the algebra structure we must assume that f transforms under the adjoint representation:

$$f \to g^{-1} f g.$$

In this case the appropriate derivative of an element f is of the form

$$D_i f = e_i f + [\omega_i, f], \tag{8.2.11}$$

where we have written $\omega = \omega_i \theta^i$. If we replace the ordinary derivative by this one, the action (8.2.9) becomes gauge invariant. The mass spectrum (8.2.10) of the broken phase however remains the same since D_i reduces to the ordinary derivative in the absence of external fields. This is not the case in the symmetric phase. In the symmetric phase, in the absence of all external fields, D_i is given by

$$D_a = 0, \qquad D_\alpha = e_\alpha.$$

The first equation is (the unorthodox) (3.5.12) with $\phi = 0$. Therefore $\Delta_v \equiv 0$ and the masses of all the scalar modes vanish. □

Example 8.4 We shall now consider in more detail electromagnetism in the minimal case $n = 2$ and examine the resulting classical mass spectrum. The fermions are Dirac fermions which take their values in the space $M_2 \otimes \mathbb{C}^2$ and the gauge group is U_2. There are therefore four U_2 doublets. From (8.2.5) and the definition of Ω_{ab} we see that the vacuum configurations are given by the values μ_a of ϕ_a which satisfy the equation

$$[\mu_a, \mu_b] - C^c{}_{ab}\mu_c = 0.$$

For general n a solution to this equation is an n-dimensional representation of $\underline{su}_2$. The number of solutions is given by the partition function $p(n)$ we mentioned in Example 3.31. If $n = 2$ there are two solutions. Matter can exist then in two phases. There is a symmetric phase in which all the gauge bosons are massless and three of them are gluon-like. The U_1 component of the gauge fields can be interpreted as a photon and the fermions are all charged. In units of $\mu/(2\sqrt{2})$ there are two doublets of mass 3 and two of mass 5. In this phase the theory resembles an SU_2 version of the standard model for the strong interactions. In the broken phase, three of the gauge bosons become massive but the U_1 component remains massless. The fermions are again all charged and massive but with different masses. There are now two doublets of mass 5, a doublet of mass 7 and a doublet of split mass 5 and 7 units. In this phase the model possesses certain similarities with the standard electroweak model. □

Example 8.5 A more radical departure from standard electromagnetism would consist of considering it from the point of view of the algebra of forms. Functions as such play a minor role in the theory as do the derivations. The most important operation is the exterior derivative. Consider then not the algebra of functions (3.3.4) but the algebra of forms (3.3.7). A connection form will now be an odd form in the sense of the $\mathbb{Z}_2$ grading. Consider the decomposition (3.3.9) of this space. Since the horizontal component of a connection will in fact have an ordinary grading even if the theory ignores it, we can add as an extra hypothesis that it be a 1-form. That is we suppress the possible 3-form contribution. Consider the minimal model of Example 3.9. Then the connection can be written in the form of a matrix

$$\omega = \begin{pmatrix} A_0 + A_3 & \omega_v \\ \omega_v^* & A_0 - A_3 \end{pmatrix}$$

where the A_0, A_3 are U_1 gauge potentials. As in the previous case, the Higgs field ϕ is the difference $\omega_v - \eta$ between the odd form ω_v and the special odd form η introduced in Section 3.2. Apart from the different gauge group, the main difference of this model with the previous one lies in the term of the total action which corresponds to the Higgs potential. It is now the integral of the norm in the sense of (3.4.16) of the curvature (3.5.15). This is exactly the Higgs potential used in the standard model of the electroweak interactions. □

Example 8.6 A more interesting example from a phenomenological point of view can be constructed using the algebra of forms of Example 3.10 with the anti-hermitian operator η defined by Equation (3.2.9). Consider as even elements not the entire M_3^+ but the subalgebra

$$\mathbb{H} \oplus \mathbb{C} \subset M_3^+$$

where $\mathbb{H}$ is the ring of quaternions. Then the associated potential A is an $SU_2 \times U_1$ potential. The complex numbers which appear in the definition of η are parameters of the theory. The total algebra is chosen to be $\mathcal{A} = \mathcal{C}(V) \otimes M_3^+$ and the module of 1-forms is given by

$$\Omega^1(\mathcal{A}) = \mathcal{C}(V) \otimes \Omega_\eta^1 \oplus \Omega^1(\mathcal{C}(V)) \otimes M_3^+$$

where $\Omega^1(\mathcal{C}(V))$ is the module of de Rham forms and Ω_η^1 was defined in Example 3.10. The methods of Section 6.1 yield in this case a differential calculus which is larger than the tensor product of the de Rham calculus $\Omega^*(V)$ over V and the calculus Ω_η^* of Example 3.10. This possibility was evoked already in Section 3.3. The resulting model is known as the Connes-Lott model. $\square$

Example 8.7 Consider now an electromagnetic field F on a manifold Σ of dimension p. Suppose that Σ is parallelizable and let θ^a be a moving frame. We choose F with values in $i\mathbb{R}$ and we write

$$F = \frac{1}{2} F_{ab} \theta^a \theta^b, \qquad P^{ab}{}_{cd} F_{ab} = F_{cd}$$

with $P^{ab}{}_{cd}$ given by (3.1.44). The *Born-Infeld action* $S_{BI}(g + B, F)$ on Σ is given by

$$k^2 S_{BI}(g + B, F) = I(g + B, F) - I(g + B, 0)$$

with

$$I(g + B, F) = \int_\Sigma \sqrt{\det |g + B + kF|} \; \eta_\Sigma. \tag{8.2.12}$$

The matrix g_{ab} is inverse to the matrix g^{ab} of components of the metric. The parameter k could be chosen, for example, equal to the string coupling constant.

In the noncommutative case recall that the components of $\hat{g}$ are to be written in the form $\hat{g} = \hat{\eta} + \hat{B}$ as discussed in Example 3.35. Recall also the definition of the integral (3.4.10) as well as the notation of Example 3.33. Let $\hat{\mathcal{A}}_\Sigma$ be a noncommutative deformation of an algebra of functions on a manifold Σ. Consider also the action $S_{BI}(\hat{g}, \hat{F})$ on $\hat{\mathcal{A}}_\Sigma$ given by

$$k^2 S_{BI}(\hat{g}, \hat{F}) = I(\hat{g}, \hat{F}) - I(\hat{g}, 0)$$

for a potential $A \in \Omega^1(\hat{\mathcal{A}}_\Sigma)$ with

$$I(\hat{B}, \hat{F}) = \int \sqrt{\det |\hat{g} + k\hat{F}|} \; \hat{\eta}_\Sigma. \tag{8.2.13}$$

The ordinary Born-Infeld action depends on $\hat{B}$; the noncommutative version depends on a set of commutation relation defined by quantities J^{ij} which depend also on $\hat{B}$. It has been shown that there is a map (3.5.16) such that the two actions can be identified:

$$S_{BI}(g + B, F) = S_{BI}(\hat{g}, \hat{F}). \tag{8.2.14}$$

If the algebra contains a matrix factor as was considered in Section 3.3 then the expansion of the determinant is ambiguous because of the ordering ambiguity in the λ_a.   □

8.3   Modified Kaluza-Klein theory

The goal of Kaluza-Klein theory is to reduce all of bosonic field theory to the theory of gravity in a higher-dimensional manifold, a manifold which is most conveniently chosen to be a principal fibre bundle P over space-time. For simplicity suppose P to be the trivial bundle $P = V \times SU_n$, the product of space-time V and an *internal manifold* SU_n. Let $\mathcal{C}(V)$ be the algebra of smooth functions on space-time and $\mathcal{A}$ the algebra of smooth functions on P. There is therefore an embedding of associated algebras

$$0 \to \mathcal{C}(V) \to \mathcal{A},$$

where

$$\mathcal{A} = \mathcal{C}(V) \otimes \mathcal{C}(SU_n). \tag{8.3.1}$$

Kaluza-Klein theory in the usual sense can be described equally well by the usual projection of P onto space-time or by the above embedding. The internal structure is described by the manifold SU_n or by the algebra of functions $\mathcal{C}(SU_n)$. One can extend Kaluza-Klein theory by replacing $\mathcal{C}(SU_n)$ by an algebra of functions on an arbitrary compact manifold. One can also modify it by replacing $\mathcal{C}(SU_n)$ by an associative algebra which is not necessarily an algebra of functions. One can then no longer refer to the projection as there is no internal manifold, but the embedding remains.

We shall consider here a matrix algebra M_n as possible *internal structure*. The total algebra $\mathcal{A}$ is given then by Equation (3.3.4):

$$\mathcal{A} = \mathcal{C}(V) \otimes M_n.$$

This choice has the two advantages which we have already mentioned in the Introduction. There is no valid notion of a point in the associated matrix geometry and since the algebra of derivations is of finite dimension, the associated particle spectrum is finite. Also the total momentum space remains of dimension 4. This is important for the renormalizability of the theory. The price to be paid for this however is a reduction of the symmetry group as well. Formally, the invariance group of the complete structure is $SO(3 + n^2 - 1, 1)$. If we restrict the local rotations to those which do not mix the ordinary θ^α with the algebraic θ^a, this group reduces to $SO(3,1) \times SO_{n^2-1}$. Pure gravity in dimension 5 is 1-loop renormalizable. However, the theory in dimension 4 obtained by a truncation to the lowest-order modes is no longer renormalizable even at the 1-loop level. In this case the advantage gained by the reduction in dimension does not compensate the loss of symmetry. Finally, a third advantage of interest to us is the fact that on the space of derivations of the total algebra $\mathcal{A}$ we can still define a linear connection although $\mathcal{A}$ is noncommutative.

The identification of the internal structure is the first step of the Kaluza-Klein construction; the second step is the construction of a linear connection on $\Omega^1(\mathcal{A})$. We shall follow the usual prescription, using the connection ω of the previous section, the linear connection (3.6.31) on M_n and a linear connection on V.

The gauge group U_n acts on the algebraic structure as in (3.1.20). If the integer m introduced in Section 3.1 is less than n the set of generators does not remain invariant. We must choose then $m = n$. In this case the group acts through the adjoint representation

$$U_n \to SO_{n^2-1}.$$

Only the group $SU_n/\mathbb{Z}_n$ acts nontrivially and we have an embedding

$$SU_n/\mathbb{Z}_n \hookrightarrow SO_{n^2-1}.$$

We shall be forced then also to suppose that

$$\phi_r^0 = 0, \qquad A_\alpha^0 = 0,$$

and that $g \in \mathcal{SU}_n$, the group of local SU_n gauge transformations. From (8.2.1) and (8.2.2) the connection ω can be therefore written in components as

$$\omega = (A_\alpha^a \theta^\alpha + \omega_r^a \theta^r)\lambda_a, \qquad \omega_r^a = -\delta_r^a + \phi_r^a. \qquad (8.3.2)$$

As we have defined them the components A_α^a and ϕ_r^a are dimensionless. In Section 3.1 we considered a change of basis $\lambda^a \to \Lambda_b^a \lambda^b$. Under a change of basis the components ϕ_b^a in the expansion $\phi = \phi_b^a \lambda_a \theta^b$ transform as

$$\phi_b^a \to \Lambda_c^a \Lambda^{-1d}_{\ b} \phi_d^c.$$

Under a gauge transformation

$$\phi' = \phi_b^a (g^{-1}\lambda_a g)\theta^b = \phi'^a_{\ b}\lambda_a \theta^b$$

and so $\phi'^a_b = \Lambda^{-1a}_{\ c}\phi_b^c$. Only the upper index changes. Gauge invariance is a stronger requirement than simple invariance under the change of basis. We shall adopt in this section the convention of using the letters $(r, s, \dots)$ for those indices which are unaffected by gauge transformations.

The gauge group $\mathcal{SU}_n$ acts on $\mathcal{A}$ and the ordinary coordinate transformations act on $C(V)$ and therefore also on $\mathcal{A}$. We suppose that there is no correlation between these two actions but it is, however, possible to correlate them. For example consider only the Lorentz transformations acting on $C(V)$. We could require that the algebra $\mathcal{A}$ be a representation of the Lorentz group, and that the gauge group be also replaced by the Lorentz group. This requirement would lead necessarily to algebras with an infinite number of generators. The internal structure instead of representing internal-symmetry degrees of freedom, would represent the spin as in Section 8.1.

We first apply the derivative (8.2.11) to an element $f \in \mathcal{A}$:

$$D_i f = e_i f + [\omega_i, f] = (e_i + \omega_i^a e_a) f.$$

This suggests that we introduce the derivations

$$\tilde{e}_\alpha = e_\alpha + A_\alpha^a e_a, \quad \tilde{e}_r = e_r + \omega_r^a e_a = \phi_r^a e_a \tag{8.3.3}$$

of the algebra $\mathcal{A}$. They constitute also part of the covariant derivative which we shall later apply to spinor fields. When the Higgs fields vanish one sees from (8.3.2) that $\omega(\tilde{e}_\alpha) = 0$; the components $\tilde{e}_\alpha$ are annihilated by the form ω. They correspond to the horizontal lift of a vector field on V to a vector field on P in the case (8.3.1). Under a gauge transformation the $\tilde{e}_i$ transform as

$$\tilde{e}_i \rightarrow g_* \tilde{e}_i \equiv g^{-1} \tilde{e}_i g.$$

We have then

$$g_* \tilde{e}_r = \phi_r^a g_* e_a = \phi_r^a \Lambda^{-1b}{}_a e_b, \quad g_* \tilde{e}_\alpha = g_*(e_\alpha + A_\alpha^a e_a) = e_\alpha + A'^a_\alpha e_a.$$

The A'^a_α are the components of the gauge-transformed potential.

Dual to the derivations $\tilde{e}_i$ are the 1-forms $\tilde{\theta}^i$ given by

$$\tilde{\theta}^\alpha = \theta^\alpha, \quad \tilde{\theta}^r = \chi_a^r (\theta^a - A_\alpha^a \theta^\alpha). \tag{8.3.4}$$

We have here used the inverse χ_a^r to the matrix ϕ_r^a:

$$\chi_a^r \phi_s^a = \delta_s^r.$$

Under a gauge transformation χ_b^r transforms only with respect to the lower index. By definition the gauge transformations $g^* \tilde{\theta}^i$ of the $\tilde{\theta}^i$ are determined by the equations $g^* \tilde{\theta}^i(g_* \tilde{e}_j) = \delta_j^i$. We have then

$$g^* \tilde{\theta}^\alpha = \theta^\alpha, \quad g^* \tilde{\theta}^r = \chi_a^r \Lambda_b^a (\theta^b - A'^b_\alpha \theta^\alpha).$$

The $\tilde{\theta}^r$ is a mixed combination of a moving frame on V and the Stehbein on the algebraic structure. The $\tilde{\theta}^i$ is not the most general frame on the total structure since the components $\tilde{\theta}^{\alpha}$ have no algebraic contribution. However, an arbitrary metric on the total structure can be written using the line element $ds^2 = g_{ij}\tilde{\theta}^i\tilde{\theta}^j$ with g_{ij} the Lorentz metric in dimension $4 + n^2 - 1$ and with $\tilde{\theta}^i$ of the form (8.3.4). The metric determines then θ^{α} as well as the coefficients ϕ_r^a and A_{α}^a.

We restrict first our considerations to that special class of connections for which the curvature of the algebraic structure vanishes:

$$\Omega_{rs} = 0. \tag{8.3.5}$$

This is the case which most resembles ordinary Kaluza-Klein theory. Either ϕ_r^a vanishes then or ϕ_r^a belongs to the gauge orbit of δ_r^a. We have seen that these values correspond to the stable vacua of the theory. The first gives rise to a singular set of 1-forms $\tilde{\theta}^r$. Consider the second value, which corresponds to the physical vacuum. It yields a $\tilde{\theta}^i$ which is formally very similar to the usual moving frame constructed on a principal SU_n-bundle. On the other hand if we compare with (8.2.2) we see that the vertical component of the connection vanishes. It is in the other vacuum $\phi = 0$ that the connection ω most resembles a connection in a trivial SU_n-bundle.

Let $\omega^{\alpha}{}_{\beta}$ now be a linear connection on space-time, an $\underline{so}(3,1)$-valued 1-form satisfying the structure equations

$$d\theta^{\alpha} + \omega^{\alpha}{}_{\beta}\theta^{\beta} = 0,$$

$$\Omega^{\alpha}{}_{\beta} = d\omega^{\alpha}{}_{\beta} + \omega^{\alpha}{}_{\gamma}\omega^{\gamma}{}_{\beta}.$$

These are Equations (2.2.1) and (2.2.2) with vanishing torsion. We must construct an $\underline{so}(3+n^2-1,1)$-valued 1-form $\tilde{\omega}^i{}_j$ on $\Omega^1(\mathcal{A})$ satisfying the first structure equations

$$d\tilde{\theta}^i + \tilde{\omega}^i{}_j\tilde{\theta}^j = 0. \tag{8.3.6}$$

Under the condition (8.3.5) the solution to these equations is given by

$$\tilde{\omega}^{\alpha}{}_{\beta} = \omega^{\alpha}{}_{\beta} + \frac{1}{2}F_a{}^{\alpha}{}_{\beta}\tilde{\theta}^a,$$

$$\tilde{\omega}^{\alpha}{}_a = \frac{1}{2}F_a{}^{\alpha}{}_{\beta}\tilde{\theta}^{\beta}, \tag{8.3.7}$$

$$\tilde{\omega}^a{}_b = -\frac{1}{2}C^a{}_{bc}\tilde{\theta}^c + C^a{}_{cb}A^c_{\alpha}\tilde{\theta}^{\alpha}.$$

Except for an additional term on the right-hand side of the equation for $\tilde{\omega}^a{}_b$, this connection is of the same form as the usual one (2.2.33) which one constructs on an SU_n-bundle. The extra term, we shall see, is what remains of the covariant derivative of the Higgs fields. The solution (8.3.7) is not gauge invariant; it is given in the particular gauge $\phi_r^a = \delta_r^a$. To construct (8.3.7) we have implicitly used the linear connection (3.6.31) on the matrix factor of (3.3.4).

Consider now a general SU_n connection with an arbitrary Higgs field. The matrix χ_a^r can be considered in (8.3.4) as a transformation of the frame θ^a away from its value in the physical vacuum. Set

$$\Omega_{ij} = \Omega^a{}_{ij}\lambda_a.$$

and define the gauge-invariant quantities

$$\tilde{\Omega}^r{}_{ij} = \chi_a^r \Omega^a{}_{ij}.$$

If we define $C^k{}_{ij}$ by the equation $[e_i, e_j] = C^k{}_{ij} e_k$ then the commutator of the $\tilde{e}_i$ is given by

$$[\tilde{e}_i, \tilde{e}_j] = \tilde{\Omega}^r{}_{ij}\tilde{e}_r + C^k{}_{ij}\tilde{e}_k.$$

The gauge invariance of the coefficients on the right-hand side is evident from the transformation properties of the $\tilde{e}_i$.

Define

$$\tilde{C}^r{}_{st} = \chi_a^r C^a{}_{bc}\phi_s^b\phi_t^c = \tilde{\Omega}^r{}_{st} + C^r{}_{st}.$$

Then the solution to (8.3.6) is given by

$$\tilde{\omega}^\alpha{}_\beta = \omega^\alpha{}_\beta + \frac{1}{2}\tilde{\Omega}_r{}^\alpha{}_\beta\tilde{\theta}^r,$$

$$\tilde{\omega}^\alpha{}_r = \frac{1}{2}\tilde{\Omega}_r{}^\alpha{}_\beta\tilde{\theta}^\beta + \frac{1}{2}(\tilde{\Omega}_s{}^\alpha{}_r + \tilde{\Omega}_r{}^\alpha{}_s)\tilde{\theta}^s, \tag{8.3.8}$$

$$\tilde{\omega}^r{}_s = \frac{1}{2}(\tilde{\Omega}^r{}_{\alpha s} - \tilde{\Omega}_{s\alpha}{}^r)\tilde{\theta}^\alpha - \frac{1}{2}(\tilde{C}^r{}_{st} - \tilde{C}_t{}^r{}_s + \tilde{C}_{st}{}^r)\tilde{\theta}^t.$$

The geometry of the algebraic structure, determined by the structure constants $\tilde{C}^r{}_{st}$, is equal to that of SU_n with the Killing metric but deformed by the components ϕ_r^a of the Higgs field. If we set $A_\alpha = 0$, $\omega^\alpha{}_\beta = 0$ and suppose that ϕ_a is a constant then we have from (8.3.4)

$$d\tilde{\theta}^r = -\frac{1}{2}\tilde{C}^r{}_{st}\tilde{\theta}^s\tilde{\theta}^t.$$

In the sense which we have defined the expression in Section 3.6 the $\tilde{\omega}^i{}_j$ defines a linear connection. Because of (3.6.32) the generalized flip σ is the ordinary flip, $\sigma(\tilde{\theta}^i \otimes \tilde{\theta}^j) = \tilde{\theta}^j \otimes \tilde{\theta}^i$, and the covariant derivative is given by

$$D\tilde{\theta}^i = -\tilde{\omega}^i{}_j \otimes \tilde{\theta}^j.$$

From the expression for the torsion map (3.6.4) one sees that (8.3.6) is indeed the condition that it vanish. A general 1-form can be written $\xi = \xi_i\tilde{\theta}^i = \tilde{\theta}^i\xi_i$ and the covariant derivative can be extended by either of the Leibniz rules

$$D\xi = D(\xi_i\tilde{\theta}^i) = d\xi_i \otimes \tilde{\theta}^i + \xi_i D\tilde{\theta}^i,$$

$$D\xi = D(\tilde{\theta}^i\xi_i) = \sigma(\tilde{\theta}^i \otimes d\xi_i) + (D\tilde{\theta}^i)\xi_i.$$

These two expressions are consistent because the coefficients of $\tilde{\omega}^i{}_j$ are all central elements and because of the form of σ. The general Leibniz rules of Section 3.6 follow directly from the ordinary Leibniz rule for the differential.

Example 8.8 The construction of the frame $\tilde{\theta}^i$ is straightforward because it was based on the derivations $\tilde{e}_i$ which are a basis of the complete set of all derivations of an algebra which has a simple product structure of a commutative algebra and a matrix algebra. An interesting question concerns the possible extension to more general frames, involving for example quantum-group symmetries. For this one can consider an algebra which is the tensor product of an algebra of functions on space-time (or on any other smooth manifold) and a noncommutative version $\mathcal{K}$ of an algebra of functions on a space which is in some sense compact. If one supposes that on both factors there is a differential calculus with a free module of 1-forms then a frame can be constructed as well as linear connections using the frame formalism of Section 3.6. The difficult part of the construction would be the identification of a (quantum) group of automorphisms of $\mathcal{K}$ which leaves the frame invariant and which one could consider as a gauge symmetry.   □

To complete the Kaluza-Klein construction it is necessary to consider the second structure equations

$$\tilde{\Omega}^i{}_j = d\tilde{\omega}^i{}_j + \tilde{\omega}^i{}_k\tilde{\omega}^k{}_j$$

and the equations of motion which follow from a suitable action. That this be the most general invariant which yields second-order field equations is the only condition one can impose in the absence of any criterion of renormalizability. Invariance under local $SO(3,1) \times SU_n$ transformations permits an infinite formal sum of terms involving arbitrary powers of the components $\Omega_{\alpha\beta}$, $\Omega_{\alpha a}$, Ω_{ab} of Ω as well as of the components ϕ_a of the Higgs field. If on the other hand one imposes a rather formal $SO(3 + n^2 - 1, 1)$-invariance which mixes the space-time θ^α with the algebraic θ^a, then the most general invariant can be constructed using only the components of the curvature form $\tilde{\Omega}^i{}_j$. Using the construction (2.2.8) and (3.4.11) we can define

$$S_p = \int \tilde{\Omega}^{i_1 i_2} \cdots \tilde{\Omega}^{i_{2p-1} i_{2p}} * \tilde{\theta}_{i_1 \ldots i_{2p}}$$

where the integral is as defined in Equation (3.4.17). It consists of an integration over the algebraic structure as well as over space-time. In an ordinary compact manifold the S_p would be proportional to the generators of the Euler classes in even dimensions up to and including $4 + n^2 - 1$. The action is then a finite sum

$$S = \sum_{p=0}^{N+1} \alpha_p S_p, \qquad N = \left[\frac{n^2 - 1}{2}\right] \tag{8.3.9}$$

with arbitrary coefficients α_i.

It has been shown in the usual Kaluza-Klein case with a manifold as internal structure that a consistent classical theory with a reasonable stable vacuum can be only based on a lagrangian which includes some of the higher-order terms in the above expansion. We shall restrict however our attention here to the Einstein-Hilbert term (2.2.41), which when properly normalized can be written as

$$S_1 = \frac{1}{n^2 - 1}\mu_P^{n^2+1}\int \tilde{\Omega}^{ij} * \tilde{\theta}_{ij}.$$

The integral over the algebraic factor is

$$\int \tilde{\theta}^1 \cdots \tilde{\theta}^{n^2-1} = (n^2 - 1)\mu_P^{1-n^2}\chi.$$

We have here defined $\chi = \det \chi_b^r$. It is gauge invariant because the transformation Λ_b^a is orthogonal. By convention we have chosen it to be positive. The action can be therefore written in the form of an integral over space-time,

$$S_1 = \int \mathcal{L}\,\theta^0\theta^1\theta^2\theta^3$$

with

$$\mathcal{L} = -\mu_P^2\chi\tilde{R}.$$

A straightforward calculation yields then that the Ricci scalar $\tilde{R}$ is given by

$$\tilde{R} = R + \frac{1}{4}\tilde{\Omega}_{r\alpha\beta}\tilde{\Omega}^{r\alpha\beta}$$
$$+ \frac{1}{2}\tilde{\Omega}_{ras}\tilde{\Omega}^{ras} + \frac{1}{2}\tilde{\Omega}_{ras}\tilde{\Omega}^{sar} + \tilde{\Omega}_{r\alpha}{}^r\tilde{\Omega}_s{}^{\alpha s} - 2D_\alpha\tilde{\Omega}_r{}^{\alpha r}$$
$$+ \frac{1}{4}(\tilde{C}_{rst}\tilde{C}^{rst} - 2\tilde{C}_{rst}\tilde{C}^{trs}). \qquad (8.3.10)$$

All of the terms on the right-hand side are gauge invariant. The extended gauge invariance is no longer manifest however.

The second term in (8.3.10) is a modified version of the gauge-boson lagrangian which we used in the previous section. It includes also an infinite sum of terms with insertions of Higgs fields which come from the inverse χ_a^r of ϕ_r^a near the physical vacuum. The next three terms are the Higgs kinetic terms, also modified by insertions. The last term gives rise to an effective cosmological constant.

The Higgs potential which comes from (8.3.10) is given by

$$V(\phi) = \frac{1}{4}\mu_P^2\chi(\tilde{C}_{rst}\tilde{C}^{rst} - 2\tilde{C}_{rst}\tilde{C}^{trs}).$$

It has the correct physical dimensions since $\tilde{C}^r{}_{st}$ has units of mass. The potential $V(\phi)$ is singular at $\phi = 0$. As is to be expected it is quite different from the potential (8.2.5) which comes from the generalized electromagnetic action (8.2.3).

If we introduce the variables

$$\xi^{ab} = g^{rs}\phi_r^a\phi_s^b$$

then $\tilde{R}$ can be written in a more familiar form:

$$
\begin{aligned}
\tilde{R} = R + R_{(i)} &+ \frac{1}{4}\xi_{ab}^{-1}\Omega^a{}_{\alpha\beta}\Omega^{ba\beta} \\
&+ \frac{1}{4}\xi_{ab}^{-1}\xi_{cd}^{-1}(D_\alpha\xi^{ac}D^\alpha\xi^{bd} + D_\alpha\xi^{ab}D^\alpha\xi^{cd}) \\
&- D_\alpha(\xi_{ab}^{-1}D^\alpha\xi^{ab}).
\end{aligned}
$$

(8.3.11)

The scalar curvature $R_{(i)}$ of the algebraic structure is given by

$$R_{(i)} = \frac{1}{4}\xi_{ab}^{-1}\xi^{cd}\xi^{ef}C^a{}_{ce}C^b{}_{df} - n\mu_P^2 g_{ab}\xi^{ab}.$$

If we define $\xi = \det\xi^{ab}$ then the Higgs potential can be rewritten as

$$V(\phi) = \mu_P^2\xi^{-1/2}R_{(i)}.$$

A fermion ψ is a function which takes its values in the space

$$\mathcal{H}' = \mathcal{H}\otimes\mathbb{C}^N\otimes\mathbb{C}^4, \qquad N = 2^{[(n^2-1)/2]},$$

where $\mathcal{H}$ is an M_n-module. The covariant derivative of ψ is given by

$$\tilde{D}_i\psi = \tilde{e}_i\psi + \frac{1}{4}\tilde{\omega}_i{}^j{}_k\gamma_j\gamma^k\psi, \qquad \gamma^k \equiv (1\otimes\gamma^\alpha, \gamma^r\otimes\gamma^5).$$

Written out in detail the operator $\tilde{D}_i$ is given by

$$
\begin{aligned}
\tilde{D}_\alpha &= D_\alpha - \frac{1}{4}\tilde{\Omega}_{r\alpha\beta}\gamma^r\gamma^\beta\gamma^5 + \frac{1}{8}(\tilde{\Omega}_{r\alpha s} - \tilde{\Omega}_{s\alpha r})\gamma^r\gamma^s, \\
\tilde{D}_r &= D_r + \frac{1}{8}\tilde{\Omega}_{r\alpha\beta}\gamma^\alpha\gamma^\beta + \frac{1}{4}(\tilde{\Omega}_{r\alpha s} - \tilde{\Omega}_{s\alpha r})\gamma^s\gamma^\alpha\gamma^5 \\
&\quad - \frac{1}{8}(\tilde{C}_{str} - \tilde{C}_{rst} + \tilde{C}_{trs})\gamma^s\gamma^t.
\end{aligned}
$$

The expression for D_α was given in (8.2.6). In the particular case $\phi_r^a = \delta_r^a$, the operator $\tilde{D}_r$ reduces to the one given in (8.2.7).

Example 8.9 Consider the simple Ansatz

$$\phi_r^a = e^\sigma\delta_r^a, \qquad A_\lambda^a = 0.$$

Then the connection (8.3.8) is given by

$$\tilde{\omega}^\alpha{}_\beta = \omega^\alpha{}_\beta, \qquad \tilde{\omega}^\alpha{}_r = \partial^\alpha\sigma\tilde{\theta}_r, \qquad \tilde{\omega}^r{}_s = -\frac{1}{2}C^r{}_{st}\tilde{\theta}^t e^\sigma.$$

The Ricci tensor derived from this connection is

$$\tilde{R}_{\alpha\beta} = R_{\alpha\beta} - (n^2 - 1)(D_\alpha D_\beta \sigma - \partial_\alpha \sigma \partial_\beta \sigma),$$
$$\tilde{R}_{\alpha r} = 0,$$
$$\tilde{R}_{rs} = \left[(n^2 - 1)\partial_\alpha \sigma \partial^\alpha \sigma - D_\alpha D^\alpha \sigma - \frac{n}{2} e^{2\sigma} \mu_P^2 \right] g_{rs}.$$

The field equations $\tilde{R}_{ij} = 0$ yield the same equations as traditional Kaluza-Klein theory using as internal metric the conformal factor $e^{-2\sigma}$ times the metric of a space of constant curvature. The expression for the potential $V(\phi)$ becomes

$$V(\sigma) = -\frac{1}{2} n(n^2 - 1) \mu_P^4 e^{-(n^2 - 3)\sigma}.$$

There is no stable solution with constant σ.   □

Example 8.10 Matrices can also be used to give a finite 'fuzzy' description of the space complementary to a Dirichlet p-brane, a description which will allow one perhaps to include the reasonable property that points should be intrinsically 'fuzzy' at the Planck scale. This has much in common with the noncommutative version of Kaluza-Klein theory which we have just described. Strings naturally play a special role here since they have a world surface of dimension two and an arbitrary matrix can always be written as a polynomial in two given matrices.   □

We have learned that some bosonic field theories on the commutative geometry of space-time can be re-expressed as abelian-gauge theory in an appropriate noncommutative geometry. This fact is quite the analogue of the dichotomy in general relativity between the components of a metric considered as external fields in a flat background and the same components considered as defining the metric and therefore a non-flat geometry. A certain number of examples have been considered which exhibit the property of an external field which can be incorporated into a redefinition of the basic geometry. The noncommutative structure can be considered as containing extra modes all of whose dynamics are given by the one abelian action. This was discussed in some detail in the previous two sections.

Example 8.11 Some of the most illuminating examples are taken from the field of simple hamiltonian mechanics. Complicated non-local non-polynomial hamiltonians can be considered as the free-particle hamiltonian in appropriately chosen geometries. A free particle in motion in a curved space-time can be considered as a particle in a flat space-time moving under the influence of an external field. There is an analogous example in noncommutative geometry. Consider an interaction hamiltonian $H = H_0 + V$ on the real line $\mathbb{R}$ with time added or not. Then for appropriate V these hamiltonians are equivalent to free hamiltonians acting on often exotic noncommutative structures. We discussed one of these in Example 7.32. An important dynamical variable which can also be considered as part of the space-time geometry is classical spin; we saw in Section 8.1 that a relativistic spinning particle can be described as an ordinary particle in a noncommutative geometry.   □

Example 8.12 A further example was discussed in Example 8.7 which suggests that to describe certain fields one is lead naturally to consider noncommutative geometries. It involves an external field B which can be absorbed into an appropriate redefinition of the commutation relations of a noncommutative geometry. We touched on this point in Example 7.39. When considered as part of the geometry the field B changes the structure of the gauge group, indirectly because of the way the commutation relations of the algebra depend on it. A Yang-Mills potential A has one gauge group in the presence of a B field considered as external field and its noncommutative counterpart $\hat{A}$ has another. Since the physics cannot depend on the interpretation of the field there must be a well-defined map $\hat{A} = \hat{A}(A, B)$, the Seiberg-Witten map, which reduces to the identity when $B = 0$. The set of noncommutative structures over space-time is in many aspects similar to a Kaluza-Klein extension. The B field acts then as a set of extra coordinates which parametrize the extra dimensions. The map is a change of gauge which depends on the extra parameters. We have in fact identified the gauge group as the unitary elements of the algebra. When we change the structure of the algebra this entails necessarily a change in the structure of the gauge group and hence of the Lie algebra. In certain cases the change involves a finite number of parameters in the commutation relations. As an example of this one can consider (4.2.6) with the θ^{ij} real numbers. A gauge transformation which depends on these extra parameters is equivalent to a local gauge transformation in a Kaluza-Klein extension of the theory with the θ^{ij} as the local coordinates of the extra dimensions. The variation described in Equation (3.5.20) is however for fixed 'Kaluza-Klein' parameters and gives only the variation of Γ under change of gauge. Having found the solution explicitly in terms of the extra parameters one could calculate also their variation.   □

In Examples 8.11 and 8.12 we have presented an extension to noncommutative geometry of the point of view defended by Wheeler (1957) that it is of interest to consider space-time as a geometric structure and the metric thereon as a special dynamical variable, not simply a classical spin-2 field. The interest of Example 8.12 is that the relation between the classical field (the B-field) and the geometry (the commutation relations) is more subtle. Both however constitute the antisymmetric part of a σ-symmetric metric.

Notes

For further details of the relation mentioned in Section 8.1 between the spinning particle and the conformal group we refer to Duval & Fliche (1978). An introductory discussion of the problems involved in defining the position operator of a spinning particle is given in Bacry (1988). The content of Section 8.1 is closely related to the idea of Markov (1940) and Snyder (1947b, 1947a). Example 8.1 is taken from the articles of Snyder (1947b, 1947a) and Yang (1947). See also Madore (1989b) and Borchers (1993). Example 8.2 is a version proposed by Doplicher et al. (1995).

The double-bundle structure we considered in Section 8.2, which gives rise to a quartic Higgs potential, was first investigated by Forgács & Manton (1980) and by Manton (1979). The lagrangian (8.2.4) is the standard lagrangian chosen for all gauge theories which use the Higgs mechanism. Given a gauge group the theories differ according to the representation in which the Higgs particles lie and the form of the Higgs potential. The particular expression to which we have been led has been also found by slightly different, group theoretical, considerations in the context of dimensional reduction by Fairlie (1979), Harnad et al. (1980), and Chapline & Manton (1980). For further details and a review of the previous literature we refer to Kubyshin et al. (1989) or to Kapetanakis & Zoupanos (1992). What our formalism shows is that the Higgs potential is itself the action of a gauge potential on a purely algebraic structure. Different restrictions from those introduced in Section 5.2 result in different physical models. See, for example, Dubois-Violette et al. (1989b, 1989a, 1990b, 1990a, 1991), Balakrishna et al. (1991) and Madore (1991b, 1993a, 1993b). If, as in Example 8.5, one uses the exterior derivative (3.2.1) one obtains another set of models which have been studied by Connes (1990), Connes & Lott (1991), and by Coquereaux et al. (1991). See also Paschke et al. (1999). The main difference with the former lies in the form of the Higgs potential which is in fact identical in form to that used in the standard electroweak model. In this case the $\mathbb{Z}_2$ grading of the algebra of forms suggests a comparison with supersymmetry. In fact, as pointed out by Hussain & Thompson (1991), the noncommutative models based on (3.2.1) are similar in structure to a 'supersymmetric' model proposed by Ne'eman (1979), Fairlie (1979) and others (Dondi & Jarvis 1979). The grading now is due to parity and not to a boson-fermion symmetry. We refer to Hwang et al. (1996) for a review of this and its relation to the Connes-Lott model and the work of Coqueraux et al. (1991). A detailed exposition of all the model-building possibilities based on the idea of Example 8.6 is given in the review articles by Kastler (1993) and by Várilly & Gracia-Bondía (1993). A shorter exposition is to be found in Iochum & Schücker (1994), in Kastler (2000) and in Schücker (2000). For longer, self-contained, expositions we refer to the books by Landi (1997) or Figueroa et al. (2000).

The lagrange function of the Born-Infeld action (Born & Infeld 1934) is the volume element of a nonsymmetric metric (Einstein & Straus 1946) in which the nonsymmetric part is interpreted as a supplementary field (Kunstatter et al. 1983). Because of its conjectured relevance to the study of brane dynamics it has been the subject of considerable interest (Gibbons 1998; Callan Jr. & Maldacena 1998; Okawa 1999; Seiberg & Witten 1999; Alekseev et al. 1999; Alekseev et al. 2000; Seiberg &

Witten 1999; Tseytlin 2000).

For a recent review of Kaluza-Klein theory we refer to Appelquist *et al.* (1987), Bailin & Love (1987) or to Coquereaux & Jadczyk (1988). A detailed comparison at the 1-loop level of the divergences in pure gravity in dimension 5 with those of the corresponding truncated field theory in dimension 4 has been given by Coquereaux & Esposito-Farèse (1990). The expressions (8.3.3) which lift a set of vector fields on V to the complete set of all derivations of $\mathcal{A}$ have been reconsidered recently by Dubois-Violette & Masson (1998). The Equation (8.3.8) for the connection as well as (8.3.11) for the Ricci scalar coincide with the usual expression one obtains on a higher-dimensional manifold by imposing a transitive invariance group on the geometry of the hidden dimensions. Compare, for example, Cho (1985) with Madore & Mourad (1993). The generalized action (8.3.9) is due to Lovelock (1971). Its use in producing a stable vacuum was pointed out by Deruelle & Madore (1986). We have restricted our attention to the case of space-time with a finite additional algebraic structure and with no topological twist (Madore 1989a; Madore 1990; Madore & Mourad 1993). For a discussion of a Kaluza-Klein theory with a $\mathbb{Z}_2$ fibre we refer, for example, to Chamseddine *et al.* (1993) or to Sitarz (1994a). See also on this subject Landi *et al.* (1994) and Kehagias *et al.* (1995). A possible connection between the quantum fluctuations of the gravitational field and Kaluza-Klein theory has been explored by Madore & Mourad (1995).

We refer to the recent literature for a more detailed description of Example 8.10, in general (Polchinski 1998) and within the context of M(atrix)-theory (Banks et al. 1997; Ganor et al. 1997; Bonora & Chu 1997; Ho & Wu 1997; Taylor 1988; Connes et al. 1998; Landi et al. 1999b). The action of the matrix description of the complementary space is conjectured to be associated to the action in the infinite-momentum frame of a super-membrane (de Wit et al. 1988) of dimension p. Since quite generally the compactified factors of the surfaces normal to the p-branes are of the Planck scale we conclude (Madore & Saeger 1998) that they have ill-defined topology and that a matrix description will include a sum over many topologies. Attempts have been made to endow them with a smooth differential structure (Madore 1996; Grosse et al. 1997b). More details of Example 8.11 and Example 8.12 can be found in the literature (Schwenk & Wess 1992), (Schomerus 1999; Wess 2000; Madore et al. 2000b; Madore et al. 2000a; Jurco & Schupp 2000; Jurco et al. 2000).

References

Abdesselam, B., A. Chakrabarti, and R. Chakrabarti (1998). Towards a general construction of nonstandard R_h-matrices as contraction limits of R_q-matrices: the $U_h(SL(N))$ algebra case. *Mod. Phys. Lett. A* **13** 779.

Ackermann, T. and J. Tolksdorf (1996). A generalized Lichnerowicz formula, the Wodzicki residue and gravity. *J. Geom. Phys.* **19** 143.

Aghamohammadi, A. (1993). The two-parametric extension of h-deformation of $GL(2)$ and the differential calculus on its quantum plane. *Mod. Phys. Lett. A* **8** 2607.

Aghamohammadi, A., M. Khorrami, and A. Shariati (1995). h-deformation as a contraction of q-deformation. *J. Phys. A: Math. Gen.* **28** L225.

— (1997). $SL_h(2)$-symmetric torsionless connections. *Lett. Math. Phys.* **40** 95.

Ajduk, Z., S. Pokorski, and A. Trautman (Eds.) (1989). Proceedings of the XI Warsaw symposium on elementary particle physics. *New Theories in Physics*, Kazimierz, May 1988. World Scientific Publishing.

Alekseev, A. Y., A. Recknagel, and V. Schomerus (1999). Non-commutative world-volume geometries: Branes on $SU(2)$ and fuzzy spheres. *J. High Energy Phys.* **09** 023. hep-th/9908040.

— (2000). Brane dynamics in background fluxes and non-commutative geometry. *J. High Energy Phys.* **05** 010. hep-th/0003187.

Ali, S. T., A. Odzijewicz, M. Schlichenmaier, and A. Strasburger (Eds.) (1999). Proceedings of the XVI workshop on geometric methods in physics. *Coherent states, differential and quantum geometry*, Bialowieza, July 1997. *Rep. on Math. Phys.* **43** Number 1-2, 231–238.

Almeida, P. (Ed.) (2000). Proceedings of the European Mathematical Society Summer school. *Noncommutative Geometry and Applications*, Monsaraz and Lisbon, September 1997.

Ambjørn, J. and T. Jonsson (1997). *Quantum Geometry, a statistical field theory approach*. Cambridge University Press.

Aneva, B., D. Arnaudon, A. Chakrabarti, V. Dobrev, and S. Mihov (2000). On combined standard-nonstandard or hybrid (q, h)-deformations. *(to appear)*. math.QA/0006206.

Applequist, T., A. Chodos, and P. Freund (1987). *Modern Kaluza-Klein Theory and Applications*. Number 65 in Frontiers in Physics. Benjamin/Cummings.

Araki, H. (1987). Bogoliubov automorphisms and Fock representations of canonical anticommutation relations. *Contemp. Math.* **62** 23.

Artin, M., J. Tate, and M. V. den Berge (1990). Some algebras associated to automorphisms of elliptic curves. *Progress in Math. (Birkhäuser)* **88** 33.

Aschieri, P. and L. Castellani (1996). Bicovariant calculus on twisted $ISO(N)$, quantum Poincaré group and quantum Minkowski space. *Int. J. Mod. Phys. A* **11** 4513.

Aschieri, P., L. Castellani, and A. M. Scarfone (1999). Quantum orthogonal planes: $ISO_{q,r}(n+1, n-1)$ and $SO_{q,r}(n+1, n-1)$ bicovariant calculi. *Euro. Phys. Jour. C* **7** 159. q-alg/9709032.

Aschieri, P. and P. Schupp (1996). Vector fields on quantum groups. *Int. J. Mod. Phys. A* **11** 1077–1100. q-alg/9505023.

Ashtekar, A., A. Corichi, and J. Zapata (1998). Quantum theory of geometry III: Non-commutativity of Riemannian structures. *Class. and Quant. Grav.* **15** 2955.

Atiyah, M. (1967). *K-Theory.* W.A. Benjamin, New York.

— (1969). Global theory of elliptic operators. In *Proceedings of the Int. Symp. on Funct. Analysis, Tokyo.* Volume 21.

Bacry, H. (1988). *Localizability and Space in Quantum Physics.* Number 308 in Lect. Notes in Phys. Springer-Verlag.

Baez, J. and J. Muniain (1994). *Gauge Fields, Knots, and Gravity.* Number 4 in Series on Knots and Everything. World Scientific Publishing.

Bailin, D. and A. Love (1987). Kaluza-Klein theories. *Rep. Prog. Phys.* **50** 1087–1170.

Balachandran, A. (1996). Bringing up a quantum baby. In *Proceedings of the Workshop on Frontiers in Field Theory, Quantum Gravity and String Theory.* Puri, December, 1996.

Balachandran, A., G. Bimonte, G. Landi, F. Lizzi, and P. Teotonio-Sobrinho (1998). Lattice gauge fields and noncommutative geometry. *J. Geom. Phys.* **24** 353.

Balakrishna, B., F. Gürsey, and K. Wali (1991). Towards a unified treatment of Yang-Mills and Higgs fields. *Phys. Rev.* **D44** 3313.

Banks, T., W. Fischler, S. Shenker, and L. Susskind (1997). M theory as a matrix model: a conjecture. *Phys. Rev.* **D55** 5112.

Baum, P. and R. Douglas (1982). K-homology and index theory. *Proc. Symp. Pure Math.* no. 38, 117.

Bautista, R., A. Criscuolo, M. Durdević, M. Rosenbaum, and J. Vergara (1996). Quantum Clifford algebras from spinor representations. *J. Math. Phys.* **37** 5747.

Bayen, F., M. Flato, C. Fronsdal, A. Lichnerowicz, and D. Sternheimer (1978). Deformation theory and quantization. *Ann. Phys.* **111** 61.

Bellissard, J., A. van Elst, and H. Schulz-Baldes (1994). The noncommutative geometry of the quantum Hall effect. *J. Math. Phys.* **35** 5373.

Berezin, F. (1975). General concept of quantization. *Commun. Math. Phys.* **40** 153.

Berline, N., E. Getzler, and M. Vergne (1991). *Heat Kernels and Dirac operators.* Springer-Verlag.

Biedenharn, L. (1989). The quantum group $SU_q(2)$ and a q-analogue of the boson operators. *J. Phys. A: Math. Gen.* **22** L873.

Bimonte, G., E. Ercolessi, G. Landi, F. Lizzi, and G. Sparano (1996). Lattices and their continuum limits. *J. Math. Phys.* **20** 318.

Bimonte, G., F. Lizzi, and G. Sparano (1994). Distances on a lattice from noncommutative geometry. *Phys. Lett.* **B341** 139.

Birmingham, D., M. Blau, M. Rakowski, and G. Thompson (1991). Topological field theory. *Phys. Rep.* **209** 129.

Blackadar, B. (1998). *K-Theory for operator algebras* (Second ed.). Number 5 in Mathematical Sciences Research Institute Publications. Cambridge University Press.

Blokhintsev, D. (1970). *Prostranstvo i vremya v mikromire (Space and time in the Microworld).* Izdatelstvo, Moskwa.

Bohr, N. and L. Rosenfeld (1933). Zur Frage der Meßbarkeit der elektromagnetischen Feldgrößen. *Kgl. Danske. Vidensk. Selskab. Mat-fyz. Meddelser* **12** no. 8, 1.

Bonora, L. and C. Chu (1997). On the string interpretation of M(atrix) theory. *Phys. Lett.* **B410** 142.

Booss, B. and D. Bleecker (1985). *Topology and Analysis.* Springer-Verlag.

Borchers, H. (1993). A non-commuting realization of Minkowski space. *Quantum and Non-Commutative Analysis* **11**.

Bordemann, M., J. Hoppe, P. Schaller, and M. Schlichenmaier (1991). $gl(\infty)$ and geometric quantization. *Commun. Math. Phys.* **138** 209.

Born, M. and L. Infeld (1934). Foundations of the new field theory. *Proc. Roy. Soc.* **A144** 425–451.

Bourgeois, F. and M. Cahen (1999). Can one define a preferred symplectic connection? See Ali, Odzijewicz, Schlichenmaier, & Strasburger (1999), pp. 35–42.

Boutet de Monvel, L., G. Zampieri, and A. D'Agnolo (1993). *D-Modules, Representation Theory, and Quantum Groups.* Number 1565 in Lect. Notes in Math. Springer-Verlag.

Bratteli, O. and D. Robinson (1987). *Operator Algebras and Quantum Statistical Mechanics.* Springer-Verlag.

Bredon, G. (1993). *Topology and Geometry.* Springer-Verlag.

Bresser, K., F. Müller-Hoissen, A. Dimakis, and A. Sitarz (1996). Noncommutative geometry of finite groups. *J. Phys. A: Math. Gen.* **29** 2705.

Brzeziński, T. and S. Majid (1993). Quantum group gauge theory on quantum spaces. *Commun. Math. Phys.* **157** 591.

Cahen, M., S. Gutt, and J. Rawnsley (1990). Quantization of Kähler manifolds I. *J. Geom. Phys.* **7** 45.

Callan Jr., C. G. and J. Maldacena (1998). Brane dynamics from the Born-Infeld action. *Nucl. Phys.* **B513** 198. hep-th/9708147.

Carey, A. and S. Ruijsenaars (1987). On fermion gauge groups, current algebras and Kac-Moody algebras. *Acta Appl. Math.* **10** 1.

Carminati, L., B. Iochum, D. Kastler, and T. Schücker (1997). On Connes' new principle of general relativity. can spinors hear the forces of space-time? In *Operator Algebras and Quantum Field Theory*. International Press.

Carow-Watamura, U., M. Schlieker, M. Scholl, and S. Watamura (1990). Tensor representations of the quantum group $SL_q(2,\mathbb{C})$ and quantum Minkowksi space. *Z. Physik C - Particles and Fields* **48** 159.

Carow-Watamura, U., M. Schlieker, and S. Watamura (1991). $SO_q(N)$ covariant differential calculus on quantum space and quantum deformation of Schroedinger equation. *Z. Physik C - Particles and Fields* **49** 439.

Carow-Watamura, U., M. Schlieker, S. Watamura, and M. Weich (1991). Bicovariant differential calculus on quantum groups $SU_q(N)$ and $SO_q(N)$. *Commun. Math. Phys.* **142** 605.

Carow-Watamura, U. and S. Watamura (1997). Chirality and Dirac operator on noncommutative sphere. *Commun. Math. Phys.* **183** 365.

— (2000). Noncommutative geometry and gauge theory on fuzzy sphere. *Commun. Math. Phys.* **212** 395. hep-th/9801195.

Cartier, P. (1984). Homologie cyclique: rapport sur des travaux récents de Connes, Karoubi, Loday, Quillen ... In *Séminaire Bourbaki*, Number 621 in Séminaire Bourbaki. Soc. Math. Francais.

Castellani, L. (1995). Differential calculus on $ISO_q(N)$, quantum Poincaré algebra and q-gravity. *Commun. Math. Phys.* **171** 383.

Castellani, L., R. D'Auria, and P. Fré (1991). *Supergravity and Superstrings, a Geometric Perspective*. World Scientific Publishing.

Cerchiai, B. L., G. Fiore, and J. Madore (2000). Geometrical tools for quantum euclidean spaces. *Commun. Math. Phys.* math.QA/0002007.

Cerchiai, B. L., R. Hinterding, J. Madore, and J. Wess (1999a). A calculus based on a q-deformed Heisenberg algebra. *Euro. Phys. Jour. C* **8** 547–558. math.QA/9809160.

— (1999b). The geometry of a q-deformed phase space. *Euro. Phys. Jour. C* **8** 533–546. math.QA/9807123.

Cerchiai, B. L. and J. Wess (1998). q-deformed Minkowski space based on a q-Lorentz algebra. *Euro. Phys. Jour. C* **5** 553.

Chaichian, M., A. Demichev, and P. Prešnajder (2000). Quantum field theory on noncommutative space-times and the persistence of ultraviolet divergences. *Nucl. Phys.* **B567** 360. hep-th/9812180.

Chamseddine, A. and A. Connes (1996). Universal formula for noncommutative geometry actions: Unification of gravity and the standard model. *Phys. Rev. Lett.* **77** 4868.

Chamseddine, A., G. Felder, and J. Fröhlich (1993). Gravity in non-commutative geometry. *Commun. Math. Phys.* **155** 205.

Chapline, G. and N. Manton (1980). The geometrical significance of certain Higgs potentials: An approach to grand unification. *Nucl. Phys.* **B184** 391.

Chari, V. and A. Pressley (1994). *A Guide to Quantum Groups*. Cambridge University Press.

Cho, S., R. Hinterding, J. Madore, and H. Steinacker (2000). Finite field theory on noncommutative geometries. *Int. J. Mod. Phys. D* **9** no. 2, 161. hep-th/9903239.

Cho, S., J. Madore, and K. Park (1998). Noncommutative geometry of the h-deformed quantum plane. *J. Phys. A: Math. Gen.* **31** no. 11, 2639–2654. q-alg/9709007.

Cho, Y. (1985). Geometric symmetry breaking. *Phys. Rev. Lett.* **55** 2932.

Choquet-Bruhat, Y. and C. DeWitt-Morette (1989). *Analysis, Manifolds and Physics*. North-Holland Publishing, Amsterdam.

Chu, C.-S. and P.-M. Ho (1997). Poisson algebra of differential forms. *Int. J. Mod. Phys.* **12** 5573–5587. q-alg/9612031.

Chu, C.-S., P.-M. Ho, and Y.-C. Kao (1999). Worldvolume uncertainty relations for D-branes. Neuchâtel Preprint NEIP-99-006. hep-th/9904133.

Collins, P. A. and R. W. Tucker (1976). Classical and quantum mechanics of free relativistic membranes. *Nucl. Phys.* **B112** 150.

Connes, A. (1986). Non-commutative differential geometry. *Publications of the I.H.E.S.* **62** 257.

— (1988). The action functional in non-commutative geometry. *Commun. Math. Phys.* **117** 673.

— (1990). Essay on physics and non-commutative geometry. In *The Interface of Mathematics and Particle Physics*. Oxford, September 1988. Clarendon Press, Oxford.

— (1994). *Noncommutative Geometry.* Academic Press.

— (1996). Gravity coupled with matter and the foundation of non-commutative geometry. *Commun. Math. Phys.* **192** 155.

— (2000). A short survey of noncommutative geometry. hep-th/0003006.

Connes, A., M. R. Douglas, and A. S. Schwarz (1998). Noncommutative geometry and matrix theory: Compactification on tori. *J. High Energy Phys.* **02** 003.

Connes, A. and J. Lott (1991). Particle models and noncommutative geometry. In P. Binétruy, G. Girardi, & P. Sorba (Eds.), *Recent Advances in Field Theory.* Annecy-le-Vieux, March 1990. *Nucl. Phys. (Proc. Suppl.)* **18B** 29.

— (1992). The metric aspect of non-commutative geometry. In *Proceedings of the Cargèse Summer School, July 1991.* Plenum Press, New York.

Connes, A. and M. Rieffel (1987). Yang-Mills for non-commutative two-tori. *Contemp. Math.* **62** 237.

Constantinescu, F. and H. de Groote (1994). *Geometrische und algebraische Methoden der Physik: Supermannigfaltigkeiten und Virasoro-Algebren.* B.G. Teubner, Stuttgart.

Coquereaux, R. (1989). Noncommutative geometry and theoretical physics. *J. Geom. Phys.* **6** 425.

Coquereaux, R. and G. Esposito-Farèse (1990). 1-loop divergences in quantum gravity: the Einstein-Maxwell-Kaluza-Klein system. *Class. and Quant. Grav.* **7** 1583.

Coquereaux, R., G. Esposito-Farèse, and G. Vaillant (1991). Higgs fields as Yang-Mills fields and discrete symmetries. *Nucl. Phys.* **B353** 689.

Coquereaux, R. and A. Jadczyk (1988). *Riemannian Geometry Fiber Bundles Kaluza-Klein Theories and all that....* Number 16 in Lect. Notes in Phys. World Scientific Publishing.

Cornwell, J. (1992). *Group Theory in Physics : Supersymmetries and Infinite-Dimensional Algebras*, Volume 10 of *Techniques of Physics.* Academic Press.

Cotta-Ramusino, P. and M. Rinaldi (1992). Link-diagrams, Yang-Baxter equation and quantum holonomy. In M. Gerstenhaber & J. Stasheff (Eds.), *Quantum Groups with applications to Physic. Contemp. Math.* **134** 19–44.

Cuntz, A. and D. Quillen (1995). Algebra extensions and nonsingularity. *J. Amer. Math. Soc.* **8** 251.

Curtright, T. and C. Zachos (1990). Deforming maps for quantum algebras. *Phys. Lett.* **B243** 237–244.

Dabrowski, L., H. Grosse, and P. Hajac (1999). Strong connections and Chern-Connes pairing in the Hopf-Galois theory. math.QA/9912239.

Dabrowski, L., P. M. Hajac, G. Landi, and P. Siniscalco (1996). Metrics and pairs of left and right connections on bimodules. *J. Math. Phys.* **37** 4635–4646. q-alg/9602035.

Dabrowski, L., T. Krajewski, and G. Landi (2000). Some properties of non-linear sigma models in noncommutative geometry. In L. Castellani, F. Lizzi, & G. Landi (Eds.), *Noncommutative geometry and Hopf algebras in Field Theory and Particle Physics.* Torino, July 1999. World Scientific Publishing. hep-th/0003099.

Davidson, K. R. (1996). *C*-Algebras by Example.* Number 6 in Fields Institute Monographs. Amer. Math. Soc., Providence, Rhode Island.

de Azcárraga, J., P. Kulish, and F. Rodenas (1997). Twisted *h*-spacetimes and invariant equations. *Z. Physik C - Particles and Fields* **76** 567.

de Wit, B., J. Hoppe, and H. Nicolai (1988). On the quantum mechanics of supermembranes. *Nucl. Phys.* **B305** 545.

Demidov, E., Y. Manin, E. Mukhin, and D. Zhdanovich (1990). Non-standard quantum deformations of $GL(n)$ and constant solutions of the Yang-Baxter equation. *Prog. Theor. Phys. (Suppl.)* **102** 203.

Deruelle, N. and J. Madore (1986). On the vanishing of the cosmological constant. *Phys. Lett.* **A114** 185.

DeWitt, B. (1984). *Supermanifolds.* Cambridge University Press.

Dimakis, A. and J. Madore (1996). Differential calculi and linear connections. *J. Math. Phys.* **37** no. 9, 4647–4661. q-alg/9601023.

Dimakis, A. and F. Müller-Hoissen (1993). Noncommutative symplectic geometry and quantum mechanics. *Int. J. Mod. Phys. A (Proc. Suppl.)* **3** 214.

— (1994). Discrete differential calculus, graphs, topologies and gauge theory. *J. Math. Phys.* **35** 6703.

Dirac, P. (1926a). The fundamental equations of quantum mechanics. *Proc. Roy. Soc.* **A109** 642.

— (1926b). On quantum algebras. *Proc. Camb. Phil. Soc.* **23** 412.

Dobrev, V. (1992). Duality for the matrix quantum group $GL_{p,q}(2, \mathbb{C})$. *J. Math. Phys.* 3419–3430.

Doebner, H.-D. and J.-D. Hennig (Eds.) (1991). Proceedings of the Clausthal meeting. *Quantum Groups,* Clausthal, July 1989. Number 370 in Lect. Notes in Phys. Springer-Verlag.

Doebner, H.-D., W. Scherer, and C. Schulte (Eds.) (1997). Group 21. *Physical Applications and Mathematical Aspects of Geometry, Groups and Algebras,* Goslar, July 1996. World Scientific Publishing.

Dondi, P. and P. Jarvis (1979). A supersymmetric Weinberg-Salam model. *Phys. Lett.* **B84** 75.

Doplicher, S., K. Fredenhagen, and J. Roberts (1995). The quantum structure of spacetime at the Planck scale and quantum fields. *Commun. Math. Phys.* **172** 187.

Drinfeld, V. (1988). Quantum groups. *Sov. Math. Dokl.* 212.

Dubois-Violette, M. (1988). Dérivations et calcul différentiel non-commutatif. *C. R. Acad. Sci. Paris* **307** 403.

— (1991). Noncommutative differential geometry, quantum mechanics and gauge theory. In C. Bartocci, U. Bruzzo, & R. Cianci (Eds.), *Differential geometric methods in theoretical physics*. Rapallo, June 1990. Number 375 in Lect. Notes in Phys., pp. 13–24. Springer-Verlag.

— (1993). Complex structures and the Elie Cartan approach to the theory of spinors. See Oziewicz, Jancewicz, & Borowiec (1993).

Dubois-Violette, M., R. Kerner, and J. Madore (1989a). Classical bosons in a noncommutative geometry. *Class. and Quant. Grav.* **6** no. 11, 1709–1724.

— (1989b). Gauge bosons in a noncommutative geometry. *Phys. Lett.* **B217** 485–488.

— (1990a). Noncommutative differential geometry and new models of gauge theory. *J. Math. Phys.* **31** no. 2, 323–330.

— (1990b). Noncommutative differential geometry of matrix algebras. *J. Math. Phys.* **31** no. 2, 316–322.

— (1991). Super matrix geometry. *Class. and Quant. Grav.* **8** no. 6, 1077–1089.

Dubois-Violette, M. and G. Launer (1990). The quantum group of a non-degenerate bilinear form. *Phys. Lett.* **B245** 175.

Dubois-Violette, M. and J. Madore (1987). Conservation laws and integrability conditions for gravitational and Yang-Mills field equations. *Commun. Math. Phys.* **108** 213–223.

Dubois-Violette, M., J. Madore, T. Masson, and J. Mourad (1995). Linear connections on the quantum plane. *Lett. Math. Phys.* **35** no. 4, 351–358. hep-th/9410199.

— (1996). On curvature in noncommutative geometry. *J. Math. Phys.* **37** no. 8, 4089–4102. q-alg/9512004.

Dubois-Violette, M. and T. Masson (1998). $SU(n)$-connections and noncommutative differential geometry. *J. Geom. Phys.* **25** 104.

Dubois-Violette, M. and P.-W. Michor (1996). Connections on central bimodules. *J. Geom. Phys.* **20** 218.

Dunford, N. and J. T. Schwartz (1994). *Linear Operators : General Theory*. John Wiley & Son.

Durdević, M. (1996). Geometry of quantum principal bundles I. *Commun. Math. Phys.* **175** 457.

— (1998). Differential structures on quantum principal bundles. *Rep. on Math. Phys.* **41** 91.

Duval, C. and H. Fliche (1978). A conformal invariant model of localized spinning test particles. *J. Math. Phys.* **19** 749.

Eguchi, T., P. Gilkey, and A. Hansen (1980). Gravitation, gauge theories and differential geometry. *Phys. Rep.* **66** 213.

Eguchi, T. and H. Kawai (1982). Recent developments in the theory of large-N gauge fields. In K. Kikkawa, N. Naskanishi, & H. Nariai (Eds.), *Gauge Theory and Gravitation.* Nara, August, 1982. Volume 176 of *Lect. Notes in Phys.*, pp. 133–140. Springer-Verlag.

Ehrenfest, P. (1927). Bemerkung über die angenäherte Gültigkeit der klassischen Mechanik innerhalb der Quantenmechanik. *Z. für Physik* **45** 455–457.

Einstein, A. and P. Bergmann (1938). On a generalization of Kaluza's theory of electricity. *Ann. of Math.* **39** 683.

Einstein, A. and E. Straus (1946). Generalization of the relativistic theory of gravitation, II. *Ann. of Math.* **47** no. 4, 731–741.

Eisenbud, E. (1995). *Commutative Algebra with a view Toward Algebraic Geometry*, Volume 150 of *Graduate Texts in Mathematics.* Springer-Verlag.

Emch, G., H. Narnhofer, W. Thirring, and G. Sewell (1994). Anosov actions on non-commutative algebras. *J. Math. Phys.* **35** 5582.

Ewen, H. and O. Ogievetsky (1994). Classification of the $GL(3)$ quantum matrix groups. q-alg/9412009.

Faddeev, L., N. Reshetikhin, and L. Takhtajan (1990). Quantization of Lie groups and Lie algebras. *Lenin. Math. Jour.* **1** 193.

Fairlie, D. (1979). The interpretation of Higgs fields as Yang Mills fields. In *Geometrical and Topological Methods in Gauge Theories*, Montreal. Number 129 in Lect. Notes in Phys. Springer-Verlag.

Fairlie, D. and C. Zachos (1992). Quantized planes and multiparameter deformations of Heisenberg and $GL(n)$ algebras. In T. Curtright, L. Mezincescu, & R. Nepomechie (Eds.), *Quantum Field Theory, Statistical Mechanics, Quantum Groups and Topology.* Coral Gables, January 1991. pp. 81–92. World Scientific Publishing.

Fairlie, D. B., P. Fletcher, and C. K. Zachos (1989). Trigonometric structure constants for new infinite algebras. *Phys. Lett.* **B218** 203–206.

Fairlie, D. B. and C. K. Zachos (1989). Infinite dimensional algebras, sine brackets and $SU(\infty)$. *Phys. Lett.* **B224** 101–107.

— (1991). Multiparameter associative generalizations of canonical commutation relations and quantized planes. *Phys. Lett.* **B256** 43–49.

Fichtmüller, M., A. Lorek, and J. Wess (1996). q-deformed phase space and its lattice structure. *Z. Physik C - Particles and Fields* **71** 533.

Figueroa, H., J. Gracia-Bondía, and J. Váilly (2000). *Elements of Noncommutative Geometry.* Birkhauser Advanced Texts. Birkhäuser Verlag, Basel.

Filk, T. (1996). Divergencies in a field theory on quantum space. *Phys. Lett.* **376** 53.

Fiore, G. (1994). Quantum groups $SO_q(N)$, $Sp_q(n)$ have q-determinant, too. *J. Phys. A: Math. Gen.* **27** 1.

— (1996). The q-euclidean algebra $U_q(e^N)$ and the corresponding q-euclidean lattice. *Int. J. Mod. Phys. A* **11** 863.

Fiore, G., M. Maceda, and J. Madore (2000). Metrics on the real quantum plane. *(to be published).*

Fiore, G. and J. Madore (1998). Leibniz rules and reality conditions. *Euro. Phys. Jour. C.* math/9806071.

— (2000). The geometry of quantum euclidean spaces. *J. Geom. Phys.* **33** 257–287. math/9904027.

Floratos, E., J. Iliopoulos, and G. Tiktopoulos (1989). A note on $SU(\infty)$ classical Yang-Mills theories. *Phys. Lett.* **B217** 285.

Forgács, P. and N. Manton (1980). Space-time symmetries in gauge theories. *Commun. Math. Phys.* **72** 15.

Fröhlich, J. and T. Kerler (1993). *Quantum Groups, Quantum Categories and Quantum Field Theory.* Number 1542 in Lect. Notes in Math. Springer-Verlag.

Fuchs, J. (1995). *Affine Lie Algebras and Quantum Groups: An Introduction, With Applications in Conformal Field Theory.* Cambridge Monographs in Math. Phys. Cambridge University Press.

Gambini, R., J. Pullin, and A. Ashtekar (1996). *Loops, Knots, Gauge Theories and Quantum Gravity.* Cambridge University Press.

Gamkrelidze, R. (Ed.) (1991). Encyclopædia of mathematical sciences. *Geometry I, Encyclopædia of Mathematical Sciences.* Volume 28. Springer-Verlag.

Ganor, O., S. Ramgoolam, and W. Taylor (1997). Branes, fluxes and duality in M(atrix)-theory. *Nucl. Phys.* **B492** 191.

Garay, L. (1995). Quantum gravity and minimum length. *Int. J. Mod. Phys. A* **10** 145.

Ge, M.-L. (Ed.) (1992). *Quantum Group and Quantum Integrable Systems.* World Scientific Publishing.

Georgelin, Y., J. Madore, T. Masson, and J. Mourad (1997). On the noncommutative riemannian geometry of $GL_q(n)$. *J. Math. Phys.* **38** no. 6, 3263–3277. q-alg/9507002.

Gerstenhaber, M. (1964). On the deformation of rings and algebras. *Ann. of Math.* **79** no. 1, 59–103.

Gibbons, G. (1998). Born-Infeld particles and Dirichlet p-branes. *Nucl. Phys.* **B514** 603. hep-th/9709027.

Gilkey, P. (1984). *Invariance Theory, the Heat Equation and the Atiyah-Singer Index Theorem*, Volume 11 of *Math. Lecture Series*. Publish or Perish Inc.

Ginibre, J. and G. Velo (1980). The classical field limit of nonrelativistic bosons. I. Borel summability for bounded potentials. *Ann. Phys.* **128** 243–285.

Göckeler, M. and T. Schücker (1987). *Differential geometry, gauge theories and gravity*. Cambridge University Press.

Gómez, C., M. Ruiz-Altaba, and G. Sierra (1996). *Quantum Groups in Two-Dimensional Physics*. Cambridge University Press.

Gonzalez-Arroyo, A. and C. Korthals-Altes (1983). Reduced model for large-N continuum field theories. *Phys. Lett.* **B131** 396.

— (1988). The spectrum of Yang-Mills theory in a small twisted box. *Nucl. Phys.* **B311** 433.

Gonzalez-Arroyo, A. and M. Okawa (1983). A twisted model for large-N lattice gauge theory. *Phys. Lett.* **B120** 174.

González-López, A., N. Kamran, and P. Olver (1994). Quasi-exact solvability. *Contemp. Math.* **160** 113.

Gopakumar, R., S. Minwalla, and A. Strominger (2000). Noncommutative solitons. *J. High Energy Phys.* **05** 020. hep-th/0003160.

Greub, W., S. Halperin, and R. Vanstone (1976). *Connections, Curvature, and Cohomology*. Academic Press.

Gross, D. J. and N. A. Nekrasov (2000). Monopoles and strings in noncommutative gauge theory. *J. High Energy Phys.* **07** 034. hep-th/0005204.

Grosse, H. (1988). *Models in Mathematical Physics and Quantum Field Theory*. Trieste Notes in Physics. Springer-Verlag.

Grosse, H., C. Klimčík, and P. Prešnajder (1997a). Field theory on a supersymmetric lattice. *Commun. Math. Phys.* **185** 155.

— (1997b). On $4D$ field theory in non-commutative geometry. *Commun. Math. Phys.* **180** 429.

— (1997c). Topological nontrivial field configurations in noncommutative geometry. *Commun. Math. Phys.* **178** 507.

Grosse, H., W. Maderner, and C. Reitberger (1993). Schwinger terms and cyclic cohomology for massive $1 + 1$-dimensional fermions and virasoro algebras. *J. Math. Phys.* **34** 4469.

Grosse, H. and J. Madore (1992). A noncommutative version of the Schwinger model. *Phys. Lett.* **B283** no. 3-4, 218–222.

Grosse, H., J. Madore, and H. Steinacker (2000). Field theory on the q-deformed fuzzy sphere I. *Preprint.* `hep-th/0005273`.

Grosse, H. and P. Prešnajder (1993). The construction of noncommutative manifolds using coherent states. *Lett. Math. Phys.* **28** 239.

Guadagnini, E. (1993). *The Link Invariants of the Chern-Simons Field Theory.* Walter de Gruyter, Berlin.

Haag, R. (1992). *Local Quantum Physics.* Springer-Verlag.

Hajac, P. (1996). Strong connections on quantum principal bundles. *Commun. Math. Phys.* **182** 579.

Harnad, J., S. Shnider, and J. Tafel (1980). Group actions on principal bundles and dimensional reduction. *Lett. Math. Phys.* **4** 107.

Harvey, J. A., P. Kraus, F. Larsen, and E. J. Martinec (2000). D-branes and strings as non-commutative solitons. *J. High Energy Phys.* **07** 042. `hep-th/0005031`.

Heckenberger, H. and K. Schmüdgen (1997). Levi-Civita connections on the quantum groups $SL_q(N)$, $O_q(N)$ and $SP_q(N)$. *Commun. Math. Phys.* **185** 177.

Heisenberg, W. (1938). The universal length appearing in the theory of elementary particles. *Ann. Phys.* **32** 20–33.

Hellund, E. and K. Tanaka (1954). Quantized space-time. *Phys. Rev.* **94** 192.

Henneaux, M. (1985). Hamiltonian form of the path integral for theories with a gauge freedom. *Phys. Rep.* **126** 1.

Ho, P.-M. and Y.-S. Wu (1997). Noncommutative geometry and D-branes. *Phys. Lett.* **B398** 52.

Holstein, T. and H. Primakoff (1940). Field dependence of the intrinsic domain magnetization of a ferromagnet. *Phys. Rev.* **58** 1098.

Hoppe, J. (1982). Quantum theory of a massless relativistic surface. MIT Ph. D. Thesis.

Hoppe, J. (1987). Quantum theory of a relativistic surface. In G. Longhi & L. Lusanna (Eds.), *Constraint's Theory and Relativistic Dynamics.* Florence, May 1986. pp. 267–276. World Scientific Publishing.

— (1989). Diffeomorphism groups, quantization and $SU(\infty)$. *Int. J. Mod. Phys. A* **4** 5235.

Husemoller, D. (1974). *Fibre Bundles* (Second ed.). Springer-Verlag.

Hussain, F. and G. Thompson (1991). Non-commutative geometry and supersymmetry. *Phys. Lett.* **B260** 359.

Hwang, D. S., C.-Y. Lee, and Y. Ne'eman (1996). Quantum action for electroweak $SU(2/1)$ with noncommutative geometry. See Doebner, Scherer, & Schulte (1997), pp. 553–563.

Iochum, B. and T. Schücker (1994). A left-right symmetric model à la Connes-Lott. *Lett. Math. Phys.* **32** 153.

Isham, C. (1997). Structural issues in quantum gravity. In *General Relativity and Gravitation: GR14*. World Scientific Publishing.

— (1999). *Modern Differential Geometry for Physicists* (Second ed.). Number 61 in Lect. Notes in Phys. World Scientific Publishing.

Jimbo, M. (1985). A q-difference analogue of $U(\mathfrak{g})$ and the Yang-Baxter equation. *Lett. Math. Phys.* **10** 63.

Jones, V. (1995). Three lectures on knots and von Neumann algbras. In R. Coquereaux, M. Dubois-Violette, & P. Flad (Eds.), *Infinite Dimensional Geometry, Noncommutative Geometry, Operator Algebras, Fundamental Interactions*. St. Francois-Guadeloupe, May 1993. pp. 96–113. World Scientific Publishing.

Jones, V. and V. Sunder (1997). *Introduction to Subfactors*. Cambridge University Press.

Jordan, P. (1968). Über das Verhältnis der Theorie der Elementarlänge zur Quantentheorie. *Commun. Math. Phys.* **9** 279–292.

Jurco, B., S. Schraml, P. Schupp, and J. Wess (2000). Enveloping algebra valued gauge transformations for non- abelian gauge groups on non-commutative spaces. hep-th/0006246.

Jurco, B. and P. Schupp (2000). Noncommutative Yang-Mills from equivalence of star products. *Euro. Phys. Jour. C* **14** 367. hep-th/0001032.

Kadison, R. and J. Ringrose (1997). *Fundamentals of the Theory of Operator Algebras*. Amer. Math. Soc., Providence, Rhode Island.

Kadyshevsky, V. G. (1978). Fundamental length hypothesis and new concept of gauge vector field. *Nucl. Phys.* **B141** 477.

Kalau, W., N. Papadopoulos, J. Plass, and J.-M. Warzecha (1995). Differential algebras in non-commutative geometry. *J. Geom. Phys.* **16** 149.

Kalau, W. and M. Walze (1995). Gravity, noncommutative geometry and the wodzicki residue. *J. Geom. Phys.* **16** 327.

Kaluza, T. (1921). Zum Unitätsproblem der Physik. *Sitzungber. Preuss. Akad. Wiss. Phys.-Math. Kl.* **K1** 966.

Kapetanakis, D. and G. Zoupanos (1992). Coset-space dimensional reduction of gauge theories. *Phys. Rep.* **219** 1.

Karimipour, V. (1995). Bicovariant differential geometry of the quantum group $SL_h(2)$. *Lett. Math. Phys.* **35** 303–311. hep-th/9404105.

Karoubi, M. (1978). *K-Theory: An Introduction.* Springer-Verlag.

— (1982). Connections, courbures et classes caractéristiques en K-théorie algébrique. In *Current trends in algebraic topology, Part I.* London, Ontario 1981. Amer. Math. Soc., Providence, Rhode Island.

Kastler, D. (1988). *Cyclic cohomology within the differential envelope.* Hermann, Paris.

— (1993). A detailed account of Alain Connes' version of the standard model in non-commutative geometry.

— (1995). The Dirac operator and gravitation. *Commun. Math. Phys.* **166** 633.

— (1998). Spectral triples attached to quantum groups at roots of unity; I a first example. In *Proceedings of the ISI Guccia Workshop on Quantum groups and Fundamental Physical Interactions, Palermo, December 1997.* Nova Science Publishers, USA.

— (2000). Noncommutative geometry and fundamental interactions. See Almeida (2000).

Kastler, D., J. Madore, and T. Masson (1997). On finite differential calculi. See Nencka & Bourguignon (1997), pp. 135–144.

Kastler, D., J. Madore, and D. Testard (1997). Connections of bimodules in non-commutative geometry. See Nencka & Bourguignon (1997), pp. 159–164.

Kehagias, A., J. Madore, J. Mourad, and G. Zoupanos (1995). Linear connections on extended space-time. *J. Math. Phys.* **36** no. 10, 5855–5867. hep-th/9502017.

Kehagias, A., P. Meessen, and G. Zoupanos (1995). Deformed Poincaré algebra and field theory. *Phys. Lett.* **B346** 262.

Kempf, A. and G. Mangano (1997). Minimal length uncertainty relation and ultraviolet regularization. *Phys. Rev.* **D55** 7909.

Kempf, A., G. Mangano, and R. Mann (1995). Hilbert space representation of the minimal length uncertainty relation. *Phys. Rev.* **D52** 1108.

Kittel, C. (1996). *An Introduction to Solid-State Physics* (Seventh ed.). John Wiley & Son.

Klein, O. (1926). Quantentheorie und funfdimensionaler relativitätstheorie. *Z. für Physik* **37** 895–906.

Klimek, S. and A. Lesniewski (1992a). Quantum Riemann surfaces I. the unit disk. *Commun. Math. Phys.* **146** 108.

— (1992b). Quantum Riemann surfaces II. the the discrete series. *Lett. Math. Phys.* **24** 125.

Klymik, A. and K. Schmüdgen (1997). *Quantum groups and their Representations.* Springer-Verlag.

Kobayashi, S. and K. Nomizu (1969). *Foundations of Differential Geometry, I, II.* John Wiley & Son.

Kontsevich, M. (1997). Deformation quantization of Poisson manifolds, I. IHES Preprint. q-alg/9709040.

Kosinski, P., J. Lukierski, and P. Maslanka (2000). Local $D = 4$ field theory on κ-deformed Minkowski space. *Phys. Rev.* **D62** 025004. hep-th/9902037.

Kosmann-Schwarzbach, Y. (1997). Lie bialgebras, Poisson Lie groups and dressing transformations. In Y. Kosmann-Schwarzbach, B. Grammaticos, & K. Tamizhmani (Eds.), *Integrability of nonlinear systems.* Pondicherry, 1996. Number 495 in Lect. Notes in Phys, pp. 104–170. Springer-Verlag.

Koszul, J. (1960). *Lectures on Fibre Bundles and Differential Geometry.* Tata Institute of Fundamental Research, Bombay.

Krajewski, T. (1998). Classification of finite spectral triples. *J. Geom. Phys.* **28** 1.

Krajewski, T. and R. Wulkenhaar (2000). Perturbative quantum gauge fields on the noncommutative torus. *Int. J. Mod. Phys. A* **15** 1011–1030. hep-th/9903187.

Kubyshin, Y., J. Mourao, G. Rudolph, and I. Volobujev (1989). *Dimensional Reduction of Gauge Theories, Spontaneous Compactification and Model Building.* Number 349 in Lect. Notes in Phys. Springer-Verlag.

Kulish, P. (Ed.) (1991). *Quantum Groups.* Number 1510 in Lect. Notes in Math. Springer-Verlag.

Kulish, P. (1993). *On recent progress in quantum groups: an introductory review.* Friedr. Vieweg & Sohn Verlag, Braunschweig/Wiesbaden.

Kulish, P. and E. Damaskinsky (1990). On the q-oscillator and the quantum algebra $SU_q(1,1)$. *J. Phys. A: Math. Gen.* **23** L415.

Kulish, P. and A. Mudrov (1999). Twist-related geometries on q-Minkowski space. *Proc. Steklov Inst. of Math. (Trudy Mat. Inst. imeni V.A. Steklova)* **226** 97–111. math.QA/9901019.

Kulish, P. and N. Reshetikhin (1983). Quantum linear problem for the Sine-Gordon equation and higher representations. *J. Soviet Math.* **23** 2435.

Kunstatter, G., J. Malzan, and J. W. Moffat (1983). Geometrical interpretation of a generalized theory of gravitation. *J. Math. Phys.* **24** 886.

Kupershmit, B. (1992). The quantum group $GL_h(2)$. *J. Phys. A: Math. Gen.* **25** L1239.

Lam, T. (1991). *A First Course in Noncommutative Rings*, Volume 131 of *Graduate Texts in Mathematics*. Springer-Verlag.

Landi, G. (1997). *An Introduction to Noncommutative Spaces and their Geometries*, Volume 51 of *Lecture Notes in Physics. New Series M, Monographs*. Springer-Verlag.

Landi, G., F. Lizzi, and R. J. Szabo (1999a). From large N matrices to the noncommutative torus. hep-th/9912130.

— (1999b). String geometry and the noncommutative torus. *Commun. Math. Phys.* **206** 603. hep-th/9806099.

Landi, G., A. V. Nguyen, and K. Wali (1994). Gravity and electromagnetism in noncommutative geometry. *Phys. Lett.* **B326** 45.

Langmann, E. (1994). Fermion current algebras and Schwinger terms in $(3+1)$ dimensions. *Commun. Math. Phys.* **162** 1.

Langmann, E. and J. Mickelsson (1994). $(3+1)$-dimensional Schwinger terms and noncommutative geometry. *Phys. Lett.* **B338** 241.

Leitenberger, F. (1996). Quantum Lobachevsky planes. *J. Math. Phys.* **37** 3131.

Loday, J.-L. (1992). *Cyclic Homology*. Springer-Verlag.

Lorek, A., A. Ruffing, and J. Wess (1997). A q-deformation of the harmonic oscillator. *Z. Physik C - Particles and Fields* **74** 369.

Lorek, A., W. Weich, and J. Wess (1997). Non-commutative Euclidean and Minkowski structures. *Z. Physik C - Particles and Fields* **76** 375.

Lovelock, D. (1971). The Einstein tensor and its generalizations. *J. Math. Phys.* **12** 498.

Lukierski, J., P. Minnaert, and M. Mozrzymas (1996). Quantum deformations of conformal algebras introducing fundamental mass parameters. *Phys. Lett.* **B371** 215.

Lukierski, J., A. Nowicki, and H. Ruegg (1992). New quantum Poincaré, algebra and κ-deformed field theory. *Phys. Lett.* **B293** 344.

Lukierski, J., H. Ruegg, and W. Ruhl (1993). From κ-Poincaré algebra to κ-Lorentz quasigroup: a deformation of relativistic symmetry. *Phys. Lett.* **B313** 357.

Macfarlane, A. (1989). On q-analogues of the quantum harmonic oscillator and the quantum group $SU_q(2)$. *J. Phys. A: Math. Gen.* **22** 4581.

Macfarlane, A., A. Sudbery, and P. Weisz (1968). On Gell-Mann's λ-matrices, d- and f-tensors, octets, and parametrizations of $SU(3)$. *Commun. Math. Phys.* **11** 77;

MacLane, S. (1963). *Homology*. Springer-Verlag.

Madore, J. (1975). The characteristic surface of a classical spin-3/2 field in an Einstein-Maxwell background. *Phys. Lett.* **B55** 217.

— (1981). Geometric methods in classical field theory. *Phys. Rep.* **75** no. 3, 125–204.

— (1989a). Kaluza-Klein aspects of noncommutative geometry. In A. I. Solomon (Ed.), *Differential Geometric Methods in Theoretical Physics*. Chester, August 1988. pp. 243–252. World Scientific Publishing.

— (1989b). Non-commutative geometry and the spinning particle. See Ajduk, Pokorski, & Trautman (1989), pp. 524–533.

— (1990). Modification of Kaluza-Klein theory. *Phys. Rev.* **D41** no. 12, 3709–3719.

— (1991a). The commutative limit of a matrix geometry. *J. Math. Phys.* **32** no. 2, 332–335.

— (1991b). Quantum mechnics on a fuzzy sphere. *Phys. Lett.* **B263** 245.

— (1992). Fuzzy physics. *Ann. Phys.* **219** 187–198.

— (1993a). On a lepton-quark duality. *Phys. Lett.* **B305** 84–89.

— (1993b). On a noncommutative extension of electrodynamics. `hep-ph/9209226`. See Oziewicz, Jancewicz, & Borowiec (1993), pp. 285–298.

— (1996). Linear connections on fuzzy manifolds. *Class. and Quant. Grav.* **13** no. 8, 2109–2119.

— (1997). Fuzzy surfaces of genus zero. *Class. and Quant. Grav.* **14** no. 12, 3303–3312. Erratum: ibid 15:479,1998. `gr-qc/9706047`.

— (1999). On Poisson structure and curvature. `gr-qc/9705083`. See Ali, Odzijewicz, Schlichenmaier, & Strasburger (1999), pp. 231–238.

Madore, J., T. Masson, and J. Mourad (1995). Linear connections on matrix geometries. *Class. and Quant. Grav.* **12** no. 6, 1429–1440. `hep-th/9411127`.

Madore, J. and J. Mourad (1993). Algebraic Kaluza-Klein cosmology. *Class. and Quant. Grav.* **10** no. 10, 2157–2170.

— (1995). On the origin of Kaluza-Klein structure. *Phys. Lett.* **B359** 43–48. `hep-th/9506041`.

— (1998). Quantum space-time and classical gravity. *J. Math. Phys.* **39** no. 1, 423–442.

Madore, J., J. Mourad, and A. Sitarz (1997). Deformations of differential calculi. *Mod. Phys. Lett.* A **12** 975–986. `hep-th/9601120`.

Madore, J. and L. Saeger (1998). Topology at the Planck length. *Class. and Quant. Grav.* **15** 811–826. gr-qc/9708053.

Madore, J., S. Schraml, P. Schupp, and J. Wess (2000a). External fields as intrinsic geometry. *Euro. Phys. Jour. C.* hep-th/0009230.

— (2000b). Gauge theory on noncommutative spaces. *Euro. Phys. Jour. C* **16** 161–167. hep-th/0001203.

Madore, J. and H. Steinacker (2000). Propagators on the *h*-deformed Lobachevsky plane. *J. Phys. A: Math. Gen.* **33** 327–342. q-alg/9907023.

Maes, A. and A. V. Daele (1998). Notes on compact quantm groups. *Neuw Archief voor Wiskunde, Vierde serie* **12** no. 1-2, 73–112. math/9803122.

Maggiore, M. (1994). Black holes as quantum membranes. *Nucl. Phys.* **B429** 205–228. gr-qc/9401027.

Majid, S. (1988). Hopf algebras for physics at the Planck scale. *Class. and Quant. Grav.* **5** 1587.

— (1993). Braided momentum in the *q*-Poincaré group. *J. Math. Phys.* **34** 2045.

— (1995). *Foundations of Quantum Group Theory*. Cambridge University Press.

— (1997). Quantum geometry and the Planck scale. See Doebner, Scherer, & Schulte (1997).

— (1999). Quantum and braided group Riemannian geometry. *J. Geom. Phys.* **30** 113.

Maltsiniotis, G. (1993). Le langage des espaces et des groupes quantiques. *Commun. Math. Phys.* **151** 275.

Mangano, G. (1998). Path integral approach to noncommutative space-time. *J. Math. Phys.* **39** 2584.

Manin, Y. (1988). *Quantum Groups and Non-commutative Geometry*. Les Publications du Centre de Recherche Mathématiques, Montréal.

— (1989). Multiparametric quantum deformations of the general linear supergroup. *Commun. Math. Phys.* **123** 163.

Manton, N. (1979). A new six-dimensional approach to the Weinberg-Salam model. *Nucl. Phys.* **B158** 141.

Markov, M. A. (1940). Über das "vierdimensional-ausgedehnte" Elektron in dem relativistischen Quantengebiet. *J. Phys, U.S.S.R.* **2** no. 6, 453–476.

Martín, C. and D. Sánchez-Ruiz (1999). The one-loop UV-divergent structure of *U*(1)-Yang-Mills theory on noncommutative $\mathbb{R}^4$. *Phys. Rev. Lett.* **83** 476–479. hep-th/9903077.

Meissner, G. (1993). Incompressible quantum liquid versus quasi-two-dimensional electron solid. *Physica B* **184** 66.

Minwalla, S., M. V. Raamsdonk, and N. Seiberg (1999). Noncommutative perturbative dynamics. hep-th/9912072.

Misner, C., K. Thorne, and J. Wheeler (1973). *Gravitation*. W.H. Freeman and Company.

Mleko, M. and I. Sterling (1993). Application of soliton theory to the construction of pseudospherical surfaces in $\mathbb{R}^3$. *Ann. Global Anal. Geom.* **11** 65.

Moscovici, H. (1997). Eigenvalue inequalities and Poincaré duality in noncommutative geometry. *Commun. Math. Phys.* **194** 619.

Mourad, J. (1995). Linear connections in non-commutative geometry. *Class. and Quant. Grav.* **12** 965.

Moyel, J. E. (1949). Quantum mechanics as a statistical theory. *Proc. Camb. Phil. Soc.* **45** 99.

Murphy, G. (1990). *C*-algebras and Operator Theory*. Academic Press.

Nakahara, M. (1989). *Geometry, Topology and Physics*. Adam Hilger, Bristol.

Nash, C. and S. Sen (1982). *Topology and Geometry for Physicists*. Academic Press.

Ne'eman, Y. (1979). Irreducible gauge theory of a consolidated Weinberg-Salam model. *Phys. Lett.* **B81** 190.

Nencka, H. and J.-P. Bourguignon (Eds.) (1997). A conference on new trends in geometrical and topological methods. *Geometry and nature: in memory of W.K. Clifford*, Madeira, August 1995. *Contemp. Math.* **203**.

Oeckl, R. (1999). Braided quantum field theory. hep-th/9906225.

Ogievetsky, O., M. Pillin, W. Schmidke, J. Wess, and B. Zumino (1992). q deformed Minkowski space. In B. Dorfel & E. Wieczorek (Eds.), *International Symposium on the Theory of Elementary Particles*. Wendisch-Rietz, September 1992.

Ogievetsky, O., W. Schmidke, J. Wess, and B. Zumino (1992). q-deformed Poincaré algebra. *Commun. Math. Phys.* **150** 495.

Ogievetsky, O. and B. Zumino (1992). Reality in the differential calculus on q-euclidean spaces. *Lett. Math. Phys.* **25** 121.

Ohn, C. (1992). A *-product on $SL(2)$ and the corresponding nonstandard quantum-$U(\mathfrak{sl}(2))$. *Lett. Math. Phys.* **25** 85.

— (1998). Quantum $SL(3,\mathbb{C})$'s with classical representation theory. *Jour. of Alg.* **213** no. 2, 721–756.

Okawa, Y. (1999). Derivative corrections to Dirac-Born-Infeld Lagrangian and noncommutative gauge theory. hep-th/9909132.

Oziewicz, Z., B. Jancewicz, and A. Borowiec (Eds.) (1993). Proceedings of the second Max Born symposium. *Spinors, Twistors, Clifford Algebras and Quantum Deformations*, Wroclaw, September 1992. Volume 52 of *Fundamental Theories of Physics*. Kluwer Academic Publisher.

Pais, A. and G. Uhlenbeck (1950). On field theories with non-localized action. *Phys. Rev.* **79** 145.

Paschke, M., F. Scheck, and A. Sitarz (1999). Can (noncommutative) geometry accommodate leptoquarks? *Phys. Rev.* **D59** 035003. hep-th/9709009.

Paschke, M. and A. Sitarz (1998). Discrete spectral triples and their symmetries. *J. Math. Phys.* **39** 6191.

Pauli, W. (1956). Relativitätstheorie und Wissenschaft. In A. Mercier & M. Kervaire (Eds.), *Funfzig Jahre Relativitätstheorie*. Bern, July 1955. *Helv. Phys. Acta* **29 Suppl IV** 282–286.

Perelomov, A. (1986). *Generalized Coherent States and their Applications*. Springer-Verlag.

Peskin, M. E. and D. V. Schroeder (1995). *An Introduction to Quantum Field Theory*. Addison-Wesley Publishing Co.

Pittner, L. (1998). *Algebraic Foundations of Non-Commutative Differential Geometry and Quantum Groups*. Number 39 in Lect. Notes in Phys. Springer-Verlag.

Podleś, P. (1987). Quantum spheres. *Lett. Math. Phys.* **14** 193.

— (1996). Solutions of the Klein-Gordon and Dirac equations on quantum Minkowski spaces. *Commun. Math. Phys.* **181** 569.

Podleś, P. and S. Woronowicz (1990). Quantum deformation of Lorentz group. *Commun. Math. Phys.* **130** 381.

— (1996). On the classification of quantum Poincaré groups. *Commun. Math. Phys.* **178** 61.

Polchinski, J. (1998). *String Theory: An Introduction to the Bosonic String; String Theory: Superstring Theory and Beyond*. Cambridge Monographs on Mathematical Physics. Cambridge University Press.

Prugovečki, E. (1995). *Principles of Quantum General Relativity*. World Scientific Publishing.

Pusz, W. and S. Woronowicz (1989). Twisted second quantization. *Rev. Mod. Phys.* **27** 231.

Ramond, P. (1989). *Field Theory: A Modern Primer*. Addison-Wesley Publishing Co.

Reinhart, B. (1983). *Differential Geometry of Foliations*. Springer-Verlag.

Renault, J. (1980). *A Groupoid Approach to C*-algebras*. Number 793 in Lect. Notes in Math. Springer-Verlag.

Rideau, G. (1981). Covariant quantization of the Maxwell field. In E. Tirapegui (Ed.), *Field Theory, Quantization and Statistical Physics*. Dordrecht, 1980. Number 6 in Mathematical Physics and Applied Mathematics, pp. 201–226. D. Reidel Publishing Company.

Rieffel, M. (1981). C*-algebras associated with irrational rotation. *Pacific J. Math.* **93** 415.

— (1990). Non-commutative tori - a case study of non-commutative differentiable manifolds. *Contemp. Math.* **105** 191.

Riginos, A. (1976). Platonica: The anecdotes concerning the life and writing of Plato. In *Columbia Studies in the Classical Tradition, III, Anecdote 98*. E.J. Brill, Leiden.

Ruegg, H. (1990). A simple derivation of the quantum Clebsch-Gordon coefficients for $SU(2)_q$. *J. Math. Phys.* **31** 1085.

Ruijsenaars, S. (1977). On Bogoliubov transformations for systems of relativistic charged particles. *J. Math. Phys.* **18** 517.

Ryder, L. (1985). *Quantum Field Theory*. Cambridge University Press.

Sakharov, A. (1967). Vacuum quantum fluctuations in curved space and the theory of gravitation. *Doklady Akad. Nauk. S.S.S.R.* **177** 70.

— (1975). Spectral density of eigenvalues of the wave equation and vacuum polarization. *Teor. i Mat. Fiz.* **23** 178.

Schlieker, M., W. Weich, and R. Weixler (1992). Inhomogeneous quantum groups. *Z. Physik C - Particles and Fields* **53** 79.

Schmidke, W., J. Wess, and B. Zumino (1991). A q-deformed Lorentz algebra. *Z. Physik C - Particles and Fields* **52** 471.

Schmüdgen, K. (1998). Operator representations of a q-deformed Heisenberg algebra. Leipzig Preprint. `math/9805131`.

— (1999a). Commutator representations of differential calculi on quantum group $SU_q(2)$. *J. Geom. Phys.* **31** no. 4, 241–264.

— (1999b). On the construction of covariant differential calculi on quantum homogeneous spaces. *J. Geom. Phys.* **30** 23–47.

Schmüdgen, K. and A. Schüler (1993). Covariant differential calculi on quantum spaces and on quantum groups. *C. R. Acad. Sci. Paris* **316** 1155.

Schomerus, V. (1999). *D*-branes and deformation quantization. *J. High Energy Phys.* **06** 030.

Schücker, T. (2000). Geometries and forces. See Almeida (2000).

Schupp, P., P. Watts, and B. Zumino (1992). Differential geometry on linear quantum groups. *Lett. Math. Phys.* **25** 139–148. hep-th/9206029.

Schwenk, J. and J. Wess (1992). A q-deformed quantum mechanical toy model. *Phys. Lett.* **B291** 273.

Schwinger, J. (1960). Unitary operator bases. *Proc. Nat. Acad. Sci. (USA)* **46** 570.

Segal, I. (1968). Quantized differential forms. *Topology* **7** 147.

Seiberg, N. and E. Witten (1999). String theory and noncommutative geometry. *J. High Energy Phys.* **09** 032.

Seibt, P. (1987). *Cyclic Homology of Algebras.* World Scientific Publishing.

Shanahan, P. (1978). *The Atiyah-Singer index theorem: an introduction.* Springer-Verlag.

Sitarz, A. (1994a). Gravity from non-commutative geometry. *Class. and Quant. Grav.* **11** 2127.

— (1994b). Metric on quantum spaces. *Lett. Math. Phys.* **31** 35–40.

Snyder, H. (1947a). The electromagnetic field in quantized space-time. *Phys. Rev.* **72** 68.

— (1947b). Quantized space-time. *Phys. Rev.* **71** 38.

Steinacker, H. (1998). Finite dimensional unitary representations of quantum anti-de Sitter groups at roots of unity. *Commun. Math. Phys.* **192** 687.

Sternheimer, D. (1999). Deformation quantization: Twenty years after. In J. Rembieliński (Ed.), *Particles, Fields and Gravitation.* Lodz, April 1998. AIP Conference Proceedings, pp. 107. American Institute of Physics.

Stone, M. (Ed.) (1992). *Quantum Hall Effect.* World Scientific Publishing.

Sudbery, A. (1990). Noncommuting coordinates and differential operators. *J. Phys. A: Math. Gen.* **23** L697.

Sunder, V. (1987). *An Invitation to von Neumann Algebras.* Springer-Verlag.

Sylvester, J. (1884). Lectures on the principles of universal algebra. *Amer. J. Math.* **6** 271.

Szabó, L. (1989). Geometry of quantum space-time. See Ajduk, Pokorski, & Trautman (1989), pp. 517.

't Hooft, G. (1974). A planar diagram theory for strong interactions. *Nucl. Phys.* **B72** 461.

— (1996). Quantization of point particles in (2+1)-dimensional gravity and spacetime discreteness. *Class. and Quant. Grav.* **13** 1023.

Takeuchi, M. (1990). Matrix bialgebras and quantum groups. *Israel J. Math.* **72** 232.

Taylor, W. (1988). Lectures on D-branes, gauge theory and M(atrices). In *Second Trieste Conference on Duality in String Theory, June 1997.* hep-th/9801182.

Thirring, W. (1979). *A Course in Mathematical Physics*, Volume 2. Springer-Verlag.

Todorov, I. (1990). Quantum groups as symmetries of chiral conformal algebras. In *Proceedings of the 8th International Workshop on Mathematical Physics, Clausthal.* Number 370 in Lect. Notes in Phys. Springer-Verlag.

Tseytlin, A. A. (2000). Born-Infeld action, supersymmetry and string theory. hep-th/9908105. In M. Shifman (Ed.), *Yuri Golfand Memorial Volume.* World Scientific Publishing.

Tsygan, B. (1983). Homology of matrix Lie algebras over rings and Hochschild homology. *Uspekhi Math. Nauk.* **38** 217.

Turbiner, A. (1988). Quasi-exactly-solvable problems and $sl(2, \mathbb{R})$ algebra. *Commun. Math. Phys.* **118** 467–474.

— (1994). Lie-algebras and linear operators with invariant subspaces. *Contemp. Math.* **160** 263.

— (2000). Lie algebras in Fock space. Cuernavaca, 1996. q-alg/9710012. In *Complex analysis and related topics*, Number 114 in Operator Theory: Advances and Applications, pp. 265–284. Birkhäuser Verlag, Basel.

Vainerman, L. (1995). On the Gel'fand pair associated with the quantum group of motions of the plane and q-bessel functions. *Rep. on Math. Phys.* **35** 303–326.

Várilly, J. C. (2000). Introduction to noncommutative geometry. physics/9709045. See Almeida (2000).

Várilly, J. C. and J. M. Gracia-Bondía (1993). Connes' noncommutative differential geometry and the standard model. *J. Geom. Phys.* **12** 223.

— (1999). On the ultraviolet behaviour of quantum fields over noncommutative manifolds. *Int. J. Mod. Phys. A* **14** 1305. hep-th/9804001.

Vilenkin, N. (1968). *Special Functions and the Theory of Group Representations.* Number 22 in Translations of Mathematical Monographs. Amer. Math. Soc., Providence, Rhode Island.

von Neumann, J. (1955). *Mathematical Foundations of Quantum Mechanics.* Princeton University Press.

Wassermann, A. (1995). Operator algebras and conformal field theory. In *Proceedings of the International Congress of Mathematicians, Vol. 1, 2 (Zürich, 1994)*, Basel. pp. 966–979. Birkhäuser Verlag, Basel.

— (1998). Operator algebras and conformal field theory. III. Fusion of positive energy representations of LSU(N) using bounded operators. *Invent. Math.* **133** no. 3, 467–538.

Wegge-Olsen, N. (1993). *K-Theory and C*-Algebras*. Oxford University Press.

Wess, J. (2000). q-deformed Heisenberg algebras. In H. Gausterer, H. Grosse, & L. Pittner (Eds.), *Geometry and Qcpsuantumphysics.*ꞌSchladming, January 1999. Number 543 in Lect. Notes in Phys., pp. 311–382. Springer-Verlag.

Wess, J. and J. Bagger (1983). *Supersymmetry and Supergravity*. Princeton University Press.

Wess, J. and B. Zumino (1990). Covariant differential calculus on the quantum hyperplane. *Nucl. Phys. (Proc. Suppl.)* **18B** 302.

West, P. (1990). *Introduction to supersymmetry and Supergravity* (Second ed.). World Scientific Publishing.

Weyl, H. (1950). *The Theory of Groups and Quantum Mechanics*. Dover, New York.

Wheeler, J. (1957). On the nature of quantum geometrodynamics. *Ann. Phys.* **2** 604.

Wodzicki, M. (1984). Local invariants of spectral asymmetry. *Invent. Math.* **75** 143.

Woronowicz, S. (1980). Pseudospaces, pseudogroups and Pontriagin duality. In K. Osterwalder (Ed.), *Mathematical problems in theoretical physics*. Lausanne, August 1979. Number 116 in Lect. Notes in Phys., pp. 407–412. Springer-Verlag.

— (1987a). Compact matrix pseudogroups. *Commun. Math. Phys.* **111** 613.

— (1987b). Twisted $SU(2)$ group, an example of a non-commutative differential calculus. *Publ. RIMS, Kyoto Univ.* **23** 117.

— (1991). Unbounded elements affiliated with C^*-algebras and non-compact quantum groups. *Commun. Math. Phys.* **136** 399.

Yang, C. (1947). On quantized space-time. *Phys. Rev.* **72** 874.

Yukawa, H. (1949). On the radius of the elementary particle. *Phys. Rev.* **76** 300.

Zachos, C., D. Fairlie, and T. Curtright (1997). Matrix membranes and integrability. In H. Aratyn, T. Imbo, W.-Y. Keung, & U. Sukhatme (Eds.), *Supersymmetry and integrable models*. Chicago, June 1997. Number 502 in Lect. Notes in Phys., pp. 183–196. Springer-Verlag. `hep-th/9709042`.

Zinn-Justin, J. (1996). *Quantum Field Theory and Critical Phenomena* (Third ed.). Number 92 in International Series of Monographs on Physics. Oxford University Press.

Notation Index

Subject Index

Printed in the United States
By Bookmasters